Advanced civil infrastructure materials

Related titles:

The deformation and processing of structural materials
(ISBN-13: 978-1-85573-738-9; ISBN-10:1-85573-738-8)
This new study focuses on the latest research in the performance of a wide range of materials used in the construction of structures, particularly structural steels. It considers each material's processing and its deformation behaviour in structural applications. This book will help engineers by providing them with a better understanding of the performance of the major structural materials (especially metals) under different conditions in order to select the right type of material for a job and for setting design specifications. It also shows how the microstructural composition of materials is affected by processing and what influence this has on its subsequent *in situ* performance. This book will be the first to give such comprehensive coverage to the deformation and processing of all types of structural materials and will be a valuable resource for researchers in mechanical, civil and structural engineering.

Analysis and design of plated structures, Volume 1: Stability
(ISBN-13: 978-1-85573-967-3; ISBN-10: 1-85573-967-4)
Steel plated structures are used in a variety of marine and land-based applications, such as ships, off-shore platforms, power and chemical plants, box-girder cranes and bridges. The basic strength members in steel plated structures include support members, plates, stiffened panels and girders. During their lifetime structures constructed using these members are subjected to various types of loading. For example, stiffened panels are prone to buckling, resulting in progressive collapse and failure. For more effective service life it is essential to predict the behaviour and ultimate load-bearing capacity of these structures. This major new book will provide the first authoritative survey on the stability, analysis and design of plated structures particularly laminated plates, tapered steel members, concrete filled tubular columns, and stiffened composite structures. It will be a valuable source of reference for all those in mechanical, civil and structural engineering.

Inspection and monitoring techniques for bridges and civil structures
(ISBN-13: 978-1-85573-939-0; ISBN-10: 1-85573-939-9)
With the current emphasis on infrastructure rehabilitation and renewal, advanced techniques for bridge monitoring and assessment are of great interest to civil engineers and those researchers involved in the testing, inspection, management, planning, design, construction, maintenance and life extension of civil structures. The integration of these techniques can lead towards more realistic predictions of the time-dependent performance of bridges and civil structures. The potential penalties for ineffective inspection can be catastrophic. The overall emphasis now is on a life extension approach which requires the effective use of NDE techniques and reliability monitoring. This comprehensive new book discusses those techniques requiring physical measurement for inspection and/or monitoring of structures.

Details of these and other Woodhead Publishing materials books and journals, as well as materials books from Maney Publishing, can be obtained by:

- visiting www.woodheadpublishing.com
- contacting Customer Services (e-mail: sales@woodhead-publishing.com; fax: +44 (0) 1223 893694; tel.: +44 (0) 1223 891358 ext. 30; address: Woodhead Publishing Ltd, Abington Hall, Abington, Cambridge CB1 6AH, England)

If you would like to receive information on forthcoming titles, please send your address details to: Francis Dodds (address, tel. and fax as above; email: francisd@woodhead-publishing.com). Please confirm which subject areas you are interested in.

Maney currently publishes 16 peer-reviewed materials science and engineering journals. For further information visit www.maney.co.uk/journals.

Advanced civil infrastructure materials

Edited by
Hwai Chung Wu

Woodhead Publishing and Maney Publishing
on behalf of
The Institute of Materials, Minerals & Mining

CRC Press
Boca Raton Boston New York Washington, DC

WOODHEAD PUBLISHING LIMITED
Cambridge England

Woodhead Publishing Limited and Maney Publishing Limited on behalf of
The Institute of Materials, Minerals & Mining

Published by Woodhead Publishing Limited, Abington Hall, Abington,
Cambridge CB1 6AH, England
www.woodheadpublishing.com

Published in North America by CRC Press LLC, 6000 Broken Sound Parkway, NW,
Suite 300, Boca Raton, FL 33487, USA

First published 2006, Woodhead Publishing Limited and CRC Press LLC

British Library Cataloguing in Publication Data
A catalogue record for this book is available from the British Library.

Library of Congress Cataloging in Publication Data
A catalog record for this book is available from the Library of Congress.

Woodhead Publishing Limited ISBN-13: 978-1-85573-943-7 (book)
Woodhead Publishing Limited ISBN-10: 1-85573-943-7 (book)
Woodhead Publishing Limited ISBN-13: 978-1-84569-117-2 (e-book)
Woodhead Publishing Limited ISBN-10: 1-84569-117-2 (e-book)
CRC Press ISBN-10: 0-8493-3477-2
CRC Press order number: WP3477

The publishers' policy is to use permanent paper from mills that operate a sustainable forestry policy, and which has been manufactured from pulp which is processed using acid-free and elementary chlorine-free practices. Furthermore, the publishers ensure that the text paper and cover board used have met acceptable environmental accreditation standards.

Project managed by Macfarlane Production Services, Dunstable, Bedfordshire, England (macfarl@aol.com)

Typeset by Replika Press Pvt Ltd, India
Printed by TJ International, Padstow, Cornwall, England

Contents

Contributor contact details

(* = main contact)

Chapter 1

Sidney Mindess
Department of Civil Engineering
University of British Columbia
6250 Applied Science Lane
Vancouver, British Columbia
V6T 1Z4
Canada

E-mail: smindess@civil.ubc.ca

Chapter 2

Professor Charles W. Roeder*
Department of Civil Engineering
University of Washington
Seattle
WA 98195-2700
USA

E-mail: croeder@u.washington.edu

Professor Masayoshi Nakashima
Disaster Prevention Research Institute
Kyoto University
Gokasho
Uji
Kyoto 611-0011
Japan

Chapter 3

Professor Hwai-Chung Wu
Department of Civil and
Environmental Engineering
Wayne State University
5050 Anthony Wayne Drive
Detroit
MI 48202
USA

Fax: 00 1 313 577 3881
E-mail: hcwu@eng.wayne.edu

Chapter 4

Dr Julio F. Davalos
Department of Civil and
Environmental Engineering
West Virginia University
Morgantown
WV 26506-6103
USA

Tel: (304) 293-3031, Ext. 2632
Fax: (304) 293-7109
E-mail: Julio.Davalos@mail.wvu.edu

Dr Pizhong Qiao*
Department of Civil Engineering
The University of Akron
Akron
OH 44325-3905
USA

Tel: (330) 972-5226
Fax: (330) 972-6020
E-mail: qiao@uakron.edu

Luyang Shan
Department of Civil Engineering
The University of Akron
Akron
OH 44325-3905
USA

Tel: (330) 972-5226
Fax: (330) 972-6020
E-mail: ls30@uakron.edu

Chapter 5

Vistasp M. Karbhari
Department of Structural
Engineering and Materials Science
and Engineering Program
University of California, San Diego
La Jolla
CA 92093-0085
USA

Tel: 858-534-6470
Fax: 858-534-6373
E-mail: vkarbhari@ucsd.edu

Chapter 6

Professor Habib J. Dagher
Director, AEWC Center
5793 AEWC Building
University of Maine
Orono
ME 04469-5793
USA

Tel: (207) 581-2138
E-mail: hd@umit.maine.edu

Chapter 7

John Harrison
TecEco Pty Ltd
Tec-Cement Pty Ltd (Sales)
497 Main Road
Glenorchy TAS 7010
Australia

Tel: 61 3 62497868 (a.m., weekends)
Tel: 61 3 62713000 (p.m., weekdays)
E-mail: john.harrison@tececo.com

Preface

Concrete and steel are by far the most widely used man-made construction materials in the world. Despite numerous successful constructions of worldwide infrastructures, many of these infrastructures are now rapidly deteriorating. It has been estimated that infrastructure rehabilitation costs, in the US alone, will reach into trillions of dollars over the next twenty years. One of the primary causes for the decay of our infrastructures is the deterioration of the materials used in the construction and repair of these structures. Intrinsically, concrete is brittle hence vulnerable to cracking, whereas structural steel is heavy and prone to corrosion. These alarming infrastructural problems indicate an urgent need for more durable construction materials ultimately to replace conventional concrete and steel, and most likely to supplement these popular materials in the near future.

In recent decades, material development in response to the call for more durable infrastructures has led to many exciting advancements, especially the paradigm design example of Fiber Reinforced Plastics Composites which enjoys tremendous success in the aerospace and automotive industries. It has inspired many composite designs involving concrete and wood. These fiber reinforced composites, with unique properties, are now being explored in many infrastructural applications. Even concrete and steel are being steadily improved to have better properties and durability. Such advancement also makes it possible to use advanced concrete or steel in an unprecedented way.

The purpose of this book is to provide an up-to-date review of several emerging construction materials that may have a significant impact on repairs of existing infrastructures and/or new constructions. Two areas take high priority in this book. First, we emphasize the materials design concept which leads to the creation of advanced composites by synergistically combining two or more constituents. Such design methodology is made possible by several key advancements in materials science and mechanics. Second, we conclude each chapter with selective examples of real-world applications using such advanced materials. This includes relevant structural design guidelines and mechanics to assist readers in comprehending the uses of these advanced materials.

Selected authors who are recognized for their expertise in each field have each contributed to certain chapters. It is hoped that this book will be of value both to graduate and undergraduate students of civil engineering, materials science, architecture and construction engineering, and will serve as a useful reference guide for researchers and practitioners in the construction industry.

H C Wu
Detroit, Michigan

1

Advanced concrete for use in civil engineering

S MINDESS, University of British Columbia, Canada

1.1 Introduction

Cementitious materials are certainly the oldest manufactured materials of construction, and today are the predominant construction materials worldwide, their production far outstripping those of steel, timber, asphalt and other building materials. Their use goes back at least 9000 years, as in a mortar floor discovered in Israel (Malinowski and Garfinkel 1991; Bentur 2002). Similar lime mortars, prepared by calcining limestone, were used throughout the Middle East. Gypsum-based mortars were later used in Egypt, for example in the construction of the Pyramid of Cheops (~3000 B.C.); these were prepared by calcining impure gypsum. However, neither of these cementing materials is hydraulic; that is, they will not harden under water.

Later, the Greeks and the Romans were able to produce hydraulic lime mortars, mixed with finely ground volcanic ash. In particular, the Romans used a material called *pozzolana*, found in the vicinity of the village of Pozzuoli, near Mt Vesuvius. This material is described by Vitruvius* writing in about 27 B.C.:

> There is also a kind of powder which from natural causes produces astonishing results. It is found in the neighborhood of Baiea and in the country belonging to the towns round about Mt Vesuvius. This substance, when mixed with lime and rubble, not only lends strength to buildings of other kinds, but even when piers of it are constructed in the sea, they set hard under water.

Materials such as these were used extensively in hydraulic structures, such as piers and sea walls. They were also used to make a form of concrete. Probably the best preserved Roman structure, the Pantheon, dating from the

*Vitruvius, *The Ten Books of Architecture,* Bk II, Ch. VI (New York, Dover, 1960), pp. 46–7.

second half of the second century A.D. was built largely of concrete. The 44 m diameter dome was constructed by pouring concrete into ribbed sections and letting it harden. In addition, to reduce the weight, a lightweight aggregate (pumice) derived by crushing a porous volcanic rock was used.

Though the quality of cementing materials declined through the Middle Ages, high quality mortars were later rediscovered after the 14th century A.D., and by the middle of the 18th century, research began which led to the development of 'modern' Portland cements. John Smeaton, who in 1756 had been commissioned to rebuild the Eddystone Lighthouse off the coast of Cornwall, England, carried out a series of experiments with different limes and pozzolans, and discovered that the best limestones for this purpose were those containing high proportions of clayey material. While this was still not quite Portland cement, it represented a distinct improvement over the slaked limes then in use.

Other developments followed rapidly. James Parker in England patented (1796) a natural hydraulic cement, produced by calcining nodules of impure limestone containing clay. In France, Louis-Joseph Vicat (1813) prepared artificial hydraulic lime by calcining synthetic mixtures of limestone and clay. Finally, in 1824, Joseph Aspdin, a Leeds builder, took out a patent on 'portland' cement (so named because of a real or fancied resemblance of the hardened cement to a naturally occurring limestone quarried on the Isle of Portland, which was then a popularly used material in buildings). This cement was prepared by calcining finely ground limestone, mixing this with finely divided clay, calcining the mixture again in a kiln, and then finely grinding the resulting clinker. This patent was followed by work in many countries, leading to improved kiln designs and methods of proportioning the raw materials.

However, work on the chemistry of Portland cement really began only late in the 19th century, with the publication in 1887 of the doctoral thesis of Henri Le Chatelier in France. He established that the main cementing phase in the cement was tricalcium silicate ($3CaO{\cdot}SiO_2$), and identified some of the other phases as well. Subsequent work by many others finally resulted in the high quality Portland cements that we use today. Finally, in the 1930s, a systematic study was begun on the role of chemical admixtures in Portland cement concrete. (This technology was not entirely new; the Romans had used animal fat, blood and milk in some of their concretes, probably to improve the workability.) Air entraining agents, water reducers, superplasticizers, set retarders and accelerators, corrosion inhibitors, and various other admixtures make possible the high performance concretes that can now be produced.

Still, to the lay person, concrete is a deceptively simple material: You mix together cement, water and aggregates, and it gets hard – what else is there to know? Indeed, it is this apparent simplicity that has helped to make concrete such a widely used material. The reality, of course, is very different. Modern concretes, despite the popular prejudice to the contrary, are truly 'high-tech' materials. We can produce concretes for virtually any purpose. However, our ability to produce advanced concretes, with controlled rheological properties, compressive strengths of several hundred MPa, tensile strength of over 50 MPa, high toughness, and excellent long-term durability, depends upon our understanding of the fundamental mechanisms governing the chemical and mechanical properties of the concrete, and our ability to manipulate the microstructure (and sometimes even the nanostructure) of the material. Without this understanding, concrete structures such as the 452 meters tall Petronas Towers in Kuala Lumpur would not have been possible In the remainder of this chapter, some of the recent developments in the production and use of modern advanced concrete will be described.

1.2 What is modern advanced concrete?

Advanced concrete, more commonly referred to as 'high-performance concrete', may be defined as 'concrete meeting special combinations of performance and uniformity requirements that cannot always be achieved routinely using conventional constituents and normal mixing, placing and curing practices' (Russell 1999). In other words, advanced concrete is simply concrete that is significantly better in some respects than the concrete that we usually make. Clearly, the above definition is, like concrete itself, highly time-dependent; what might have seemed like very advanced concrete 100 years ago would look very ordinary today, just as today's 'advanced' concretes will seem quite primitive 100 years from now.

1.3 Materials

Before discussing the properties of some of the advanced concretes available today, it is necessary first to say a few words about the materials that are currently available. It must be emphasized that, in order to produce truly high performance concretes, it is essential that the composition and properties of *all* of the components of the concrete be carefully considered: cement chemistry and particle size distribution, the chemistry and size distribution of the supplementary cementing materials, the chemistry of the admixtures and their possible reactions with the cementing materials, and the physical, chemical and mechanical properties of the aggregates. In addition, there must be strict quality control governing every phase of the concrete-making process; the design engineer(s), the cement producer, the concrete supplier and the contractor must all work together to ensure the quality of the final product. Even with the best choice of materials, if the concrete is not placed, finished and cured properly, the result will be a substandard material.

1.3.1 Portland cements

The basic raw materials for Portland cement production are limestone (the major source of CaO) and clays or shales containing about 55 to 60% SiO_2, 15 to 25% Al_2O_3 and 5 to 10% Fe_2O_3. The production technology depends upon high-temperature chemistry. Based on the work of Taylor (1997), the high temperature reactions for the formation of cement clinker have been described as follows (Bentur 2002):

1. Decomposition of the clay minerals (~500 to 800 °C)
2. Decomposition of the calcite (~700 to 900 °C)
3. Reactions of the calcite (or lime formed from it), SiO_2 and the decomposed clays to form $2CaO{\cdot}SiO_2$ (~1000 to 1300 °C)
4. Clinkering reactions at about 1300 to 1450 °C to form $3CaO{\cdot}SiO_2$. A melt of aluminate and ferrite is formed to act as a flux to facilitate the formation of $3CaO{\cdot}SiO_2$ by the reaction between CaO and $2CaO{\cdot}SiO_2$
5. Cooling back to ambient temperature, during which time the melt crystallizes to form the ferrite and aluminate phases.

The resulting clinker is then interground with gypsum (to control the setting behavior of the cement). The ternary phase diagram of the CaO-Al_2O_3-SiO_2 system into which the various cementitious materials fit is shown in Fig. 1.1.

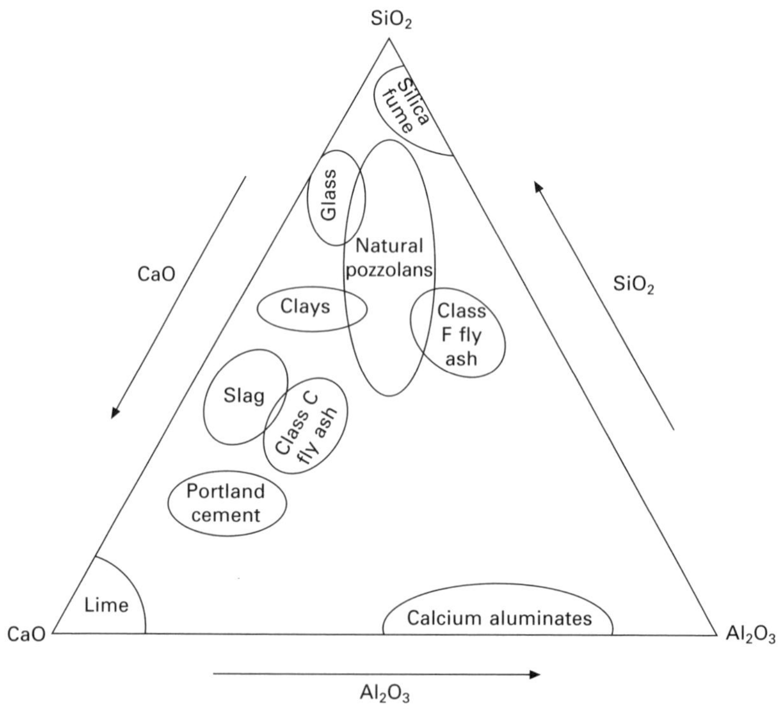

1.1 Ternary phase diagram of the CaO-Al_2O_3-SiO_2 system (after Bentur, 2002).

While this helps enormously in our understanding of clinker chemistry, it remains a considerable simplification of the true state of affairs, in that it overlooks several important factors:

1. The presence of the roughly 5 to 8% of impurities present in the clinker, the exact type and amount of which depend on the particular raw material sources. These impurities include alkalis (sodium and potassium), sulfates, magnesium, manganese, iron, potassium, titanium, and perhaps others as well.
2. The fact that the mineral phases formed are not pure, but are 'doped' with various other ions, depending on the exact chemistry of the raw feed.
3. The different mineral phases are not in the form of separate grains; each cement particle will generally contain several phases.

These simplifications are not particularly important for 'ordinary' concretes, whose properties (in both the fresh and hardened states) can be predicted reasonably well based on the aggregate grading, the cement content, and the water/cement (w/c) ratio. The strength and other concrete properties are largely independent of the details of the cement chemistry. However, for high performance concretes, these details can be of vital importance, since such concretes will invariably contain both mineral admixtures and chemical admixtures, and in particular superplasticizers (also known as high range water reducers). The behavior of these much more complex mixtures can be greatly affected by the 'minor' components of the cement, and by the cement mineralogy and composition. The problems of cement-superplasticizer incompatibility and other adverse admixture interactions can create difficulties in finding satisfactory mix designs for concretes for some special applications.

It should also be noted that some of the cements produced today lead, on the whole, to less durable concretes than the 'old' (pre-1960) concretes. This has been driven largely by the demands of the construction industry for higher rates of strength gain, since these permit more rapid form removal and lead to economies in construction costs. The higher rate of strength gain has been achieved primarily by grinding the cement more finely and increasing the tricalcium silicate content; unfortunately, both measures lead to higher heats of hydration and a greater possibility of thermal cracking. Modern concretes also appear to be somewhat more brittle than their older counterparts.

1.3.2 Aggregates

As is the case with cements, ordinary concrete is quite insensitive to the aggregate properties, as long as the aggregates fall within the normal range of particle size distributions, and are not chemically reactive with the cement (as in alkali-aggregate or alkali-carbonate reactions). Even aggregates that do not meet the specified grading limits may often be used to make satisfactory concretes, albeit at additional cost. However, for advanced concretes, the aggregate properties become increasingly important. Obviously, for the very high strength concretes which will be described later, the aggregates must be strong enough and stiff enough so that they do not become the strength limiting factor. For the new generation of self compacting concretes (SCC) which will also be described later, the aggregate grading must be very tightly controlled to achieve the appropriate rheological properties. It may also be necessary to 'engineer' in some way the interfacial region at the cement-aggregate interface, as this is often the 'weak link' in the system. Unfortunately, most concrete research over the past few decades has focused on the binder system; it is now becoming necessary to focus equal attention on the aggregates which, after all, make up 70 to 80% by volume of the concrete.

1.3.3 Chemical admixtures

For the purposes of the present discussion, only superplasticizers amongst the various chemical admixtures will be discussed here, since they are essential for the production of most advanced concretes. These types of admixtures were first developed in the 1960s. They are synthetic, high molecular weight, water soluble polymers, that adsorb onto the surfaces of the cement (and supplementary cementitious) particles and are very effective in dispersing these particles. They can permit very large reductions in the amount of water required for adequate workability through three mechanisms:

1. They build up a negative charge on the particle surfaces sufficient to cause repulsion amongst them.
2. They increase the affinity between the particles and the liquid phase.
3. Steric hindrance, which occurs when the adsorbed polymers are oriented in such a way as to weaken interparticle attraction (Haneharo and Yamada, 1999).

There are four main types of superplasticizers currently available (Aitcin 1998; Sakai *et al.* 2003). Melamine superplasticizers (polycondensates of formaldehyde and melamine sulfonate) were the first superplasticizers to be used commercially, and are still widely used in both North America and Europe, and are considered to be very reliable in terms of quality and performance. Naphthalene superplasticizers (polycondensates of formaldehyde and naphthalene sulfonate) are available as either sodium or calcium salts. While the sodium salt version is more commonly used because it is less expensive, the calcium salt version, which does not contain any chlorides, is required for nuclear applications of reinforced or prestressed concrete, to protect the steel from chloride-induced corrosion. In general, these superplasticizers tend to be less expensive than melamine superplasticizers.

Lignosulfonate-based superplasticizers are, by themselves, less effective than the other two, and are not commonly used in conjunction with either of the other two. Recently, a new family of superplasticizers, the polycarboxylates (Sakai *et al.* 2003) have been developed (sometimes referred to as 'comb-type' superplasticizers) These superplasticizers maintain the fluidity of the concrete much longer than the naphthalene and melamine superplasticizers but without excessively retarding the concrete mix. They appear to be particularly effective for producing self-compacting concretes (see section 1.4.4).

Unfortunately, not all superplasticizers interact in the same way with all cements (in large part because different cements contain different impurities), and there is no way yet of predicting which combination of cement and superplasticizer will give the best results. Unless there is prior information available, concrete producers may have to resort to a 'trial and error' procedure to find the best combination.

1.3.4 Mineral admixtures

Most concretes produced today contain one or more mineral admixtures; it would be unusual to find a pure Portland cement concrete (Malhotra, 1987). As mentioned above, this is particularly true of advanced concretes, for which not only the strength, but also the rheological properties and the durability must be tightly controlled. The mineral admixtures (also known as 'supplementary cementing materials'), are either preblended with the cement (more common in Europe), or added separately at the batching plant (more common in North America). The three most common such materials are fly ash, granulated blast furnace slag and silica fume. In ordinary concrete, fly ash or blast furnace slag are used primarily for economic reasons, as they are cheaper than the Portland cement that they replace; silica fume, being more expensive than cement, is not used. However, for very high strength, low permeability materials, the use of silica fume, commonly in conjunction with one or sometimes both of the other two materials, is essential. The physical properties of some common supplementary cementing materials are given in Table 1.1; their particle size distributions are shown in Fig. 1.2.

Silica fume

Silica fume is a waste by-product of the production of silicon and silicon alloys, and is most commonly used to make high strength concretes. It is essential for concretes with compressive strengths greater than 100 MPa, and makes it easier to produce high strength concretes in the 60 to 100 MPa range. It is best used at addition rates of about 8 to 10% by weight of cement.

Table 1.1 Physical characteristics of some supplementary cementing materials

Material	Mean size (μm)	Surface area (m^2/g)	Particle shape	Specific gravity
Portland cement	10–15	<1	Angular, irregular	~3.2
Natural pozzolans	10–15**	<1	Angular, irregular	Variable
Fly ash (F and C)	10–15	1–2	Mostly spherical	2.2–2.4
Silica fume	0.1–0.3	15–25	Spherical	2.2
Rice husk ash	10–20**	50–100	Cellular, irregular	<2.0
Calcined clay (metakaolin)	1–2	~15	Platy	2.4

** After grinding

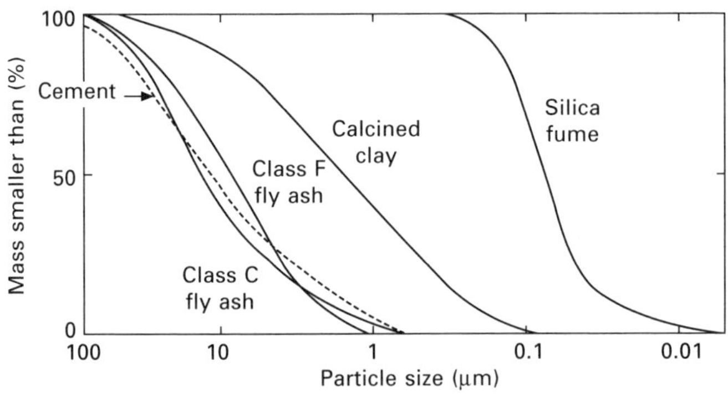

1.2 Particle size distributions of some supplementary cementing materials.

It affects concrete properties by a number of different mechanisms (Mindess *et al.* 2003):

1. It eliminates the growth of calcium hydroxide at the cement-aggregate interface, or transforms the calcium hydroxide to calcium silicate hydrate by the pozzolanic reaction between silica and lime.
2. It eliminates many of the large pores at the cement-aggregate interface, making it denser.
3. It modifies the rheological properties of the fresh concrete, in such a way that internal bleeding is reduced, and less superplasticizer is required to achieve a particular workability.
4. Because of its very fine particle size (about 1/100 the mean size of cement grains), silica fume has a marked filler effect; that is, it is able to pack between the cement grains, lowering the mean size of the capillary pores, and increasing the density of the material.

Fly ash

Fly ash is the inorganic, noncombustible residue of powdered coal after burning in power plants. While it is used in concrete (at typical replacement rates of about 15% by weight of cement) primarily because it is cheaper than cement, it also has the beneficial effect of reacting with the lime produced during cement hydration to produce more calcium silicate hydrate (i.e. the same pozzolanic reaction that silica fume undergoes). This will generally increase the long-term strength and improve the durability of the concrete. By itself, fly ash cannot be used to achieve concrete strengths higher than about 100 MPa; for higher strengths, it must be used in conjunction with silica fume.

Blast furnace slag

Granulated blast furnace slags are the ground residue from metallurgical processes, either from the production of metals from ore or from the refinement of impure metals. They are rich in lime, silica and alumina (and are thus somewhat similar to Portland cements). Pure slag cements react only very slowly, and so commercially they must be activated, most commonly by Portland cement. Typical dosage rates of slag are in the range of 15 to 39% by weight of cement, though much higher dosages are sometimes used. As with fly ash, for compressive strengths in excess of about 100 MPa, they must be used in conjunction with silica fume.

1.4 Modern advanced concretes

In the remainder of this chapter, a number of the principal advanced, or high performance, concretes that are used today will be described. The choice is an arbitrary one, and is not intended to be an exhaustive list of all advanced concretes. Rather, it is intended to describe some of the basic principles, and to illustrate the range of possibilities.

1.4.1 High strength concretes

For most brittle materials, strength is inversely proportional to porosity. A typical expression for this empirical observation is

$$f_c = f_{c0}e^{-kp} \qquad 1.1$$

where f_c is the strength, f_{c0} is the strength at zero porosity (sometimes referred to as the 'intrinsic' strength), p is the porosity, and k is a constant that depends on the particular system. Expressions of this form are also applicable to concrete, for which the porosity is largely determined by the w/c ratio.

The w/c ratio 'law' enunciated by Duff Abrams in 1918 states that 'For given materials, the strength depends only on one factor – the ratio of water to cement.' This can be expressed as

$$f_c = (A/B)^{(w/c)} \tag{1.2}$$

where A and B are constants which depend on the cement properties, age, curing conditions, and so on. This works reasonably well for ordinary concretes, and even up to strengths of about 120 MPa as shown in Fig. 1.3; indeed, it forms the basis for most mix design procedures.

However, as the w/c ratio decreases, particularly below about 0.3, this relationship becomes increasingly unreliable. The constant 'A' is frequently taken to be 96.5 MPa (14,000 psi), implying a maximum strength of 96.5 MPa, but we know that it is possible to make concretes with strengths well in excess of 600 MPa. Thus, simple reliance on the w/c ratio law does not work for ultra-high-strength concretes. This should come as no surprise, since Abram's law was formulated long before the advent of superplasticizers and supplementary cementing materials. Clearly, though, something besides

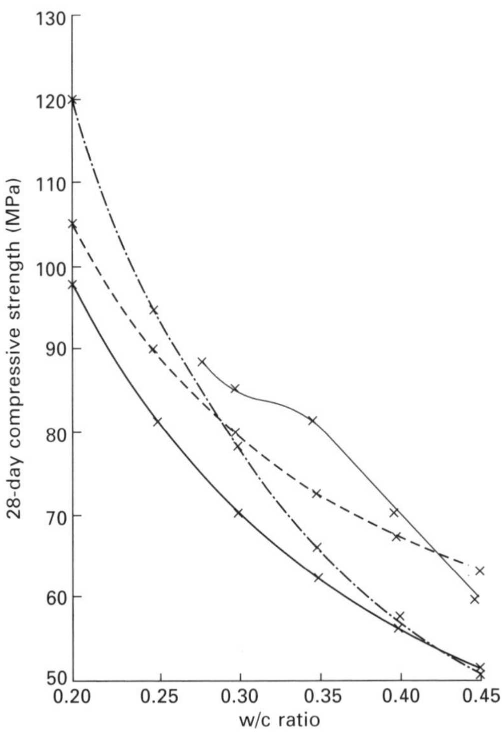

1.3 Compressive strength vs. w/c ratio for different concretes.

w/c ratio (or porosity) must play a large part in determining the strength of very low w/c ratio concretes. Unfortunately, we do not yet know what these additional factors are, though they may be related to an improved cement-aggregate bond, or to stronger bonds within the hydrated cement paste system. Table 1.2 shows some mix proportions and properties of some commercially available high strength concretes in North America (Burg and Ost 1994). These are empirically arrived at mix designs, since there are as yet no generally accepted procedures for a more rational design of such mixes.

Table 1.2 Mix proportions and properties of some commercially available high strength concretes (after Burg and Ost, 1994)

Units per m^3	Mix number					
	1	2	3	4	5	6
Cement, Type 1, kg	564	475	487	564	475	327
Silica fume, kg	–	24	47	89	74	27
Fly ash, kg	–	59	–	–	104	87
Coarse aggregate SSD (14 mm crushed limestone), kg	1068	1068	1068	1068	1068	1121
Fine aggregate SSD, kg	647	659	676	593	593	742
HRWR Type F, litres	11.6	11.6	11.22	20.11	16.44	6.3
HRWR Type G, litres	–	–	–	–	–	3.24
Retarder, Type D, litres	1.12	1.05	0.97	1.46	1.5	–
Water to cementing materials ratio	0.28	0.29	0.29	0.22	0.23	0.32
	Fresh concrete properties					
Slump, mm	197	248	216	254	235	203
Density, kg/m^3	2451	2453	2433	2486	2459	2454
Air content, %	1.6	0.7	1.3	1.1	1.4	1.2
	Compressive strength, 100 × 200-mm moist-cured cylinders					
3 days, MPa	57	54	55	72	53	43
7 days, MPa	67	71	71	92	77	63
28 days, MPa	79	92	90	117	100	85
56 days, MPa	84	94	95	122	116	–
91 days, MPa	88	105	96	124	120	–
182 days, MPa	97	105	97	128	120	–
426 days, MPa	103	118	100	133	119	–
1085 days, MPa	115	122	115	150	132	–
	Modulus of elasticity in compression, 100 × 200-mm moist-cured cylinders					
91 days, GPa	50.6	49.9	50.1	56.5	53.4	47.9

1.4.2 Ultra-high-strength concretes

While high-strength concretes in the range of about 100 to 125 MPa are now readily available, at least in some regions, there is another class of concretes that may be referred to as ultra-high-strength. They are also sometimes referred to as 'reactive powder (RPC) concretes'. With concretes of this type, compressive strengths of over 600 MPa and flexural strengths approaching 100 MPa have been achieved.

RPC was developed in France in the early 1990s (Richard and Cheyrezy 1994; Richard 1996). The quite remarkable properties of this material were achieved by careful control of the concrete mixture, in particular the particle size distribution of *all* of the solid materials. Optimization of the particle size distribution leads to a mix approaching optimum density. RPC contains no coarse aggregate; indeed, the maximum aggregate size is 0.3 mm! This permits production of a more homogeneous material. For purposes of ductility, up to 5% by volume of steel fibers are added to the mix. Typical mix proportions and mechanical properties of two different strength levels of RPC are given in Table 1.3. It must be noted, however, that to achieve the higher strength levels some degree of heat treatment may be required. RPC was used to build a pedestrian walkway at the University of Sherbrooke, Quebec, Canada, using conventional concreting equipment and techniques, as a demonstration of using this apparently exotic material in the field.

A commercial development of RPC is now being marketed under the name of DUCTAL®. For some applications heat treatment (90 °C steam for 48 hours) is applied to improve the dimensional stability and durability. With steel fibers, compressive strengths are of the order of 150 to 180 MPa, with flexural strengths of about 32 MPa. These strengths are reduced by about 25% when polypropylene fibers are used. This material was used

Table 1.3 Typical composition and properties of RPC (adapted from Richard and Cheyrezy, 1994)

	RPC 200	RPC 800
Type V Portland cement	955 kg/m³	1000 kg/m³
Fine sand (150–400 μm)	1051 kg/m³	500 kg/m³
Silica fume	229 kg/m³	230 kg/m³
Precipitated silica (35 m²/g)	10 kg/m³	–
Ground quartz (μm)	–	390 kg/m³
Superplasticizer	13 kg/m³	18 kg/m³
Steel fibers (12.5 × 180 μm)	191 kg/m³	–
Steel microfibers (< 3 mm long)	–	630 kg/m³
Total water	153 l/m³	180 l/m³
f′c (compressive strength)	170–230 MPa	490–680 MPa
E	54–60 GPa	65–75 GPa
Flexural strength	25–60 MPa	45–102 MPa

recently to build a railway station canopy in Calgary, Canada. A somewhat similar French material, BSI®-CERACEM concrete was used to construct the toll gate roofs for the new Millau viaduct in the south of France (Thibaux *et al.*, 2004).

Another version of this technology, also developed in France, has been patented under the name of $\text{CEMTEC}_{\text{multiscale}}$® (Parant and Rossi, 2004). It is characterized by much higher cement and fiber contents than DUCTAL®, though the underlying principles are the same. This material can achieve flexural strengths of about 60 MPa. It also has an extremely low permeability (Charron *et al.* 2004). For comparison, typical mix proportions for two of these materials are given in Table 1.4.

Still another family of ultra-high-strength concretes was developed in Denmark in the 1980s by Bache (1987). This material is referred to as CRC (compact reinforced composite). It too is made with a very low water/binder ratio (~0.16 or less), and contains from 2 to 6% steel fibers, providing matrix strengths of 140 to 400 MPa. It differs from the materials described above in that it is also combined with closely spaced conventional steel reinforcement. It has been used mainly in precast elements such as staircases and balcony slabs (Aarup 2004), but has also been used in cast-in-place applications.

There are other ultra-high-strength concretes that have been produced, and still others in the development stage. Their common features are a very low water/binder ratio, the use of silica fume and superplasticizers, high contents of fibers, limitations on the maximum aggregate size, and careful control of the particle size distribution. They also require very tight quality control both in their production and in their placement. Consequently, these materials are very expensive. However, their high durability and unique architectural possibilities will ensure that they will have increasingly wide applications in a variety of specialized structures.

Table 1.4 Compositions of some commercial RPCs

Material	DUCTAL® (kg/m^3)	$\text{CEMTEC}_{\text{multiscale}}$® ($kg/m^3$)
Portland cement	710	1050.1
Silica fume	230	268.1
Crushed quartz	210	–
Sand	1020	514.3
Water	140	180.3
Fibers	40–160[a]	858[b]
Superplasticizer	10	44

[a]Either steel or polypropylene fibers (13 mm × 0.20 mm)
[b]A mixture of three different geometries of steel fibers

1.4.3 Fibre reinforced concretes

The ultra-high-strength concretes described above represent 'extreme' examples of fiber reinforced concrete, with fiber contents too high for them to be economical in more normal and widespread applications. However, there is a continuing growth in the use of 'conventional' fiber reinforced concrete (FRC), at fiber volumes less than about 1%. These concretes were first developed in the 1960s (Romualdi and Batson 1963). As we all know, plain concrete is a brittle material, with low tensile strength and strain capacities. To help overcome these disadvantages, it has become increasingly common to incorporate short, discrete discontinuous fibers into the concrete matrix. Generally, the fibers at these addition rates are not added to increase the concrete strength, though modest increases in strength may occur. Rather, the principal role of the fibers is to make the concrete tougher. That is, their function is to control the cracking of the concrete, and then to modify the behavior of the composite material once the concrete matrix has cracked; by bridging across the cracks as they begin to open, the fibers provide some post-cracking ductility to the concrete. As of 2001, over 76 million m^3 of fiber concrete were produced annually (Mindess *et al.* 2003), with the principal applications being slabs on grade, fiber shotcrete (mostly for tunnel linings) and precast members. A number of different fibers are now being used commercially, the most common being steel, polypropylene, cellulose (largely as a replacement for asbestos), carbon and glass.

It must be emphasized that, in general, fiber reinforcement is not a substitute for conventional steel reinforcement. They have different functions in concrete: reinforcing bars are used to carry the tensile and shear stresses, and thus to increase the load-bearing capacity of structural members, while the fibers are there primarily for crack control. Indeed, they should be seen as complementary materials; there are many applications (for example, blast loading and seismic loading) in which the fibers can be used effectively in conjunction with the conventional reinforcement.

There is not the space here to describe all of the properties and applications of FRC. Instead, only a few of the more recent developments will be discussed.

Impact and blast protection

There is no doubt that fiber reinforcement is one of the most effective means of enhancing the impact and blast resistance of concrete (Banthia *et al.*, 2003, 2004a). Concrete is often exposed to dynamic events, at strain rates far higher than those obtained in the standard quasi-static tests. Typical strain rates for these events are:

Fast-moving traffic:	$10^{-6} - 10^{-4}\ s^{-1}$
Gas explosions:	$5 \times 10^{-5} - 5 \times 10^{-4}\ s^{-1}$

Earthquakes:	$5 \times 10^{-3} - 5 \times 10^{-1}\ s^{-1}$
Pile driving	$10^{-2} - 10^{0}\ s^{-1}$
Aircraft landing	$5 \times 10^{-2} - 2 \times 10^{0}\ s^{-1}$

Unfortunately, we are still unable to predict the behavior of concrete at high strain rates from the results of static tests. Proper impact tests are needed for this, but at the moment there is a complete absence of any standardized, agreed upon tests. It is therefore largely impossible to compare test results obtained in different laboratories using different test techniques and different methods of analysis. Until this problem is solved, it will be difficult to develop rational design equations for the use of FRC in dynamic loading situations.

However, despite the difficulties in quantifying the behavior of FRC under impact loading, we do know that fibers help enormously under this type of loading. This is an area in which further research is warranted, particularly given our need to develop protective barriers for structures and other high value assets.

High-performance fiber reinforced cement composites

Conventional fiber reinforced concretes typically display the stress vs. strain behavior shown in the upper part of Fig. 1.4 (Naaman and Reinhardt, 2003). That is, although they do exhibit some load-carrying capacity beyond the peak load, and are clearly less brittle than plain concrete, they may be classified as 'strain-softening' materials. They fail through the localization of a single major crack. On the other hand, high performance fiber reinforced cement composites (HPFRCC) display the behavior shown in the lower part of Fig. 1.4, strain-hardening accompanied by multiple cracking. These composites will therefore have high fracture energies, and should be of particular value in seismic and other forms of dynamic loading.

A special type of HPFRCC, termed engineered cementitious composites (ECC) has been developed by Li (Li 2003; Li and Stang 2004). These materials, which typically contain about 2% by volume of fibers, also exhibit multiple cracking and strain-hardening in tension, as shown in Fig. 1.5 (Li and Stang, 2004). In particular, the cracks that develop rapidly reach a maximum width; with additional strain, however, this width stabilizes at a very low value (60 μm in the example of Fig. 1.5), which is particularly important for reasons of durability.

While materials such as these are certainly expensive in terms of their first cost, they become highly competitive when life-cycle costs are considered. This is particularly true when both the social and environmental costs of premature repairs to our infrastructure are taken into account. These materials are, as yet, still primarily in the development stage, though a great deal of

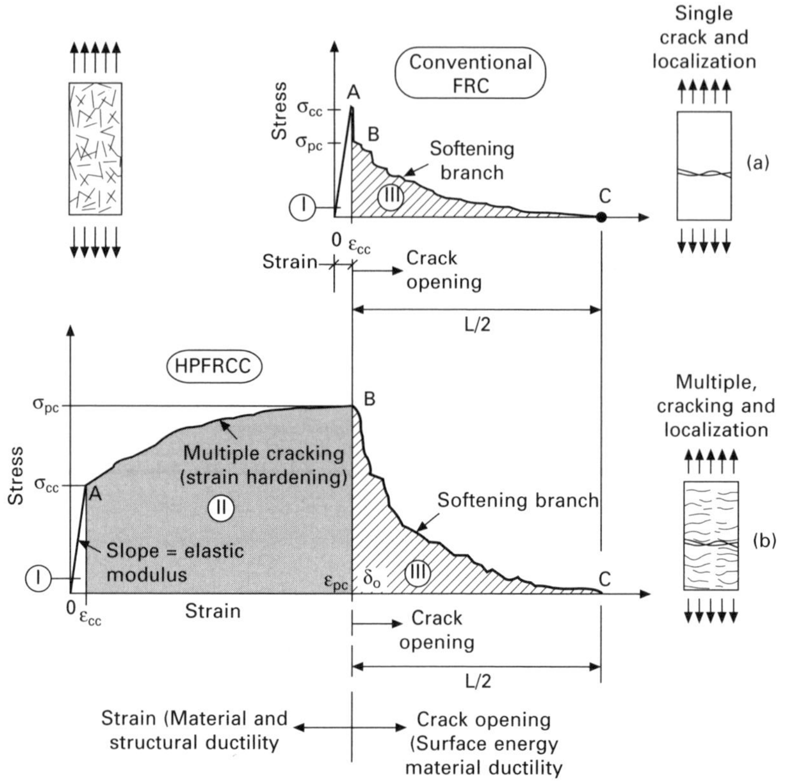

1.4 Typical stress vs. strain responses in tension of conventional FRC and high-performance fiber reinforced cement composites (from Naaman and Reinhardt (2003)).

research has already been carried out. It will take some time before they become a routine part of concrete engineering practice.

1.4.4 Self-compacting concrete

Self-compacting concrete (SCC) is a concrete that can flow on its own, without any vibration, to fill forms of virtually any shape, and to become fully compacted with no segregation or bleeding. This technology was first developed in Japan in the late 1980s, and has gained widespread acceptance, as it can be used to produce both ordinary concretes and high-strength concretes, and even fiber reinforced concretes (Banthia *et al.* 2004b). From a material point of view, SCC is considerably more expensive than ordinary concrete, but this is compensated for by the much greater ease of placement, and consequent reduction in labor requirements. It also permits proper concrete placement in areas of highly congested reinforcement, as occurs when seismic reinforcement must be used, or in other complex structures.

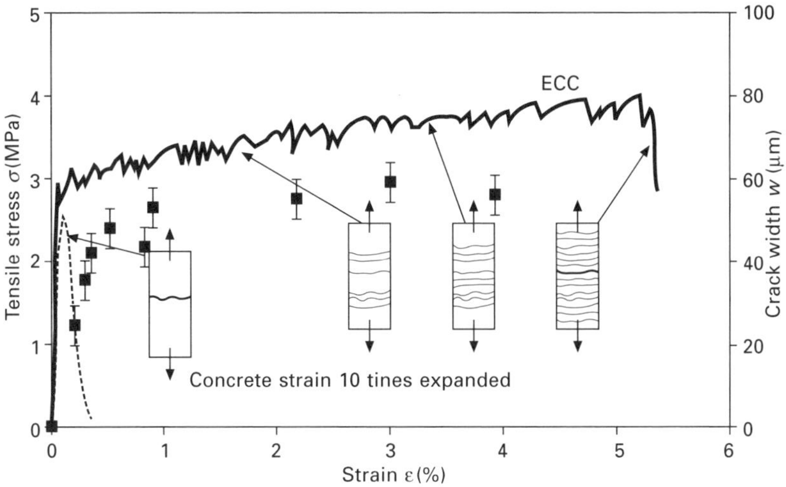

1.5 Tensile stress vs. strain response (continuous line) and crack width development (solid squares) of a typical ECC (from V. C. Li and H. Stang (2004)).

The concept behind the production of SCC is basically a simple one. The volume of fine material in the concrete is increased without particularly changing the amount of mixing water, combined with the use of relatively large amounts of superplasticizer. Typically, the ratio of fine to coarse aggregate in SCC is in the range of 1.0 to 1.3, compared to about 0.75 for ordinary concrete. In particular, there must be a relatively large amount of material with particle sizes below 0.125 mm; this is achieved by the use of appropriate supplementary cementing materials, as described above.

The fresh SCC must possess the following properties (Skarendahl and Petersson, 2000; EFNARC, 2002):

1. Filling ability, or the ability of the SCC to flow into the forms and around the reinforcing bars with no externally applied vibration, and to fill the formwork completely.
2. Passing ability, or the ability of the SCC to pass through narrow spaces in the formwork, or through and around reinforcing bars, without either blocking or segregating.
3. Segregation resistance, both while it is flowing and after it has come to rest.

These objectives can be met by the appropriate selection and proportioning of the concrete mixture. The required high fluidity can be achieved by severely limiting the coarse aggregate content, and by the use of superplasticizers; the superplasticizers also permit a reduced water/binder ratio. This reduced water/binder ratio in turn, combined with the limited coarse aggregate content, imparts a high segregation resistance. The high fluidity, combined with the segregation resistance, thus makes possible the production of SCC.

To achieve the appropriate fresh concrete properties, EFNARC (2002) suggests the following mix proportions as a first estimate:

- water/powder ratio (by volume) of 0.8 to 1.1
- total powder content of 400–600 kg/m^3
- coarse aggregate content of 28–35% of the mix volume
- water content less than 200 kg/m^3
- fine aggregate content to make up the remainder of the volume.

Of course, these values may have to be modified to meet the strength and durability requirements of the hardened concrete. If the fresh and/or hardened concrete properties are not met by this first estimate, then the mix must be adjusted, using different types or volumes of supplementary cementing materials or fine mineral fillers, using different types of superplasticizers, adding a viscosity modifying agent, and so on.

It should be noted that currently there are no well-defined or agreed upon procedures to determine the rheological properties of SCC. Thus, for mix design purposes or for quality control, a number of completely *ad hoc* or empirical tests are used, such as those suggested by EFNARC (2002).

1.4.5 High-durability concrete

Traditionally, when dealing with concrete, engineers have focused primarily on strength, on the assumption that if the concrete was strong enough, it would also be durable enough. This assumption, unfortunately, is simply not true. The concepts of 'high strength' and 'high durability' are not synonymous; they are, in fact, quite separate. There is far more distress to concrete (and, alas, far more litigation) due to poor durability than there is to low strength. In designing high-performance concrete (or any other concrete), it would be more appropriate to focus first on the environmental conditions to which the concrete will be exposed, and only then to worry about strength. Fortunately, at least some modern design codes are beginning to require such an approach. For instance, the American Concrete Institute Building Code (ACI Committee 318, 2002) is now written to indicate that durability requirements shall take precedence over strength requirements.

The first line of defense against any form of external attack (due to sulfates, deicing salts, seawater, industrial wastes and so on) is to produce a concrete with a low ratio of water to cementitious materials (w/cm). This reduces the porosity and permeability of the concrete, making the ingress of aggressive chemicals much more difficult (Gjorv, 1994). Thus, it is important to emphasize again that it is the w/cm ratio, and not the strength that governs the permeability, and hence the durability, of concrete. The lower w/cm ratio is also likely to lead to higher strengths, perhaps higher than are needed from a purely 'structural' point of view, but it is the durability requirements that must prevail. Presumably, a skilled structural designer will find a way to make use of this 'extra' strength.

Of course, permeability is not the only factor that controls durability. We must always keep in mind the other well-known durability requirements: sulfate resistant cements, air entrainment for freeze-thaw durability, low alkali cements when the aggregates are susceptible to alkali-aggregate reactions, and so on. However, even in these cases, a low w/cm ratio will help to mitigate the severity of attack. It is also essential to recognize that the concrete 'skin' (the outermost 5–10 mm of the concrete surface) generally has different properties from the 'bulk' concrete in the interior of a concrete section. Because of the 'wall effect' due to the formwork, aggregate particles are less closely packed at the concrete surface, leading to a region which tends to be richer in cement paste. If the concrete is not cured properly, this will increase the severity of surface cracking due to early drying shrinkage. This problem becomes increasingly severe as the original w/cm ratio is reduced (that is, as we enter the realm of high-performance concretes).

Abrasion resistance

The abrasion resistance of concrete is an important parameter in highway pavements, dam spillways and stilling basins, and so on. In some areas, abrasion due to ice is also of importance. Abrasion resistance is one of the few durability parameters that is in fact almost entirely proportional to the strength of the concrete, though the coarse aggregate properties and volume concentration are also of importance. Thus, high-strength concretes, particularly those made with the incorporation of silica fume, are particularly abrasion resistant. It has also been suggested that the incorporation of steel fibers into a low w/c ratio mix can improve the abrasion resistance, but there have been mixed experiences using this technique.

It has been found that at compressive strengths of about 120–150 MPa, concrete has about the same abrasion resistance as granite. In general, for the same water/binder ratios, the coarse aggregate properties are the determining factor in abrasion resistance. For instance, Laplante *et al.* (1991) found that granitic aggregates led to a more abrasion resistant concrete than did dolomite or limestone aggregates. However, there is still no standard test which can be used to predict the abrasion resistance of any particular aggregate, though tests such as the Los Angeles abrasion test may be helpful at least for screening purposes.

Freeze-thaw resistance

It is sometimes suggested that, because of their low w/cm ratios, very-high-strength concretes need not be air entrained in order to provide adequate resistance against repeated cycles of freezing and thawing. However, though the mechanisms of freezing and thawing in concrete are now pretty well understood (Pigeon and Pleau, 1995), this remains an open question, and the experimental studies to date have shown mixed performance. Some low w/c ratio mixes perform well under freeze-thaw cycling, while others do not. In addition, there is an obvious contradiction in wanting to entrain air into high-strength concretes, since every 1% of air entrainment will reduce the compressive strength by about 5%.

The experimental evidence on the necessity of using air entrainment in high-strength mixes has been summarized by Aitcin (1998). Basically, for concretes made with w/cm ratios of less than 0.25, it would appear that air entrainment is not necessary. For mixes with w/cm > 0.30, air entrainment should be used as for ordinary concretes. In the intermediate range, the necessity of air entrainment will depend on the particular cement, the presence of supplementary cementing materials, and so on, and this can be determined only by carrying out appropriate freeze-thaw tests. (In the view of the author, however, it would be prudent to conduct the appropriate freeze-thaw tests on all high-strength concretes, even those for which the w/c < 0.25).

Closely related to freeze-thaw resistance is the issue of salt scaling, that is, damage due to repeated applications of de-icing salts in freezing conditions. While the precise mechanisms underlying salt scaling are not understood, it is known that scaling is most likely to occur on surfaces that have been overvibrated, troweled too early and too long, subjected to plastic shrinkage, or where excessive bleeding has occurred. Clearly, then, the curing and finishing procedures will have a considerable influence on scaling resistance. Fortunately, care in this regard is always mandatory for high-strength concretes, and so these are generally found to be highly resistant to scaling. Scaling resistance is largely controlled by the w/c ratio, as is freeze-thaw resistance. However, it has been found (Aitcin, 1998) that it is generally easier to make a scaling resistant high-strength concrete than one that is resistant to freezing and thawing.

Corrosion of steel in concrete

The corrosion of steel in concrete is probably the single most expensive durability problem facing our concrete infrastructure. The mechanisms of steel corrosion in concrete are by now well known and understood (Bentur *et al.* 1997). Whether it occurs or not depends heavily on the quality of the concrete 'skin' as discussed earlier, and on the thickness of the concrete cover. However, a number of different 'materials' solutions to the problem of steel corrosion have been proposed over the years, all of which have both advantages and drawbacks.

Galvanized reinforcing bars have long been used to try to prevent steel corrosion. The zinc coating acts both as a barrier and as a sacrificial coating, as the zinc itself slowly oxidizes (corrodes). The effectiveness of the coating depends on its thickness, with its effective life expectancy being linearly proportional to its thickness. For instance, ASTM (1990) specifies two types of coatings, with thicknesses of about 85 μm and 150 μm. Thicknesses greater than about 200 μm may decrease the bond between the concrete and the rebar. While galvanizing is effective against corrosion induced by carbonation of the concrete cover, it is apparently less effective when the corrosion is induced by chlorides (Bentur *et al.*, 1997). As well, if the coating is too thin, then it may break when the rebars are being bent or handled at the jobsite, which can lead to very rapid localized corrosion. Finally, galvanized steel cannot be welded. Nonetheless, despite their relatively high cost, galvanized rebars seem to find a ready market.

In epoxy coated reinforcing bars, the epoxy acts as a barrier to isolate the steel from an aggressive environment. Typically, the thickness of the epoxy coating is specified (ASTM 1993) to be in the range of 130 μm to 300 μm, with a specified maximum number of defects of various kinds. The epoxy must be flexible enough to permit the bar to be bent without rupturing. If the coating is ruptured, this can lead to severe localized corrosion and failure of the rebar. There have also been some instances of the epoxy coating de-bonding from the steel when used in warm marine environments. In addition to the high cost, the major problem with epoxy coating is that the bond between the concrete and the steel is substantially reduced, increasing the possibility of cracking, and requiring the use of larger anchorage and lap lengths. Stainless steel reinforcing bars are occasionally used in extreme exposure conditions. While they are very effective in preventing corrosion, their high cost severely limits their use.

Corrosion inhibitors of various types are now becoming increasingly common. These act not as barriers to aggressive agents, but as chemicals that reduce the corrosion of the steel. There are two types of corrosion inhibitors: Anodic inhibitors stabilize and reinforce the passivating film which forms on the steel surface in the high pH environment of concrete. Cathodic inhibitors are adsorbed onto the steel surface, where they act as a barrier to the reduction of oxygen (which is the principal cathodic reaction for steel in concrete). However, it must be remembered that corrosion inhibitors are effective only if used in otherwise good concrete; they are not a panacea for corrosion in poorly designed or placed concrete.

The most common anodic corrosion inhibitor is currently calcium nitrite, which acts essentially by increasing the level of chloride necessary to initiate corrosion (Berke 1991). Cathodic inhibitors are less effective than anodic ones. They are primarily amines, phosphates, zincates and phosphonates, which have the unfortunate side effect of severe set retardation at the high dosages required for them to provide effective corrosion control.

The measures to prevent steel corrosion noted above (coatings, corrosion inhibitors) should really be seen not as absolute solutions, but as means of providing additional protection to otherwise good quality concrete. This brings us to perhaps the most sensible and efficient means of avoiding most corrosion problems, the use of the high-performance concretes that are the subject of this discussion, that is, concretes with a low w/cm ratio, and perhaps incorporating supplementary cementing materials. The low permeability of such concretes should provide adequate corrosion protection.

In particular, for more extreme exposure to chlorides, a high-performance silica fume concrete should provide good protection. There are two reasons for this. A low water/binder silica fume concrete has very low permeability, making the ingress of chloride ions difficult. As well, such concretes have high electrical resistivity (about an order of magnitude higher than that of ordinary concrete), which reduces the chloride diffusion rate. If necessary, the silica fume concrete can be combined with one of the other chemical or mechanical means of reducing corrosion described earlier.

1.4.6 Polymer modified concretes

As stated earlier, plain concrete is a brittle material. Thus, if it could be modified by the introduction of polymers, which are ductile and relatively strong, a composite material which is both strong and tough would result. As well, the incorporation of polymers should reduce the permeability, and hence improve the durability of concrete. However, this turns out to be both difficult and expensive (about three to six times as expensive as plain concrete) in practice, and so polymer-cement composites are not now widely used. However, as both durability and repair considerations become more important, these materials will find an increasing market share.

Latex-modified concrete

The most common, and easiest, way of combining a concrete with a polymer is to add the polymer latex to the concrete during the mixing process. The result is latex-modified concrete, or LMC. The most common polymers for this purpose are vinyl acetate, styrene-butadiene, vinylidine chloride and acrylic esters. The resulting materials are used primarily as bonding agents, overlays and for patching.

A schematic of how the polymer film forms in LMC is given in Fig. 1.6 (Mindess *et al.* 2003). As the mixing water evaporates, flocculation of the latex particles occurs; the particles form a continuous film as further water is removed by the cement hydration reactions. (Note that, rather than extended moist curing, the concrete should be allowed to dry out after a day or two to permit film formation.) The polymer film forms in the capillary pores and

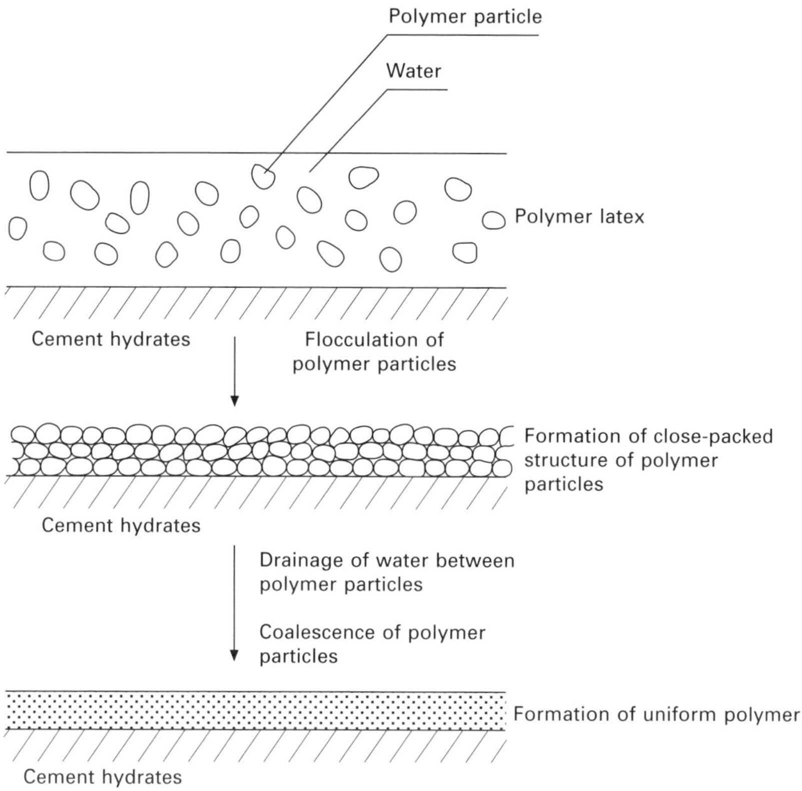

1.6 Schematic representation of polymer film formation in LMC (after Mindess *et al.*, 2003).

the cement paste-aggregate interface, thus enhancing the paste-aggregate bond, and limiting the penetration of water through the capillary pore system. The films thus formed are relatively strong and tough, and will thus increase the energy required to propagate cracks through the matrix.

The strength of LMC is higher than that of unmodified concrete under the same curing conditions in compression, and particularly in tension and flexure. LMC has a lower modulus of elasticity, but a higher strain at failure than plain concrete. Given that it is used primarily as an overlay for repair purposes, it is significant that LMC displays better bond both to old concrete and to reinforcing bars. On the other hand, it will tend to lose strength upon prolonged immersion in water. Perhaps of greater importance, LMC is more durable than plain concrete for several reasons. The polymer film reduces the permeability of the concrete. The improved resistance to tensile cracking reduces the number of cracks that can transport water into the material, and helps to keep them narrower by bridging across them (analogous to the way in which fibers work). Finally, the lower w/c ratio used in LMC concretes also leads to greater durability.

More recently, there has been an attempt to combine fibers with LMC. Work by Xu *et al.* (2004a, 2004b) has shown that such materials are superior in both strength and toughness to either LMC or FRC alone, and that they might well be used effectively both for repair purposes and where high-toughness materials are needed.

Polymer-impregnated concrete

Polymer-impregnated concrete (PIC) is prepared by impregnating the concrete with a liquid monomer, and then polymerizing the monomer *in situ* to form a solid polymer within the concrete pores. The polymerization can be induced either by using gamma radiation, or by the use of a catalyst and heat. The most common monomers used are methyl methacrylate (MMA), styrene, and the monomer of plexiglas, as they all have very low viscosities, and thus can be made fairly readily to penetrate the concrete pore system.

PIC can be much stronger (two to four times) than plain concrete, but is also more brittle. As with LMC, polymer impregnation strengthens both the hydrated cement paste and the cement-aggregate interface. More importantly, it too greatly improves the durability of the concrete, since the polymer effectively fills all of the available porosity, and prevents the ingress of aggressive chemicals. PIC is a very expensive material, and thus has seen little use in the field. Again, it will be used mostly in applications in which a long maintenance-free service life is desired, or where repairs would be difficult to undertake.

1.4.7 'Green' concrete

According to Mehta (2002), the concrete industry is currently the largest user of natural resources in the world, and thus has a considerable environmental impact. Each tonne of Portland cement requires about 1.5 tonnes of raw materials for its production, and consumes about 4000 to 7500 megajoules of energy. Portland cement production is also a major contributor of greenhouse gases, as each tonne of cement produced involves the release into the atmosphere of about one tonne of CO_2. The cement industry is now responsible for about 7% of global CO_2 emissions (Mehta 1999). Thus, for the concrete industry to remain 'sustainable', it must undergo some significant changes in the ways in which we specify, produce, use and recycle concrete (Holland 2002).

One approach to this problem is to use much greater proportions of fly ash (or granulated blast furnace slag) in concrete; this would be desirable for both economic and environmental reasons. There is thus currently a great deal of development of what is referred to as 'high-performance, high-volume fly ash concrete' (Malhotra 2002; Malhotra and Mehta 2002). Such concretes may be defined as:

- containing at least 50% fly ash by mass of the cementing materials
- having a Portland cement content of less than 200 kg/m^3
- having a water content of less than 130 kg/m^3
- having a water/cementing materials ratio of less than 0.35.

Because of their high pozzolanic content, these concretes reach their full strength potential rather more slowly than conventional concretes, which must be taken into consideration in the construction scheduling. On the other hand, the 91-day and 365-day strengths of such concretes may be 20% and 40%, respectively, higher than the 28-day strength (Malhotra, 2002), thus providing a considerable benefit for the durability of the concrete. It appears to make no particular difference whether the fly ash is added at the batching plant or is preblended with the cement. It is not clear what maximum strength can be attained with these concretes, but 28-day compressive strengths in excess of 40 MPa have certainly been obtained.

Properly produced high-volume fly ash concretes have a low permeability, and thus are highly durable. They are also more resistant to cracking than ordinary concretes of the same strength. The reasons for this may be explained by comparing the two typical mixes shown in Table 1.5 (Mehta 2002). The high-volume fly ash mix has both a lower w/cm ratio, and a substantially lower paste volume (23.4% vs. 29.5%), though both mixes have about the same 125–150 mm slump. For both of these reasons, the drying shrinkage and the heat of hydration are substantially reduced, leading to less likelihood of cracking for the fly ash concrete. However, it must be emphasized that even more than for ordinary concretes, proper curing of such high-volume fly ash mixes is essential. It is recommended (Malhotra and Mehta 2002) that there be at least seven days of moist curing, at temperatures in excess of 10 °C, for high-volume fly ash concrete to achieve its full potential.

Table 1.5 Comparison of mix proportions for 25 MPa concrete (adapted from Mehta, 2002)

	Conventional concrete		HVFA concrete	
	By mass (kg/m^3)	By volume (m^3)	By mass (kg/m^3)	By volume (m^3)
Cement	307	0.097	154	0.049
Fly ash	–	–	154	0.063
Water	178	0.178	120	0.120
Entrapped air (2%)	–	0.020	–	0.020
Coarse aggregate	1040	0.387	1210	0.449
Fine aggregate	825	0.318	775	0.299
Total	2350	1.000	2413	1.000
w/cm	0.58	–	0.38	–

While the current focus is on fly ash and blast furnace slag, which can be used as cement replacements, there are other waste materials that can be incorporated into concrete. These are primarily considered as replacements for conventional natural aggregates. (Remember that aggregates make up from 70 to 80% of the concrete volume.) There are many sources of such potential aggregates: mineral wastes, slags, incinerator residues, recycled concrete, and so on. However, there are three main considerations that must be kept in mind when evaluating such waste materials:

1. Economy: how much material is available, how far will it have to be transported, how much beneficiation will it require, and how will it affect the basic concrete mix design?
2. Compatibility with other materials: will it react adversely with other components of the mix (cement, admixtures)?
3. Concrete properties: will the material decrease the strength, the elastic modulus, or the volume stability (shrinkage, creep) of the concrete?

1.5 Conclusions

The foregoing discussion has dealt with the properties of some of the major classes of modern advanced concrete. As stated earlier, the intent was to describe some of the basic principles, and to illustrate the range of possibilities; it was not intended to be an exhaustive account of all of the concretes that can be considered to be 'high performance'. With our current understanding of basic principles, and with the materials and production technologies now available, we can produce an enormous range of Portland cement-based materials. We can largely 'tailor-make' concrete for virtually any application. The strength and durability levels that we can now achieve will almost certainly be surpassed in the future; we are limited only by our imaginations.

1.6 Sources of further information

Advanced, or high-performance, concretes are discussed in greater or lesser detail in most modern concrete textbooks, such as Mindess *et al.* (2003). The *Concrete Construction Engineering Handbook* edited by Nawy (1997) provides a broad overview of all aspects of concrete construction. By far the best single book devoted to high-performance concretes is that by Aitcin (1998). The collection of papers edited by Shah and Ahmad (1994) also contains much useful information. For fiber reinforced concretes, the most complete treatment is to be found in Bentur and Mindess (1990). For durability issues, the book by Richardson (2002) provides a comprehensive overview. Taylor's (1997) book on cement chemistry provides a detailed description of the chemistry of cement hydration, and is essential reading for a truly fundamental understanding of concrete behavior. The many publications and committee reports of the American Concrete Institute provide the most up-to-date information on concrete from a North American perspective. Similarly, the various RILEM publications are oriented more to a European perspective. Of course, any practitioner must make reference to the national codes and standards relevant to any particular country.

1.7 References

Aarup, B., (2004), 'CRC – a special fibre reinforced high performance concrete', in *Advances in concrete through science and engineering,* Bagneux, France, RILEM Publications, CD-ROM Paper 13, Hybrid-Fiber Session.

ACI Committee 318 (2002), *Building code requirements for reinforced concrete*, Farmington Hills, MI, American Concrete Institute.

Aitcin, P-C., (1998), *High-performance concrete,* London and New York, E & FN Spon.

ASTM A767-90 (1990), *Standard specification for zinc coated (galvanized) steel bars for concrete reinforcement*, West Conshohocken, PA, American Society for Testing and Materials.

ASTM A775-93 (1993), *Standard specification for epoxy-coated reinforcing steel bars,* West Conshohocken, PA, American Society for Testing and Materials.

Bache, H.H., (1987), *Compact reinforced composite, basic principles*, CBL Report No. 41, Denmark, Aalborg Portland.

Banthia, N., Bindiganavile, V. and Mindess, S., (2003), 'Impact resistance of fiber reinforced concrete: a progress report', in Naaman, A.E. and Reinhardt, H.W. *High performance fiber reinforced cement composites (HPFRCC4)*, Proceedings PRO 30, Bagneux, France, RILEM Publications, 117–131.

Banthia, N., Bindiganavile, V. and Mindess, S., (2004a), 'Impact and blast protection with fiber reinforced concrete', in di Prisco, M., Felicetti, R. and Plizzari, G.A. *Fiber-reinforced concretes BEFIB 2004*, Vol. 1, Bagneux, France, RILEM Publications, 31–44.

Banthia, N., Majdzadeh, F. and Mindess, S., (2004b), 'Hybrid fiber reinforced SCC for repairs of reinforced concrete beams', in La Tegola, A. and Nanni, A. *Proceedings of the first international conference on innovative materials for construction and restoration,* Vol. 2, Italy, Liguori Editore, 275–284.

Bentur, A., (2002), 'Cementitious materials – nine millenia and a new century: past, present, and future', *J. Mat. Civ. Eng.*, **14** (1), 2–22.

Bentur, A. and Mindess, S., (1990), *Fibre reinforced cementitious composites*, London and New York, Elsevier Applied Science.

Bentur, A., Diamond, S. and Berke, N.S., (1997), *Steel corrosion in concrete,* London and New York, E & FN Spon.

Berke, N.S., (1991), 'Corrosion inhibitors in concrete', *Concr. Int.,* **13** (7), 24–27.

Burg, R.G. and Ost, B.W., (1994) *Engineering properties of commercially available high-strength concretes (including three-year data)*, Research and development bulletin RD104, Skokie, IL, Portland Cement Association.

Charron, J.P., Denarie, E. and Bruhwiler, E., (2004), 'Permeability of UHPFRC under high stresses', in *Advances in concrete through science and engineering,* Bagneux, France, RILEM Publications, CD-ROM Paper No. 12, Hybrid-Fiber Session.

EFNARC (2002), *Specifications and guidelines for self-compacting concrete*, United Kingdom, EFNARC.

Gjorv, O.E., (1994), 'Durability', in Shah, S.P. and Ahmad, S.H. *High performance concretes and applications*, Edward Arnold, 139–160.

Haneharo, S. and Yamada, K., (1999), 'Interaction between cement and chemical admixture from the point of cement hydration, adsorption behavior of admixture and paste rheology', *Cem. Concr. Res.* **29**, 1159–1166.

Holland, T.C., (2002), 'Sustainability of the concrete industry – what should be ACI's role?', *Concr. Int.*, 24 (7), 35-40.

Laplante, P., Aitcin, P-C. and Vezina, D., (1991), 'Abrasion resistance of concrete', *J. Mat. in Civil Eng.* **3** (1), 19–28.

Li, V.C., (2003), 'On engineered cementitious composites (ECC) – a review of the material and its applications', *J. Advanced Concr. Technology* **1** (3), 215–230.

Li, V.C. and Stang, H. (2004), 'Elevating FRC material ductility to infrastructure durability', in di Prisco, M., Felicetti, R. and Plizzari, G.A. *Fibre-reinforced concretes BEFIB 2004,* Proceedings PRO 39, Bagneux, France, RILEM Publications, 171–186.

Malhotra, V.M., (ed.) (1987), *Supplementary cementing materials for concrete,* Ottawa, Minister of Supply and Services Canada.

Malhotra, V.M., (2002), 'High-performance high volume fly ash concrete', *Concr. Int.*, **24** (7), 30–34.

Malhotra, V.M. and Mehta, P.K., (2002), *High performance, high volume fly ash concrete,* Ottawa, Supplementary Cementing Materials for Sustainable Development, Inc.

Malinowski, R. and Garfinkel, Y., (1991), 'Prehistory of concrete', *Concr. Int.*, **13** (3), 62–68.

Mehta, P.K., (1999), 'Concrete technology for sustainable development', *Concr. Int.,* **21** (11), 47–53.

Mehta, P.K., (2002), 'Greening of the concrete industry for sustainable development', *Concr. Int.*, **24** (7), 23–28.

Mindess, S., Young, J.F. and Darwin, D., (2003), *Concrete*, 2nd edn, Upper Saddle River, NJ, Prentice Hall.

Naaman, A.E. and Reinhardt, H.W., (2003), 'Setting the stage: toward performance based classification of FRC composites', in Naaman, A.E. and Reinhardt, H.W. *High performance fiber reinforced cement composites (HPFRCC4)*, Proceedings PRO 30, Bagneux, France, RILEM Publications, 1–4.

Nawy, E.G., (ed.) (1997), *Concrete construction engineering handbook,* Boca Raton, CRC Press.

Parant, E. and Rossi, P., (2004), 'A new multi-scale cement composite for civil engineering and building construction fields', in *Advances in concrete through science and engineering,* Bagneux, France, RILEM Publications, CD-ROM Paper No. 14, Hybrid-Fiber Session.

Pigeon, M. and Pleau, R., (1995), *Durability of concrete in cold climates,* London and New York, E & FN Spon.

Richard, P., (1996), 'Reactive powder concrete: a new ultra-high-strength cementitious material', in de Larrard, F. and Lacroix, R. *Fourth international symposium on the utilization of high strength/high performance concrete*, Vol. 3, France, Presse de l'Ecole Nationale des Ponts et Chausées, 1343–1349.

Richard, P. and Cheyrezy, M.H., (1994), 'Reactive powder concrete with high ductility and 200–800 MPa compressive strength, in Mehta, P.K. *Concrete technology past, present and future,* SP-144, Farmington Hills, MI, American Concrete Institute, 507–518.

Richardson, M.G., (2002), *Fundamentals of durable reinforced concrete*, London and New York, Spon Press.

Romualdi, J.P. and Batson, G.B., (1963), 'Mechanics of crack arrest in concrete', *J. Eng. Mech. ASCE* **89** 147–168.

Russell, H.G., (1999), 'ACI defines high-performance concrete', *Concr. Int.* **21** (2) 56–57.

Sakai, E., Yamada, K. and Ohta, A., (2003), 'Molecular structure and dispersion-adsorption mechanisms of comb-type superplasticizers used in Japan', *J. Advanced Concr. Tech.* **1** (1) 16–25.

Shah, S.P. and Ahmad, S.H., (1994), *High performance concretes and applications*, Edward Arnold.

Skarendahl, A. and Petersson, O., (eds) (2000), *Self-compacting concrete*, RILEM Report 23, Bagneux, France, RILEM Publications.

Taylor, H.F.W., (1997), *Cement chemistry*, London, Thomas Telford.

Thibaux, T., Hajar, Z., Simon, A. and Chanut, S., (2004), 'Construction of an ultra-high-performance fibre-reinforced concrete thin-shell structure over the Millau viaduct toll gates', in di Prisco, M., Felcetti, R. and Plizzari, G.A. *Fibre-reinforced concretes BEFIB 2004,* Vol. 2, Bagneux, France, RILEM Publications, 1183–1192.

Xu, H., Mindess, S. and Fujikake, K., (2004a), 'Post crack flexural performance of high strength fiber-reinforced concrete with latex-modification', in Maultzsch *Proceedings, 11th International Congress on polymers in Concrete, ICPIC '04*, Berlin, BAM, 445–460.

Xu, H. and Mindess, S., (2004b), 'The flexural toughness of high strength fiber reinforced concrete with styrene-butadiene latex', in Zingoni A., *Progress in structural engineering mechanics and computation,* (Proceedings of the second international conference), Cape Town, A.A. Palbema Publishers, CD-ROM Chapter No. 246.

2
Advanced steel for use in civil engineering

C W ROEDER, University of Washington, USA and
M NAKASHIMA, Kyoto University, Japan

2.1 Introduction

Steel has been used as a construction material for civil engineering systems since the late 19th century. Prior to about 1890, cast and wrought iron components were used as columns and other elements in some civil engineering structures, but these structural elements generally had variable resistance, somewhat unpredictable failures, and limited ductility when compared to modern steel structures. During the 1890s, cast and wrought iron columns disappeared from common usage, and structural steel became the construction material of choice. This occurred because of improvements in the Bessemer process that permitted economical production of large quantities of steel required for building and bridge construction. Structural steel provided high strength with great ductility and consistent material behavior. These properties resulted in increased usage of steel in construction, but initial development of these steel structures followed construction practices employed with cast iron and wrought iron elements. Standard codes and specifications were not available until the 1920s (AISC 1928, ICBO 1927, AASHO 1931), and so the early structures were designed by standard practice and observations of the performance of past construction.

Experienced structural engineers designed buildings and bridges based upon simple calculations and nominal stress levels that had evolved from cast and wrought iron elements (Schneider 1905). The cost of structural steel was relatively high, while the cost of labor was relatively low, and so steel was produced as plate, bars and angles, and most structural elements were built-up members assembled from these limited shapes as depicted in Fig. 2.1(a). Built-up members and connections between members were riveted, and the construction was labor intensive. The built-up elements were light and required small amounts of material, but they were cast in concrete or encased in masonry for fire protection. This fire protection provided increased strength and stiffness to the completed structure. The increased strength and stiffness were not calculated by the structural engineer in the design, but they clearly contributed to the structural performance. Nominal design stresses in the order of 110 MPa were employed, and the nominal yield stress of the most common structural steels was 205 MPa during this period.

During the 1920s, steel design practice began a significant change in much of the world, because of increased labor costs. Local customs and economic conditions delayed this evolution in some countries, and a few countries retain remnants of the earlier practices today. The first design specifications for buildings and bridges were developed (AISC 1923, AASHO 1931, ICBO 1927), and standardized hot rolled shapes similar to the wide flanges and channels commonly used in modern construction evolved. Standard connections such as those depicted in Fig. 2.1(b) were employed, because

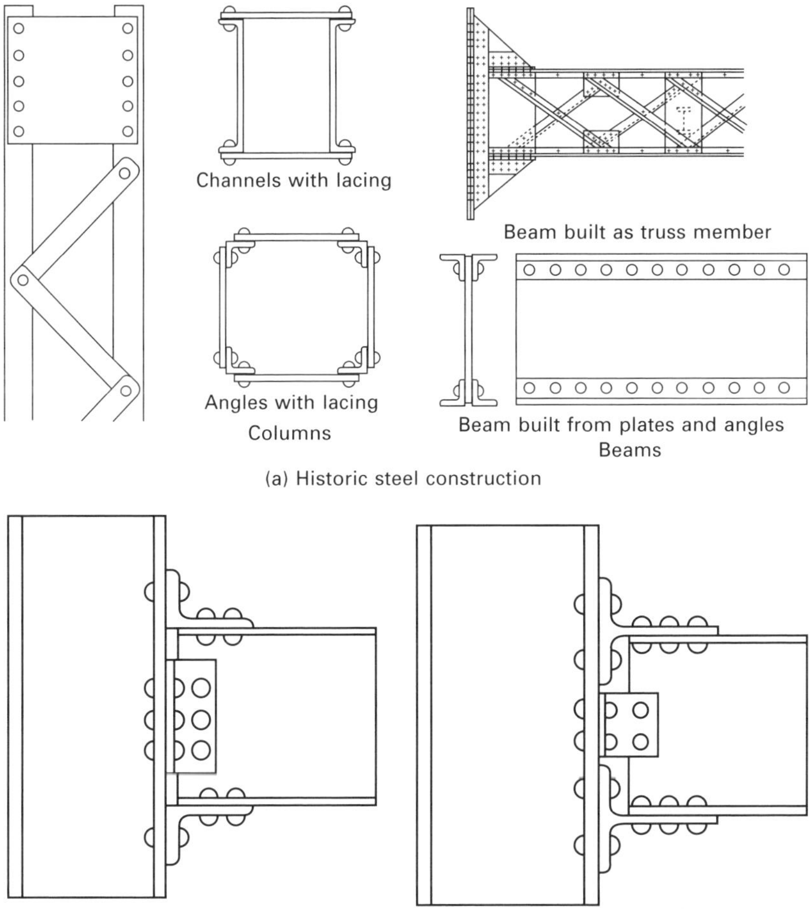

(a) Historic steel construction

(b) Riveted construction with standard members and connections

2.1 Historic steel frame construction; (a) built-up members typical of early 1900s, (b) riveted connections typical of period 1920 through 1960.

they required significantly less labor and fewer rivets than earlier construction. These early specifications employed the allowable stress method, and they generally employed permissible stress levels at approximately 125 MPa for most structural applications. Standard structural shapes such as I beams, H beams and channels reduced the labor required to assemble the structure. Further, improvements in the manufacturing process of steel produced further savings in cost and increased usage of the material.

Riveted connections and concrete encasement for fire protection were labor intensive construction practices. Increased labor costs combined with the large mass and dead load caused by the heavy concrete encasement led to further changes in engineering practice during the 1950s and 1960s. High strength bolts replaced rivets, because a much smaller number of bolts were required, and this reduced the labor costs and construction noise. Lightweight fire protection and nonstructural elements were developed to reduce the weight of the structure and further reduce labor costs. Increasing emphasis on seismic design and ultimate strength design methods led to more accurate estimates of structural resistance and ductile structural performance. Electric arc welding had been developed many years earlier, but it became commercially viable during this period, since welds could be economically completed with more uniform quality and fewer flaws. Welding permitted greater structural continuity and redundancy, and this further reduced material costs and increased the available resistance within the system. As a result, high strength bolts combined with electric arc welding became the normal construction techniques for connecting and joining steel elements. Factored load and plastic design methods evolved during this period to take advantage of the increased strength obtainable in indeterminate structures (AISC 1963). Lightweight fire protections and nonstructural elements reduced the dead loads, and increased yield stress levels (245 MPa in the 1960s) and allowable stress levels (commonly 155 MPa to 170 MPa in the 1960s) produced lighter and bigger steel structures.

These trends have continued into the current practice. Today, civil engineering construction is a continual economic competition between steel, concrete and other construction materials. As a result, steady increases in the strengths of these materials have occurred. Today the standard mild steels used in civil engineering construction have nominal yield stress values of approximately 345 MPa. The equivalent allowable stress levels for these steels are approximately 205 to 220 MPa. Operating stress levels are approximately twice as large as those used when steel was first employed. Increased operating stress levels increase the potential for stability failures, fatigue and brittle fracture. Therefore structural engineers must respond to these design issues, and building codes and specifications have generally become more complex. Labor costs have continued to increase, and engineers are today asked to create light, economical and quickly erected structures.

Some of these applications are in aggressive environments such as salt water where corrosion, corrosion protection and deterioration become increasing concerns. Nevertheless, greater material strength, higher strength to weight ratio, and greater available ductility result in steel dominating the market for bigger, taller, and longer span civil engineering structures. Steel is particularly dominant in the design and construction of bigger structures and structures with unusual geometry because of its large strength to weight ratio and the ability to economically fabricate and erect complex structures. The steel industry is continually pressed to develop improved material performance, higher strength, and more ductile materials and structural systems, because these attributes enhance this evolution.

2.2 Issues of concern

The prior discussion of the evolution of the application of steel in civil engineering practice raises several dominant issues for engineering design, research and development of steel structures. These issues include:

- *Fire protection.* Great advances have been made in lighter and more economical fire protection methods, but fire protection is still viewed as an issue of greater concern for steel structures than for other construction materials, because of the smaller members and thermal mass associated with steel structures. Fire protection adds to the structural cost, and engineers must work toward reducing these costs, while assuring adequate resistance to elevated temperatures expected during a fire.
- *Buckling.* Steel structures have smaller members and larger operating stress levels than other construction materials. Buckling and stability become more critical as higher strength materials evolve.
- *Fatigue and fracture.* Crack initiation, crack growth and propagation were not considered issues of importance in structural engineering until the last 40 or 50 years. However, the increasing yield stress, operating stress levels, emphasis on plastic and ultimate capacity, and use of welded construction have resulted in increased frequency of fatigue and fracture in bridges and industrial systems. Fatigue is presently not a major concern for building design, but fracture of steel in buildings has become a major issue for seismic design after recent earthquakes.
- *Corrosion resistance.* Steel may corrode when exposed to the environment, and this may lead to deterioration, increased maintenance, and increased construction costs. Galvanization, paint, and coatings inhibit corrosion, but they may increase the fabrication costs of the steel by several percent, Therefore, engineers are continually seeking methods of reducing these costs.
- *Weldability.* Welded construction is commonly used, because it results in stiffer, stronger structures with reduced building weight and material

requirements. Increased steel yield strength places new and continuing demands upon welding methods, because higher strength steels are usually more difficult to weld without adversely affecting the ductility and performance of the system.

- *Ductility for seismic and other loading.* Seismic design is today a requirement for many civil engineering structural systems, and steel is an ideal material for seismic design because of its material strength, stiffness, and ductility. Buckling, weldability and other issues may affect the seismic performance and ductility, and engineers are continually developing new methods to improve inelastic seismic performance of steel structures.

These factors influence the choice and consequences of using steel as a construction material, and they are often the focus of developments that are made to improve the performance of steel.

2.3 New developments

Problems and concerns of the day frequently provide the focus of future new innovations in engineering practice. As a result, many recent advances and developments in steel respond in part to the recurring issues noted above. A number of recent advances are discussed here in terms of developments in new materials, new components and new structural systems. The combination of these developments provides a more versatile and more economical construction material.

2.3.1 New materials

New steel alloys have been developed in recent years to provide greater economy and improved structural performance from steel construction. In general, these alloys address the issues noted above. Some of these material developments are discussed here.

HPS bridge steels

High performance steels (HPS) have been developed (Azizinamini *et al.* 2004) to provide higher yield strength while assuring weldability, ductile performance and good fatigue resistance for bridge construction. HPS steels are quenched and tempered low alloy plate steels manufactured under ASTM 709 (ASTM 709 2004), and they were developed to provide good ductility, Charpy Notch (CVN) toughness and weldability at relatively high yield stress. Data presented by Dexter and others (2004) suggest that HPS steels may have an upper shelf CVN toughness that may be 25% larger and a transition temperature up to 40 °C lower than achieved with the normal A572 bridge steels. Several hundred bridges have been designed or built with this material since its inception, and increased usage in the future is expected. The HPS70W alloy is commercially available from numerous suppliers and it has a nominal yield stress of 485 MPa. Other HPS alloys with yield stress up to 695 MPa have been developed, but they have not been produced in commercial quantities to date. Higher strength welds are more costly and frequently have reduced CVN toughness and greater weld flaws than achievable with lower strength steel. Further, welded splices in bridge girders typically occur at regions where fatigue rather than ultimate strength is the critical design parameter. As a result, undermatched electrodes are usually used in HPS70 bridge design practice to provide economy combined with good fatigue resistance. HPS steels are often used as flanges in hybrid plate girders with lower strength (A572 grade 50 or 345 MPa yield strength) steel used in the webs.

HPS steels permit the design of lighter bridges with longer spans and more economical bridge construction. However, the higher operating stress levels require changes in bridge design practice. Increased stress levels result in more slender flanges and webs, and therefore lower slenderness limits and stockier elements are required to develop their full plastic capacity. The strength and ductility of the HPS70 alloy has been investigated, and revised design limits have been proposed to effectively use this higher strength steel in bridge design practice while assuring economical design and construction (Barth *et al.* 2004).

Some design specifications also require deflection limits for bridge serviceability and vibration control, and research (Roeder *et al.* 2004) has shown that HPS steel bridges are more likely to be controlled by these design deflection limits than are lower strength bridge steels. This research has also shown that deflection limits are not a desirable method for controlling bridge vibrations. Serviceability and durability are important, but the impact of deflections on bridge durability is not apparent. As a result, engineers are presently moving away from these deflection based design limits for HPS steels. They are directly addressing bridge vibration and pedestrian comfort issues, but there is still ambiguity regarding the role deflections and deformations play in determining the service life of the bridge. Increased operating stress levels also increase the propensity for fatigue cracking. HPS steels reduce the risk of fatigue at these higher stress levels by using undermatched weld metals at fatigue sensitive weld locations, improved fatigue resistant details, and the increased CVN toughness of the HPS materials.

While high strength steels raise new design concerns, HPS steels offer significant advantages for bridge construction, and bridge engineers are today considering new shapes and configurations that economically use the high strength steel while circumventing the fatigue, stability and serviceability limitations that may accompany the increased stress levels. In some cases, HPS steel tubes are being used, because tubes are inherently more stable against local buckling than other structural shapes. Therefore, tubular elements may be used as chords in arches or trusses, or in other systems where they can better utilize the benefits of the higher strength material. HPS steels may also be combined with concrete in hybrid and composite systems, because this reduces deflections, vibrations and local buckling potential, while providing an economical lightweight structure.

Fire-resistant steels

As illustrated in Fig. 2.2 (AIJ 1999), Young's modulus of steel remains relatively stable with increasing temperature (from room temperature up to 700 °C), but both the yield and tensile strength decrease steadily with increasing temperature. This means that buildings are susceptible to collapse during fires (Lie 1992); hence the fire-resistance of steels and the fire protection of steel are crucial considerations of structural design. The US International Building Code (IBC) (ICC 2003) and Japanese Standard Law of Building (BSL) (Kohno 2003) stipulate that major load carrying members such as columns and beams must be able to support gravity loads for a sufficient duration of time when exposed to fire conditions. The BSL stipulates two levels of allowable stresses for major load carrying members; one is the material's yield strength (F_y) for short-term loading and the other is two-thirds of F_y for long-term loading. The BSL requires adequate fire protection so that the temperature of these critical members must be kept not higher than 350 °C during two hours of fire based on the requirement for continuous resistance of the structure after fire. The IBC requires adequate fire protection to assure a duration of time where the members will support the required gravity loads while subject to a standard fire (ASTM E119 2000), and this duration varies depending upon the building type and function.

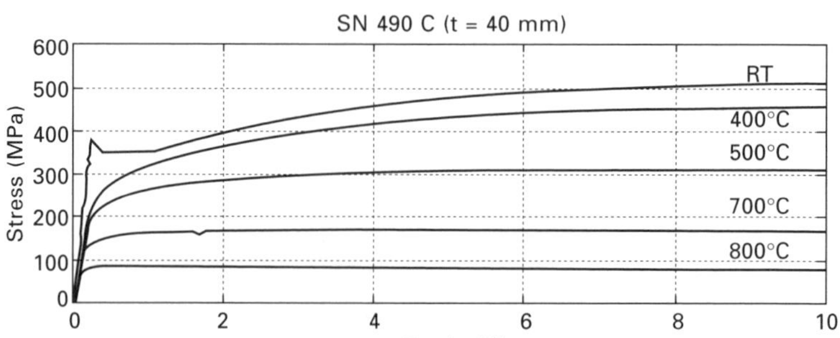

2.2 Effects of temperature on a material's stress–strain relationship (by courtesy of Japan Iron and Steel Federation).

Figure 2.3 shows a schematic of the variation of yield stress as a function of temperature for steel. The bold line is for conventional mild steel. This figure shows that conventional steels have a yield stress that is reduced to about two-thirds the yield stress in room temperature at a temperature of 350 °C. Fire-resistant steel was developed in Japan, with the aim of reducing or eliminating the amounts of costly fire-protection in some situations (Sakumoto *et al.* 1992, 1994). Addition of a few alloying elements of Mo, Nb, and Cr and strict control of heat-treatment processing was adopted to increase the yield stress in elevated temperature. The grey line of Fig. 2.3 shows the yield stress-temperature relationship of fire-resistant steel. The yield stress of this fire-resistant alloy is nearly constant up to about 400 °C, and the fire-resistant steel maintains a yield stress which is no less than two-thirds the yield stress at room temperature at temperatures up to 600 °C. Figure 2.4 shows examples of experimental relationships between the yield strength, tensile strength, and temperature for the 490 MPa grade of both conventional and fire-resistant steel (JISF 2003b). The yield strength of fire-resistant steel exceeds 210 MPA, the target value equal to two-thirds of yield stress at room temperature (i.e. 330 MPA) at this elevated temperature. The Young's modulus remains

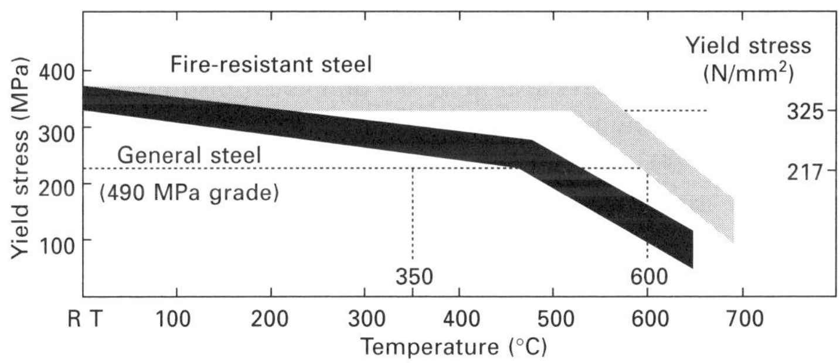

2.3 Schematic yield stress–temperature relationship (by courtesy of Japan Iron and Steel Federation).

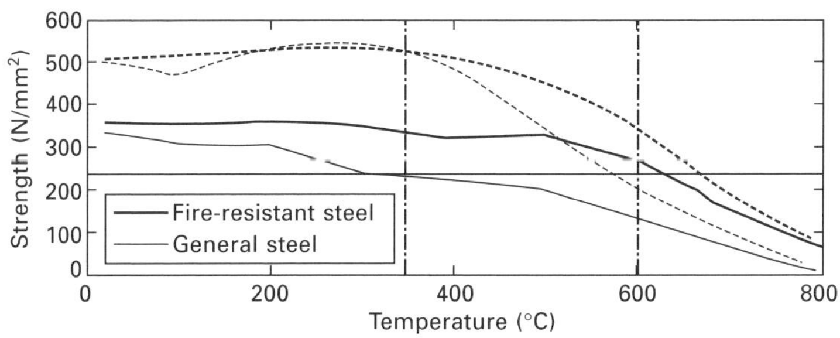

2.4 Example of variation of yield stress and tensile strength with respect to temperature (by courtesy of Japan Iron and Steel Federation).

relatively constant and maintains a value not smaller than 80% of the original Young's modulus at 600 °C.

Welding materials (for submerged arc welding and CO_2 arc welding) and high-strength bolts that conform to elevated temperature performance of fire-resistant steel were also developed. In the welding materials, the yield stress at 600 °C is no less than 250 MPa, which is larger than the target strength of 210 MPA stipulated for the base metal. The Charpy V-notch values are also higher than those obtained for conventional weld metals. High-strength bolts conforming to fire-resistant steel were developed for a common repertoire of high-strength bolts, namely torque-shear bolts, hexagon head bolts, and hot-dip galvanized bolts (JIFS 2003b, Sakumoto *et al.* 1993).

Fire-resistant steel has been commercially available for more than ten years in Japan. The steel is primarily used in parking structures, because of their open space and direct exposure to the atmosphere (JISF website). In many of these structures, fire-protection is no longer required once fire-resistant steel is adopted. While fire-resistant steels offer promise for reducing the cost of fire protection, their benefits are presently not available in many parts of the world, and their benefits may not be adequate for some applications. Therefore, economical fire protection remains a critical concern. Fire protection today is lighter and less costly than that used in past practice. Nevertheless, it is still a costly element of some steel construction, and engineers are continually searching for new methods to reduce the costs and to reduce the amount of fire protection required in the structural design (Ruddy *et al.* 2003). Reduced costs of fire protection have sometimes been achieved by using gypsum board and other materials that may serve a dual role as fire protection and an architectural element. However, another promising advance for the future appear to be in the direction of advanced analytical techniques.

Material fire ratings are commonly based upon a standard fire test such as ASTM E119 (2000), and the measured temperatures of actual fires may vary dramatically from these idealized temperatures as illustrated in Fig. 2.5. Actual fire temperature depends upon the type and available quantity of combustible materials, the available oxygen and ventilation, as well as the building application and geometry. The actual fire temperatures may exceed or be significantly less than the standard fire temperatures, because the elevated temperatures suggested by standard design fires cannot develop if there is inadequate fuel and oxygen supply to support the fire growth. Advanced computational methods permit more accurate estimates of the expected member temperatures and general structural performance during fires, and these analytical techniques show that significant reductions in the cost of fire protection may be possible in many future building designs (Milke 2002). These analyses require consideration of additional parameters from those needed for ordinary structural analysis, but the benefits of this analysis can

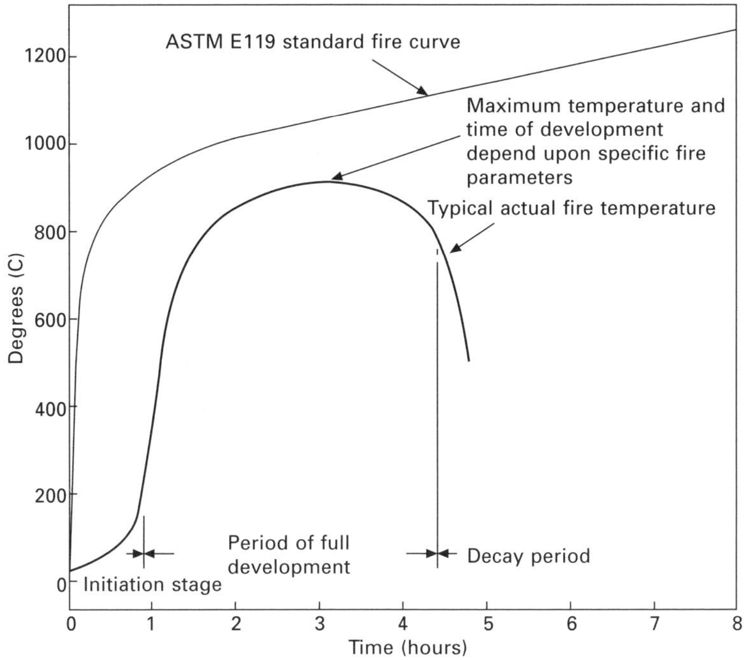

2.5 Comparison of typical measured fire temperature as function of time to Standard ASTM E119 Fire.

be improved structural safety and increased economy of steel building construction.

Steels developed for increased ductility

Steel is one of the strongest, most ductile construction materials, and light weight and ductility are particularly important for seismic design. However, the 1994 Northridge and the 1995 Hyogoken-Nanbu (Kobe) earthquakes resulted in considerable damage and nonductile performance of steel buildings despite the material ductility. These events showed that ductile materials do not automatically ensure ductile structures because the material must be used in ways that permit the ductile material properties to control system performance. As a result, the search for even greater ductility is always in progress. Comprehensive studies of this recent earthquake damage were completed, and significant changes have recently been required for seismic design of steel structures (AIJ 2001, FEMA 350 2001, AISC 2002). One contributing factor noted in these studies was that the continuing evolution toward higher yield strength steels as illustrated in Fig 2.6. This figure shows the average measured yield stress and tensile stress measured in steel test specimens reported in the literature of recent decades. It can be seen that

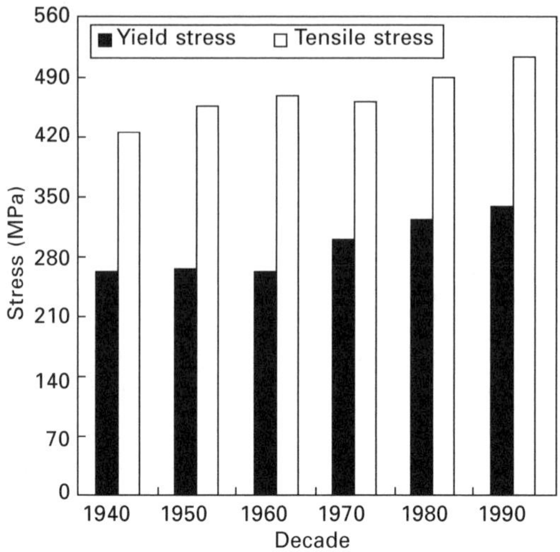

2.6 Evolution of measured material properties in mild steel in the US.

significant increases in yield stress have occurred over the past 30 to 40 years, but smaller increases in the ultimate tensile strength of steel are noted. This evolution has resulted in a steady increase in the yield to tensile strength ratio (F_y/F_u).

The ultimate tensile strength, F_u, is commonly associated with fracture or tearing of steel, and therefore an increased yield to tensile ratio reduces the amount of strain hardening reserve during seismic deformation of steel. This limits the extent of yielding during inelastic deformation and causes larger local inelastic strain demands on the locations where yield occurs (FEMA 355D 2000). As a result, greater ductility demands on welds and other critical elements may occur, and steels with higher yield to tensile strength ratios have a greater tendency to fracture during earthquake loading. Steel producers have responded to this need. In the US, A992 steel (ASTM 992 2004) has been developed to increase the nominal yield stress expected for structural steel shapes, and to limit the F_y/F_u ratio to less than 0.85.

In Japan, a higher performance steel was developed in the 1980s and adopted by the Japan Industrial Standard (JIS) in 1994 (JSA 1994). In comparison to conventional mild steel, the new steel, designated as SN steel, was intended to satisfy the following requirements.

1. To set an upper limit for the ratio of yield strength to tensile strength;
2. To reduce scatter of yield strength;
3. To lower carbon equivalent (C_{eq}) and weld crack sensitivity (P_{cm});
4. To increase through-thickness ductility;
5. To limit the tolerance of geometrical dimensions.

Requirements 1. and 2. are closely associated with ductility of steel members and frames. A small scatter of yield strengths is beneficial in securing the yielding, failure mechanism, and accordingly the ductility of the steel frame. Figure 2.7 shows comparisons in variations of yield and tensile strength between conventional mild steel and SN steel (Takanashi 2003). SN steel achieves significantly narrower bands relative to conventional steel. Further, the upper limit of the F_y/F_u ratio is set at 0.8 (for Classes B and C, as described later) for SN steel, and the allowable variation of yield strength is 120 MPa.

Requirement 3. is intended to assure good weldability, and requirement 4. is to secure strength for plates that sustain tensile stresses in the through-thickness direction. Requirement 5. aims at better constructability achieved by higher precision. Table 2.1 shows primary specifications of SN steel (JIS G 3136). Two different strengths: 400 MPa and 490 MPa (in terms of the tensile strength) and three classes, A, B, and C, are supplied. The minimum yield strength (235 MPa) and the upper and lower limit of the tensile strength are specified in Class A, while the upper and lower limits of both the yield and tensile strength and the upper limit for the F_y/F_u ratio (0.8) are specified in Classes B and C. Class C has higher performance in the through-thickness properties by controlling Sr to lower values (0.015% for Class B and 0.008% for Class C). Class B steel is used primarily for beams that sustain flexure. Class C steel is used primarily for diaphragm plates arranged in through-diaphragm connections between wide-flange beams and cold-formed box

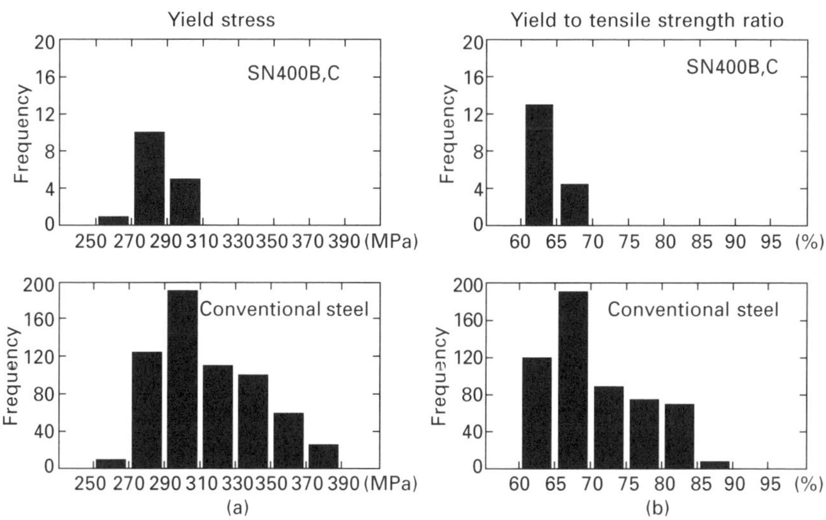

2.7 Variation of material properties for SN and conventional steel: (a) yield stress; (b) yield to tensile strength ratio (by courtesy of Japan Iron and Steel Federation).

Table 2.1 Major specifications of SN steel (JIS G 3136)

Requirement	Steel grade				
	SN400A	SN400B	SN400C	SN490B	SN490C
Yield strength (MPa)	235 min	235–355		325–445	
Tensile strength (MPa)		400–510		490–610	
Yield to tensile strength ratio (%)	–	80 max			
Charpy absorption energy (J)	–	27 min			
Though-thickness requirements (reduction of areas) (%)	–	–	25 min	–	25 min
Carbon equivalent and weld crack sensitivity (%)	C 0.24 max	C_{eq}0.36 max P_{cm}0.26 max		C_{eq} 0.46 max P_{cm} 0.29 max	

columns (as discussed later in this chapter and illustrated in Fig. 2.13(a)), while Class A steel is used for secondary members.

Development of SN steel was extensively carried out in the 1980s, and the steel was adopted in the Japanese Industrial Standard (JIS) in 1994. The Hyogoken-Nanbu (Kobe) earthquake occurred in 1995, in which many instances of welded beam-to-column connection failures were reported (Nakashima and Bruneau 1995, Nakashima *et al.* 1998). The use of SN steel accelerated significantly after the earthquake, and most major steel constructions in Japan have used SN steel in recent years.

Box sections are very commonly used in Japanese steel moment frames primarily because they have equal bending resistance in two orthogonal directions, and Japanese engineers make extensive use of three-dimensional moment framing. Unless they are very thick, box sections are cold-formed. Two strength grades of cold-formed box sections, 400 MPa and 490 MPa in tensile strength, have been available for many years. Two types of cold forming are adopted, i.e., roll-forming and press-forming. In roll-forming, the plate is initially formed and welded into a circular tube, and then the circular tube is press-formed into a box section. In press-forming, a flat plate is bent to form a C-section, and two C-sections are welded longitudinally to form a box section. Both processes involve significant cold working and local strain hardening of the steel. In the process of cold forming, corners of the cross-section sustain large plastic strains, and the ability to sustain elongation is decreased. When combined with welding, the corners become the weak spots that are susceptible to brittle fracture. Japan developed new grades of cold-form tubes (designated as BCR for roll-forming and BCP for press-forming) whose material conforms to the SN grade series. Available sizes are 200 mm (width) by 6 mm (thickness) to 550 mm by 22 mm for BCR and

200 mm by 6 mm to 1000 mm by 40 mm for BCP. The specified tensile strength for BCR is 400 MPa, while two strength grades, 400 MPa and 490 MPa, are available for BCP.

The use of energy dissipating dampers is becoming very popular in Japanese building construction. Viscous dampers, visco-elastic dampers, and hysteretic dampers are the common choices, but hysteretic dampers in which steel is used as the energy dissipating material is by far the most popular in steel building construction. According to statistics, all steel high-rise buildings constructed in Japan for the past few years have been equipped with some types of hysteretic dampers (BCJ 1995-2002). Among various types of hysteretic dampers, shear panel dampers and buckling restrained braces are most frequently adopted. Buckling restrained braces are commonly used as members that add damping and supplement stiffness in Japanese seismic design, and they are discussed in greater detail later in this chapter. The dampers are designed to take hysteretic energy even in medium earthquake events; hence they have to yield with a relatively small force. To ensure the relatively early yielding, low-yield (LYP) steel was developed in Japan. Figure 2.8 shows example stress–strain relationships of several steel grades, which include low-yield steel (designed as LYP100). Low-yield steel is characterized by a small initial yield strength (100 MPa in Fig. 2.1), a larger uniform elongation, and conspicuous strain hardening in cyclic loading condition (Nakashima 1995, Nakashima *et al.* 1994). Two grades of yield strength (100 MPa and 200 MPa) are commercially available, and they are used primarily as materials for hysteretic dampers, buckling restrained braces, and other specialty items.

Corrosion resistant steels

Steel corrodes when exposed to the environment, and the painting and protection of steels has long been a significant economic cost and occasionally a deterrent to the use of steel in exposed structural systems. A588 steel (ASTM A588

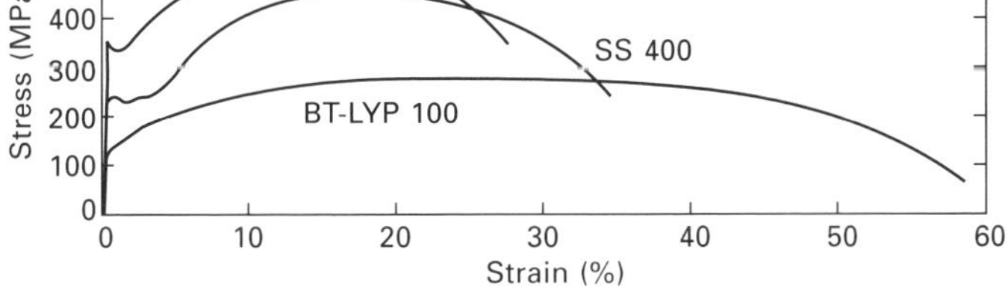

2.8 Variation of material properties for SN and conventional steel (after Nakashima *et al.* 1994).

2004) is a high-strength low-alloy structural steel that can be produced in shapes, plates, and bars for welded, riveted, or bolted construction. It is intended primarily for use in welded bridges and buildings where added durability is important. It has been used for many years in bridge construction. The atmospheric corrosion resistance of this steel in most environments is substantially better than that of carbon structural steels. When properly exposed to the atmosphere, this steel forms a patina which retards and delays further corrosion and increases the service life of most exposed structural elements. A588 steel has been shown to be particularly effective in environments where the steel remains dry for much of its service life, the patina layer is protected against wear or abrasion, and the local atmosphere does not include salt water or harsh chemical content (AISI 1995). In some cases, A588 steel is painted during fabrication to provide supplemental protection during early service life, and the patina layer forms after the initial paint coat wears away. This practice has been shown to extend the service life in most applications.

Other new steels for increased corrosion resistance have been developed by Japanese steel companies. The use of weathering steel started in the 1960s in Japan primarily for use in steel bridges. According to statistics, about 15% of existing steel bridges adopt weathering steel (Kihira *et al.* 1999, Kihira 2001). However, research has shown that basic weathering steel does not provide the intended performance when exposed to airborne salt; hence the applications of conventional weather steel are limited to environments with low deposition of airborne salt. In recent years, Japan has instituted a complete ban of studded tires in automobile production, and this forces the increased use of road de-icing salt in cold regions during the wintertime, which inevitably increases airborne salt. In consideration of these circumstances, a new type of weathering steel was developed to enhance salt corrosion resistance (JISF 2003c). The steel is nearly identical in chemical composition to conventional weathering steel except for the addition of 3% Ni. Three strength grades (i.e., 400 MPa, 490 MPa, and 570 MPa in tensile strength) and plate thickness up to 100 mm are commercially available. The new weathering steel has an enhanced capacity to inhibit the penetration of chloride ions into the steel substrate. This is achieved by means of ion exchange such that the chloride ions concentrate in the outer rust layers and the sodium ions are embedded in the inner rust layers. Table 2.2 shows examples of chemical compositions of the new weathering steel (designated as SMA400W-MOD; SMA490W-MOD; SMA570WQ-MOD) together with conventional weathering steel (designated as JIS G3114) (JISF 2003a). Welding material that contains the same level of Ni was also developed. To improve the crack sensitivity of weld metal, the carbon content of new weathering steel is reduced to 0.06% as shown in the second column of Table 2.2.

Field exposure tests have been conducted for the past ten years at several coastal locations. They verify the improved performance of the new weathering

Table 2.2 Actual chemical composition of 3%-Ni weathering steel

Grade	C	Si	Mn	P	S	Cu	Ni	Cr
SMA400 W-MOD	0.06	0.18	0.50	0.006	0.002	0.34	3.06	0.02
SMA490 W-MOD	0.06	0.20	0.85	0.005	0.002	0.32	3.05	0.02
SMA570 WQ-MOD	0.06	0.20	1.36	0.004	0.002	0.34	3.06	0.02
JIS G3114	<0.18	0.15 –0.65	<1.40	<0.035	<0.035	0.3 –0.5	0. 05 –0.3	0. 45 –0.75

steel against corrosion compared to conventional weathering steel. In one test location, the corrosion thickness reached 1.5 mm after six years for conventional weathering steel, while it remained no greater than 0.15 mm for new weathering steel. Applications of the steel for new bridge construction has increased during the past few years.

2.3.2 New components

While new steels have been developed in recent years, new ways of incorporating steel into buildings and bridges have also been employed. This subsection will describe some ways in which steel is effectively used to provide the greater benefits of the material. Often these are simply new ways of viewing and using existing materials.

Buckling restrained braces

Earthquakes induce inertial loading, and large earthquakes can cause internal forces that are much larger than those required for resistance of gravity loads. As a result, economical seismic design requires a balance of strength, stiffness and ductility, and steel structures are commonly used for seismic design. Concentrically braced frames (CBFs) are attractive for seismic design, because they provide more lateral stiffness and resistance than other structural systems with similar building cost and weight. However, the inelastic performance of most CBFs is dominated by inelastic post-buckling deformation of thc bracc. Inclastic post-buckling behavior can lead to rapid deterioration of stiffness and resistance and premature tearing or fracture of the brace or its connections under large inelastic deformations. This rapid deterioration limits the benefits and complicates the design of the CBF system.

Buckling restrained braces (BRBs) are a new component that has been developed to overcome this deficiency and significantly improve the inelastic performance of CBFs. The first BRBs were developed in Japan approximately 17 years ago (Nippon 2002), and these Japanese BRBs often employ LYP or

SN steels as described in earlier sections of this chapter. Other suppliers of BRBs have (SEAONC 2001) developed similar products in recent years. BRBs employ patented braces where the axial member yields in tension and compression without brace buckling as depicted in Fig. 2.9. This is accomplished by encasing the slender brace bar to prevent lateral deformation and buckling without bonding the bar to the encasing element. This ensures that the yield strengths of the bar develop in both tension and compression and the deterioration in stiffness and resistance due to buckling is avoided. It increases the inelastic energy dissipation, improves axial yield performance and permits development of large inelastic axial deformations. Figure 2.10 compares a typical axial force-deformation hysteresis curve for a BRB (Fig. 2.10(b)) to a typical buckling brace element (Fig. 2.10(a)). The BRB curves show a large energy dissipation capacity and no deterioration in stiffness and resistance as compared to the deterioration noted with the buckling brace element. Many engineers believe that BRBs provide superior seismic performance to that of traditional bracing systems, and they are enjoying increasing use in practice.

The seismic performance of BRBs also depends on the connection design. CBFs typically use gusset plate connections as depicted in Fig. 2.11. These gusset plate connections may have significant in-plane rotational restraint, while being relatively flexible to out-of-plane deformations. Excessive in-plane rotational stiffness and inadequate out-of-plane stiffness may be detrimental to the performance of the BRB system, and therefore the connection must be designed to control these competing requirements. BRB connections

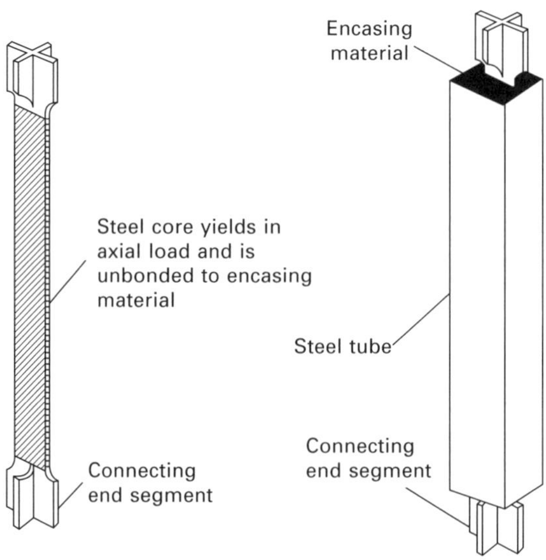

2.9 Schematic of buckling restrained brace.

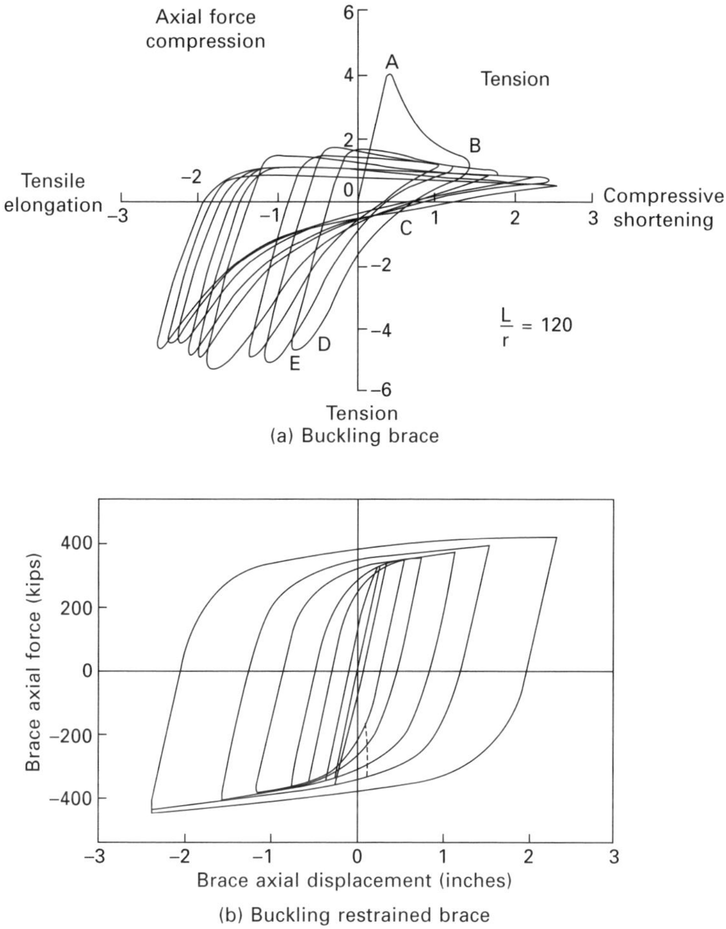

2.10 Comparison of hysteretic curves of a buckling brace and a BRB.

must support the full tensile and compressive force capacities of the brace during cyclic inelastic deformation demands. The connection must have adequate stability and lateral restraint to prevent out-of-plane deformation, and it cannot buckle or fracture prior to the development of the full resistance and ductility of the brace. Rotation or out-of-plane deformation of the buckling restrained concentrically braced frame (BRCBF) connection cannot be tolerated, because these actions may inhibit development of system resistance and ductility. Continuing work on these BRB connection issues is in progress.

BRBs are one of the most promising alternatives for both seismic design of new structures and seismic retrofitting. Additional alternative BRB elements are being developed with time (Merritt *et al.* 2003a, b), and this is reducing the cost of these patented systems. BRBs can significantly improve the seismic performance of many structures. Many structures in Japan and the

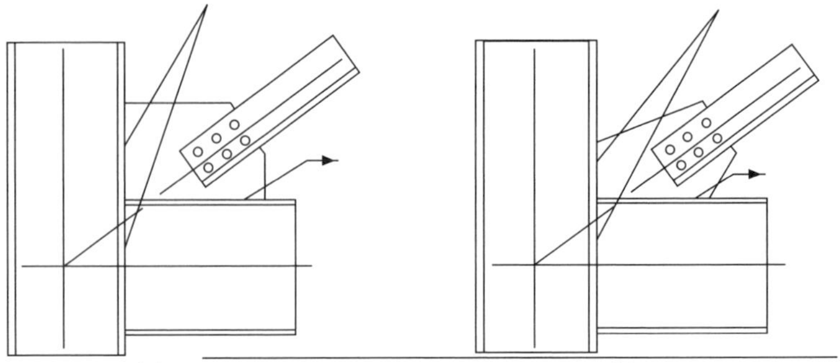

2.11 Typical gusset plate connections for a CBF.

US have employed these elements in recent years. The inelastic performance of the structural component is excellent, but they are a new system and the engineer must verify the performance that they provide. A recent guideline (SEAONC 2001) proposes some testing and acceptance criteria that can be used to verify that the BRB is appropriate for the proposed seismic design application.

2.3.3 New systems

New structural systems have also evolved to address the issues and concerns related to steel structures. This section will briefly describe several new and innovative structural systems and note the specific aspects of their construction that relate to advanced steel construction. The new systems discussed here are primarily variations of composite and hybrid elements.

Concrete filled steel tubes (CFT)

The greatest benefits of structural steel are often achieved in composite or hybrid applications, because concrete can be combined with the steel to produce significant economic and performance advantages. Concrete has relatively low compressive strength compared to structural steel, and as a result the larger volume of concrete needed to support compressive loads provides increased structural stiffness and resistance to buckling. Concrete has limited shear strength, virtually no tensile strength and limited ductility. Steel has excellent capabilities in these areas, and the resulting composite or hybrid elements are usually significantly stronger, more ductile, and more economical than steel or reinforced concrete elements acting alone.

Steel and concrete can be effectively combined in many ways, but concrete filled steel tubes (CFT) (see Fig. 2.12(a)) are enjoying increased usage for modern structural design. High strength steel tubes permit the use of smaller, lighter members with increased stiffness, stability and resistance to buckling. CFT may result in reduced construction costs, because the tube serves as both formwork and reinforcing to the concrete, and this results in a reduction of labor costs as required to reinforce the concrete element. The CFT member is much lighter than the reinforced concrete required for the same load and deformation, and this further reduces seismic design forces and design dead loads. The concrete fill adds compressive stiffness to that achieved with steel columns, and as a result the concrete fill adds axial stiffness to the column and inhibits buckling of the steel. CFT columns are particularly desirable in braced frames or as supercolumns or collector columns in taller buildings, because of their increased axial column compressive strength and stiffness. The concrete fill permits increased ductility from lighter and more slender tubes than could be achieved with steel acting alone. The confining action of the tube increases the shear strength and inelastic strain capacity of the concrete within the tube, and this further increases system ductility.

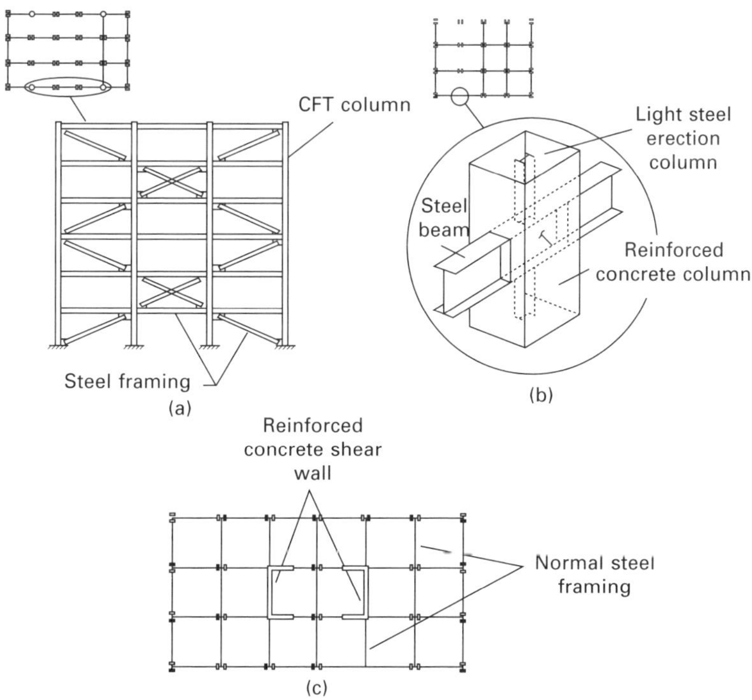

2.12 New development in composite and hybrid structural systems; (a) concrete filled steel tube; (b) RCS system; and (c) hybrid RC shear wall and steel frame system.

CFT columns require different connection details than either steel or reinforced concrete construction. A major requirement of all composite and hybrid systems is that applied loads must be appropriately shared and distributed between the two different materials, and this distribution is commonly facilitated by the connection in CFT construction. Considerable work has been done on CFT connection design (AIJ 1997, Azizinamini and Schneider 2004, MacRae *et al.* 2004). Japanese engineers commonly use internal diaphragm connections such as illustrated in Fig. 2.13(a) with CFT construction. Japanese engineers primarily use CFT columns in moment resisting frames, but variations of the internal diaphragm connection have been used in braced frames. US engineers primarily use CFT columns in braced frames of taller buildings, because of the large axial stiffness and compressive load capacity of CFT. Penetrating gusset plates, such as illustrated in Fig. 2.13(b), are commonly employed. US engineers occasionally use CFT columns in moment resisting frames of shorter buildings, and penetrating beam connections such as depicted in Fig. 2.13(c) are frequently employed. These penetrating connections are very desirable for CFT construction because they facilitate load sharing and composite action between the steel and the concrete. Research has shown that bond stress between the steel and concrete is very limited in CFT construction (Roeder *et al.* 1999), and the positive load sharing provided by the connections in Fig. 2.13 is very beneficial. CFT columns are a very promising alternative for cost effective steel construction, but steel tubes are still one of the more difficult elements to connect to other structural elements and erect in the field. As a result, additional work on new and improved connections are required to further improve this system.

This discussion of CFT has focused on CFT with circular steel tubes, but rectangular box columns such as discussed in the earlier sections on new materials are also filled with concrete to form CFT columns. These applications are quite common in Japan, because Japanese engineers make extensive use of bi-directional bending of columns and rectangular tubular sections. The benefits of CFT with rectangular box sections are significant, but somewhat different than those achieved with circular steel tubes. Rectangular tubes provide reduced confinement to the concrete and bond stress between the steel tube and concrete fill to that achieved with circular steel tubes, but the benefits of shear reinforcement and composite action are significant.

Other composite and hybrid systems

Other forms of composite or hybrid construction have recently been developed. Development of a hybrid reinforced concrete-steel reinforced concrete (RCS) construction as depicted in Fig. 2.12(b) was initially started in the US more than 15 years ago, and today it is used in both the US and Japan for intermediate sized buildings built with moment frame construction (Sheikh *et al.* 1989,

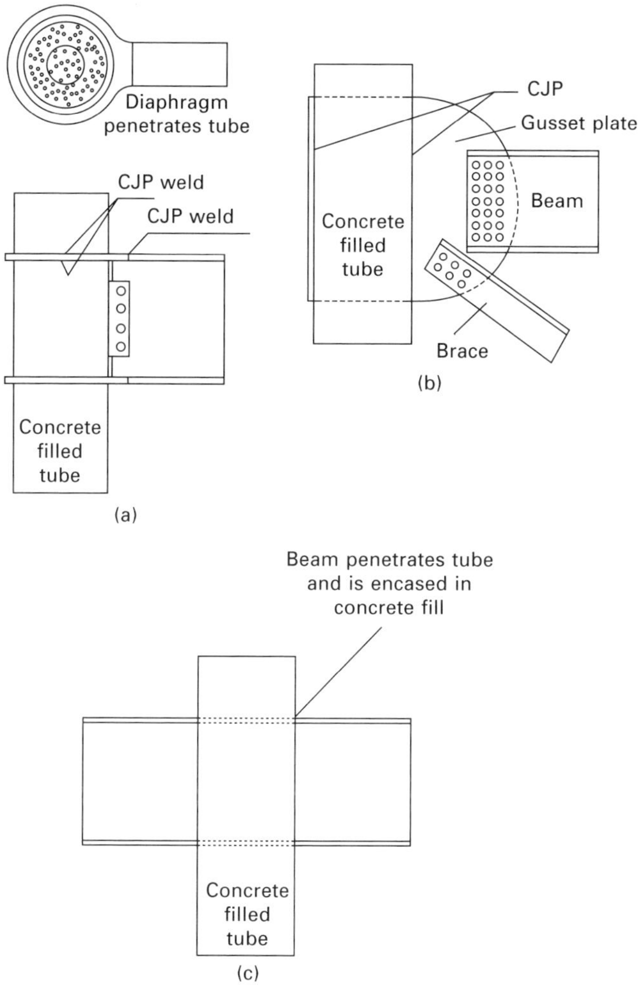

2.13 Connections used for CFT construction; (a) internal diaphragm moment frame connection; (b) penetrating gusset plate CBF connection; and (c) penetrating beam moment frame connection.

Deierlein *et al.* 1989, Nishiyama *et al.* 2004). RCS construction normally starts with erection of a steel frame with either light steel columns or widely spaced steel columns. Moment resisting frames develop resistance to lateral (wind and earthquake) load through transfer of bending moment and shear from beams to columns and through the large bending resistance and stiffness of the column. In seismic design, inelastic deformation is expected in the beams adjacent to the column. Therefore, the beam-column connection must be strong enough to fully transfer these forces and moments without adversely affecting the inelastic deformation capacity of the beam, and the column must be stiff and strong enough to control frame deflections and develop the required frame lateral resistance. Reinforced concrete (RC) columns can economically provide benefits to the steel frame in RCS construction. The RC columns are cast between widely spaced steel columns, or the RC column may encase the very light steel columns which were used to temporarily support the steel framing during erection. The RC column incorporates the steel beam to form a moment resisting frame as illustrated in the figure.

The RC column can economically be built to provide large flexural stiffness and resistance, but the connection design details are an important issue in the design of the RCS system. Connection design is important because large compressive bearing stresses are expected at the steel-concrete interfaces within the connection, and the shear design of the connection must consider the shear capacity of both the steel and concrete. Special stiffeners are frequently required in the steel beam to provide bearing surfaces between the steel and concrete as illustrated in Fig. 2.14. Column reinforcement is needed to provide shear, flexural, and axial resistance, to confine the concrete, and to ensure continued contact and interaction at the steel-concrete interfaces. RCS connections have been extensively researched, and design guidelines have been proposed for the system (ASCE 1994, Nishiyama *et al.* 2004).

Finally hybrid systems such as RC shear walls combined with steel framing as depicted in Fig. 2.12(c) have also been employed in recent years. Older variations of this alternative have been used in prior years. However, today's systems frequently result in initial construction of the shear wall by slip or jumping forming methods. The steel framing may be purchased and fabricated while the wall is being constructed. Steel erection may then proceed very quickly after the wall is complete. The wall provides great lateral resistance and stiffness to the structure, and therefore the steel frame can be very light. The steel frame provides additional ductility to the RC wall and it supports the gravity load even as the wall is damaged during or deformed during earthquake deformation.

The connection design is again an issue in obtaining good performance from the structural system. Considerable research has been performed on these connections, and guidelines and recommendations have been developed. This hybrid RC wall-steel frame system is less frequently used than the CFT system, but it offers clear benefits for a range of structural applications. The reader is referred to references noted in this section for additional information and guidance regarding connection design for these alternative composite systems.

2.4 Sample structures

Finally in reviewing recent advances in steel structures, a number of noteworthy structures can be seen. Most of these applications employ one or more of the

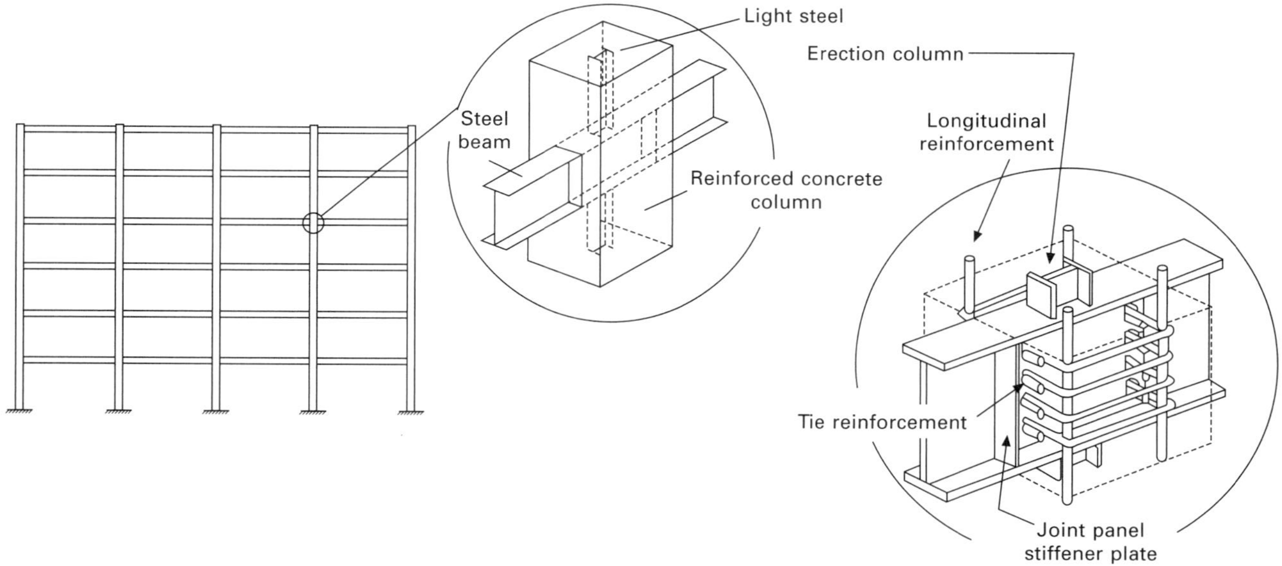

2.14 Characteristics of RCS moment frame connections.

new concepts described earlier in this chapter. A brief discussion of a few of these structures is provided.

Taipei 101 building

As noted in prior comments, steel dominates the civil engineering construction market for taller, bigger, or unusual structures. The Taipei 101 building was recently completed as a steel frame building in Taipei, Taiwan (Fig. 2.15). This building is 508 m tall with 412,500 m^2 of floor area, and it is particularly noteworthy since it is one of the tallest buildings in the world. It was designed by C.Y. Lee & Partners and Evergreen Consulting Engineering, Inc. The construction was a joint venture between Kumagai Gumi, Taiwan Kumagai, RSEA, and TaYo Wei (KTRT JV), and the total construction costs were approximately $0.8 billion (US) (*Structural Engineer* 2003). The steel was a high strength material as discussed earlier in this chapter.

Taiwan is susceptible to large earthquakes, and design for lateral loads is a major element of the structural design. The high rise portion of the building is built in a series of eight-storey modules. Each eight-storey module has internal bracing that provides shear resistance within that module and effectively transmits gravity and lateral loads within the module to a series of composite

2.15 Taipei 101 Building (Courtesy of Shaw Shieh, Evergreen Structural Engineers and Partners).

supercolumns. The composite supercolumns were made of concrete filled box steel sections. The supercolumns act together with the outriggers to resist wind and earthquake loads. The composite supercolumns of this building effectively gather a large portion of loads on the building, and they permit a relatively open structure with good visibility and exposure. These supercolumns are very stiff and strong, and they provide a relatively stiff, strong structure for lateral load and drift control. The steel was a high strength alloy (SM570M) as discussed in the new materials section, and the concrete was also a high strength material of 70 MPa.

The building also uses a tuned mass damper to further control wind vibration. The tuned mass damper is located above the 87th and suspended by cable from the 92nd floor. The mass is a 5.6 m steel sphere that was assembled on site from steel plate. It has a mass of 660,000 kg. (*Structural Engineer* 2003).

Seattle central library

Another new and very unusual steel building is the Seattle Central Library shown in Fig. 2.16. The building was designed by the joint venture OMA/LMN (Office for Metropolitan Architecture of Rotterdam and LMN Architects of Seattle) featuring the work of Dutch architect Rem Koolhaas. The structural engineering was completed by the firm of Magnusson Klemencic Associates (formerly Skilling Ward Magnusson and Barkshire) in Seattle with early cooperation of Arup (McKinlay *et al.* 2003). The general contractor was Hoffman Construction Company. The building has 12 floors with approximately 37,000 m^2 of floor area and it was built at a total cost of approximately $155 million (US). This building makes extensive use of A992 steel described earlier in this chapter, and it employs some innovative concepts of fire protection.

The architectural geometry of the building is very unusual. The architect started with the concept that libraries will have many uses beyond books, and he proposed independent but connected spaces for these diverse uses. He accomplished this by assembling the 12 stories into five overlapping platforms. These platforms can be seen through the skin of the building in Fig. 2.16(a). Each platform consisted of several floors, and the platforms had truss framing to permit large column spacings and open spaces within the building. The trusses within each platform provided the lateral load integrity and gravity load resistance within the platform. The platforms and the columns supporting the platforms were not aligned within the building as seen in the photo. Seattle is susceptible to large earthquakes, and the architects wanted to use a diamond shaped steel seismic grid around the perimeter of the building to provide full lateral resistance to the individual platforms of the structure. The seismic grid is fully exposed on the interior of the building.

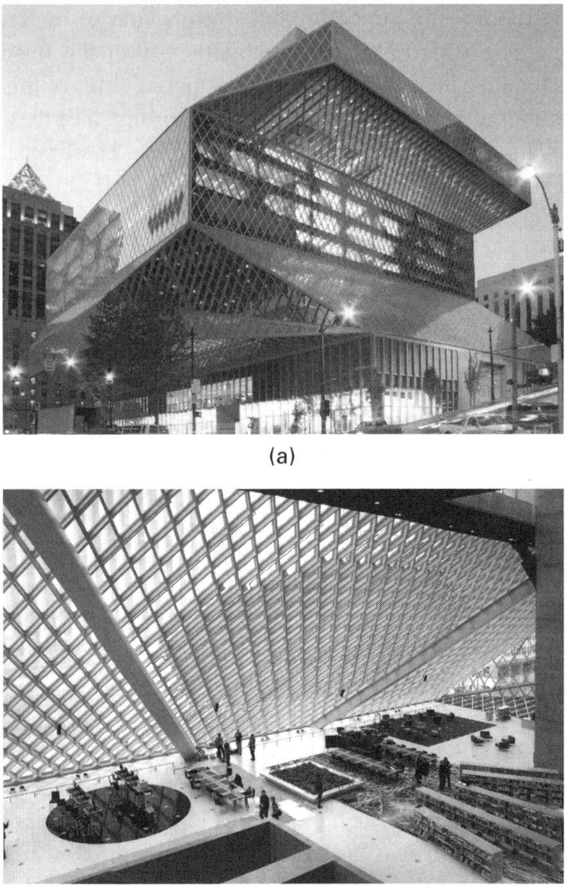

(a)

(b)

2.16 Seattle Central Library; (a) external view, (b) internal view (photos courtesy of Michael Dickter/Magnusson Klemencic Associates).

The platforms and building envelope were aligned to provide a large eight-storey central atrium. The sloped seismic grid was integrated with the exterior window-walls or envelope of the structure, and the combined interaction of the seismic grid and window mullion support system reduces the cost and complexity of the window support system over these open spaces. The diamond pattern seismic grid carries the full seismic load, but it supports no gravity load. Therefore, fire protection was avoided for the lateral load grid, since the Seattle Building Code requires fire protection only for members supporting gravity load. Sprinklers were installed in the building as a primary form of fire protection. A sophisticated fluid dynamics computer model for smoke flow was completed to limit the flow requirements and cost of this sprinkler system. In addition, structural analyses were completed for the building with and without the seismic grid to verify that the platform system was self supporting under gravity load if the seismic grid were constructed without fire protection.

2.5 Future trends

The future is always impossible to predict. However, past trends suggest broad future trends for steel bridge and building construction. Economic competition is increasing throughout the world, and so it is increasingly important to build more economical structures, which provide good system performance. Structural steel has long been the strongest and most ductile of the common construction materials. Steel has high strength in shear and tension, and if properly designed, it has good resistance in compression. Steel has a relatively heavy unit weight, but its higher strength makes steel structures relatively light (in weight per unit area) compared to most other construction materials. Further, the unit cost of steel per unit load supported, is significantly lower for steel than other new composite materials. As a result, steel is likely to continue its common usage for many years into the future. Steel is ideally suited for bigger, taller, and longer span structures, and steel is likely to continue to dominate this portion of the civil engineering structures market in the near future.

Composite and hybrid systems such as CFT offer considerable benefits in design and constraint of larger, longer and taller structures, because the benefits of composite behavior often exceed the sum of the benefits of the materials acting alone. As a result, engineers are likely to use these systems with increasing frequency in the future. The goal of increased economy in civil engineering construction is partly achieved through higher strength steel, composite action, and improved materials, but it is partly achieved through more accurate prediction of element and structural performance. Engineers today use more sophisticated analytical tools in predicting system performance, and they assume greater accuracy from these predictions. Higher strength materials increase the likelihood of many alternative failure modes such as buckling, fatigue, and fracture. As a result, structural designers must consider a wider range of behaviors, and building codes and specifications are likely to become increasingly complex. Figure 2.17 demonstrates this progression. The figure plots the page count of the American Institute of Steel Construction Specification and Commentary over the past 80 years.

There has been a dramatic increase in length and complexity in the design requirements for steel structures, and this has been accompanied by increased design stress levels and an increased number of failure modes that must be considered in design. It is likely that this trend will continue. Engineers will need to design their structures to be more and more economical with higher operating stress levels, and they will need to consider more and more adverse

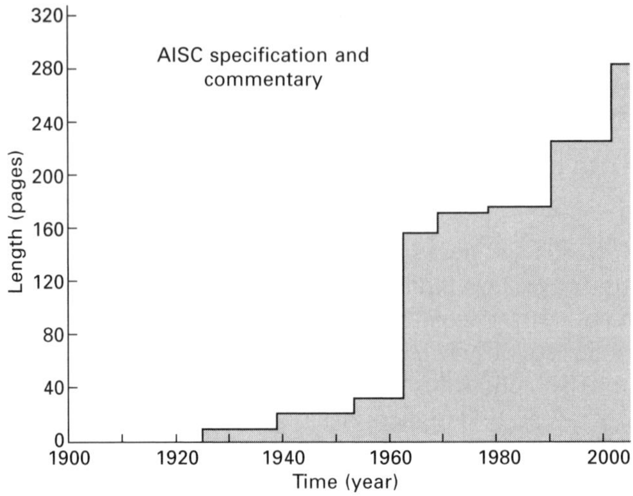

2.17 Length of AISC design specification as a function of time.

behaviors that accompany the more liberal design limits. Codes and specifications will likely become even more complex to reflect this practice. As a result, it is expected that engineers will increasingly resort to methods of ameliorating this trend. Increased use of composite and hybrid structures is one clear way this may be achieved. Increased use of hybrid and composite systems may reduce the probability of many adverse performances, without necessarily increasing the cost or complexity of design. Increased use of specialty items such as buckling restrained braces may also relieve some aspects of this trend. However, the cost of new developments can be high in our increasingly competitive environment.

2.6 Sources of further information

Throughout this chapter numerous references have been provided regarding specific issues and developments noted for steel structures. The reader is referred to these references for more detailed information on those issues.

2.7 Acknowledgments

The authors acknowledge the Japan Iron and Steel Federation for providing them with information on fire-resisting steel and SN steel and with a few figures included here. Doctor Mamoru Kono of Building Research Institute, Japan, provided the authors with information about fire-resistant steel. Mr Takeshi Katayama of Nippon Steel Co. is acknowledged for providing the authors with information about Japanese weathering steel. Mr Derek Beaman and Mr Michael Dickter of Magnusson Klemencic Associates provided information and photos regarding the Seattle Central Library. Mr Shaw Shieh of Evergreen Structural Engineers and Partners provided information and photos for the 101 Taipei building. The authors wish to thank these individuals and organizations and others who provided information for this chapter.

2.8 References

AASHO (1931), *Standard Specifications for Highway Bridges*, American Association of State Highway (and Transportation) Officials, first edition, Washington, D.C.

AIJ (1997), *Recommendations for Design and Construction of Concrete filled Steel tubular Structures*, Architectural Institute of Japan, Tokyo, 1997, 333 pp.

AIJ (1999), *Recommendations for Fire Resistant Design of Steel Structures*, Architectural Institute of Japan, Tokyo, 1999, 209pp.

AIJ (2001), *Recommendation for Design of Connections in Steel Structures*, Architectural Institute of Japan, Tokyo, Japan, November 2001, 296 ps.

AISC (1920), *Steel Construction,* American Institute of Steel Construction, New York, NY.

AISC (1923), *Standard Specification – American Institute of Steel Construction*, American Institute of Steel Construction, New York, NY. (Can also be found in *Steel Construction*, AISC, New York, NY 1928).

AISC (1963), *Standard Specification for the Design, Fabrication and Erection of Structural Steel for Buildings – Part 2*, American Institute of Steel Construction, New York, NY.

ASCE (1994), 'Guidelines for Design of Joints Between Steel Beams and Reinforced Concrete Columns,' *Journal of Structural Engineering*, ASCE, Vol 115, No. 8, pp. 2330–57.

AISI (1995), *Performance of Weathering Steel in Highway Bridges: A Third Phase Report*, American Iron and Steel Institute, Washington, D.C.

AISC (2002), *Seismic Provisions for Structural Steel Buildings*, American Institute of Steel Construction, Chicago, IL.

ASTM E119 (2000), 'E119 Standard Test Methods for Fire Tests of Building Construction and Materials,' *ASTM International*, West Conshohocken, PA.

ASTM A588 (2004), 'A588/A588M-04 Standard Specification for High-Strength Low-Alloy Structural Steel with 50 ksi [345 MPa] Minimum Yield Point to 4-in. [100-mm] Thick,' *ASTM International*, West Conshohocken, PA.

ASTM 709 (2004), 'A709/A709M-04 Standard Specification for Carbon and High-Strength Low-Alloy Structural Steel Shapes, Plates, and Bars and Quenched-and-Tempered Alloy Structural Steel Plates for Bridges,' *ASTM International*, West Conshohocken, PA.

ASTM 992 (2004), 'A992/A992M-04 Standard Specification for Structural Steel Shapes,' *ASTM International*, West Conshohocken, PA.

Azizinamini, A., and Schneider, S.P., (2004), 'Moment Connections to circular Concrete-Filled Steel Tube Columns,' ASCE, *Structural Engineering Journal*, Vol 130, No. 2, pp. 213–22.

Azizinamini, A., Barth, K., Dexter, R., and Rubeiz, C., (2004), 'High Performance Steel: Research Front – Historical Account of Research Activities,' *Journal of Bridge Engineering*, ASCE, Vol. 9, No. 3, pp. 212–17.

Barth, K.E., Hartnagel, B.A., White, D.W., and Barker, M.G., (2004), 'Recommended Procedures for Simplified Inelastic Design of Steel I-Girder Bridges,' *Journal of Bridge Engineering*, ASCE, Vol. 9, No. 3, pp. 230–42.

BCJ (1995-2002), *Building Letters*, The Building Center of Japan, 1995–2002.

Deierlein, G.G., Sheikh, T.M, Yura, J.A., and Jirsa, J.O., (1989), 'Beam-Column Connections for Composite Frames: Part 2, *Journal of Structural Engineering*, ASCE, Vol. 115, No. 11, pp. 2877–96

Dexter, R.J., Wright, W.J., and Fisher, J.W., (2004), 'Fatigue and Fracture of Steel Girders,' *Journal of Bridge Engineering*, ASCE, Vol. 9, No. 3, pp. 278–86.

FEMA 355D (2000), *State of the Art Report on Connection Performance*, Federal Emergency Management Agency, Washington. D.C.

FEMA 350 (2001), *Recommended Seismic Design Criteria for New Steel Moment-Frame Structures*, Federal Emergency Management Agency, Washington. D.C.

ICBO (1927), *Uniform Building Code*, first edition, International Conference of Building Officials, Whittier, CA.

ICC (2003), *2003 International Building Code*, International Code Council, Falls Church, VA.

JISF (2003a), 'New weathering steel for unpainted bridges,' *Steel Construction Today and Tomorrow*, No.2, March 2003, The Japan Iron and Steel Federation, Tokyo, 9–11.

JISF (2003b), 'Fire-Resistant Steel – Excellent Properties at Elevated Temperatures –,' *Steel Construction Today and Tomorrow*, Japan Iron and Steel Federation, No. 3, June 2003, pp. 9–10.

JISF (2003c), 'New Weathering Steel for Unpainted Bridges – Application Examples of 3%-Ni Steel in High Airborne Salt Environment –, *Steel Construction Today and Tomorrow*, Japan Iron and Steel Federation, No. 2, March 2003, pp. 9–12.

JISF website. http://www.jisf.or.jp/index.htm, The Japan Iron and Steel Federation (in Japanese).

JSA (1994), *JIS Handbook*, JISG3136 Steel, Japan Standards Association, 1994.

Kihira, H., (2001), 'Concept toward Durable Bridge Design by Use of Weathering Steel,' *Proceedings of the Seventeenth US-Japan Bridge Engineering Workshop*, Public Works Research Institute, Tsukuba Japan, 2001, pp. 337–350.

Kihira, H., Usami, A., Tanaba, K., Ito, M., Shigesato, G., Tomita, Y., Kusunoki, T., Tsuzuki, T., Ito, S., and Murata, T. (1999), 'Development of Weathering Steel for Coastal Atmosphere,' *Electrochemical Society Proceedings*, Vol. 99–26, 1999, pp. 127–136.

Kohno, M., (2003), Fire Resistance of Steel Structures – Its Verification Method in the Building Standard law of Japan, *Steel Construction Today and Tomorrow*, Japan Iron and Steel Federation, No. 3, June 2003, pp. 6–8.

Lie, T.T. ed. (1992), *Structural Fire Protection*, ASCE Manual 78, American Society of Civil Engineers, Reston, VA.

MacRae, G.W., Roeder, C.W., Gunderson, C. and Kimura, Y., (2004), 'Brace-Beam-Column Connections for Concentrically Braced Frames with CFT Columns,' ASCE, *Structural Engineering Journal*, Vol 130, No. 2, pp. 233–43.

McKinlay, B., Beaman, D., and Carlson, A., (2003), 'Pushing the Envelope,' *Civil Engineering*, ASCE, Reston, VA, March.

Merritt, S., Uang, C.M., and Benzoni, G., (2003A), *Subassemblage Tests of Core Brace Buckling Restrained Braces*, Report TR 2003/01, Department of Structural Engineering, University of California, San Diego.

Merritt, S., Uang, C.M., and Benzoni, G., (2003B), *Subassemblage Tests of Star Seismic Buckling Restrained Braces*, Report TR 2003/04, Department of Structural Engineering, University of California, San Diego.

Milke, J.A., (2002), 'Analytical Methods for Determining Fire Resistance of Steel Members,' *The SFPE Handbook of Fire Protection Engineering*, third edition, National Fire Protection Association, Quincy, MA.

Nakashima, M., (1995), 'Strain-Hardening Behavior of Shear Panels Made of Low Yield Steel, I: Test,' *Journal of Structural Engineering*, ASCE, Vol. 121, No. 12, December, pp. 1742–1749.

Nakashima, M., and Bruneau, M., (1995), *Preliminary Reconnaissance Report of the 1995 Hyogoken-Nanbu Earthquake*, edited by Nakashima, M. and Bruneau, M., The Architectural Institute of Japan, April, 1995, 216pp.

Nakashima, M., *et al.*, (1994), 'Energy Dissipation Behavior of Shear Panels Made of Low-Yield Stress Steel,' *Journal of Earthquake Engineering and Structural Dynamics*, Vol. 23, No. 12, pp. 1299–1313.

Nakashima, M., Inoue, K., and Tada, M., (1998), 'Classification of Damage to Steel Buildings Observed in the 1995 Hyogoken-Nanbu Earthquake,' *Engineering Structures*, Vol. 20, No. 4–6, April 1998, pp. 271–281.

Nippon, (2002), *Unbonded Brace*™, Nippon Steel Corporation, Steel Structure Division, Tokyo, Japan, pp. 6.

Nishiyama, I., Kuramoto, K., and Noguchi, H., (2004), 'Guidelines: Seismic Design of Composite Reinforced Concrete and Steel Buildings,' *Journal of Structural Engineering*, ASCE, Vol 130, No. 4, pp. 336–42.

Roeder, C.W., Cameron, B., and Brown, C.B., (1999), 'Composite action in concrete filled tubes', *Structural Engineering*, ASCE, Vol 125, No. 5, May 1999, pp. 477–84.

Roeder, C.W., Barth, K.E., and Bergman, A., (2004), 'Effect of Live-Load Deflections on Steel Bridge Performance,' *Journal of Bridge Engineering*, ASCE, Vol. 9, No. 3, pp. 259–67.

Ruddy, J.L., Mario, J.P., Ioannides, S.A., and Alfawakhiri, F., (2003), *Fire Resistance of Structural Steel Framing*, Steel Design Guide 19, American Institute of Steel Construction, Chicago, IL.

Sakumoto, Y., Yamaguchi, T., Ohashi, M., and Saito, H., (1992), 'High-Temperature Properties of Fire-Resistant Steel for Buildings,' *Journal of Structural Engineering*, ASCE, Vol. 118, No. 2, 1992, pp. 392–407.

Sakumoto, Y., Keira, K., Furumura, F., and Abe, T., (1993), 'Tests of Fire Resistant Bolts and Joints,' *Journal of Structural Engineering*, ASCE, Vol. 119, No. 11, 1993, pp. 3131–3150.

Sakumoto, Y., Yamaguchi, T., Okada, Yoshida, M., Tasaka, S., and Saito, H., (1994), 'Fire Resistance of Fire-Resistant Steel Columns,' *Journal of Structural Engineering*, ASCE, Vol. 120, No. 4, 1994, pp. 1103–1119.

Schneider, C.C., (1905), 'The Structural Design of Buildings,' *Transactions, American Society of Engineers*, Vol LIV, pp. 371–412.

SEAONC (2001), *Recommended Provisions for Buckling Restrained Braced Frames*, Structural Engineering Association of Northern California, Seismology and Structural Standards Committee, San Francisco, CA, October 2001.

Sheikh, T.M, Deierlein, G.G., Yura, J.A., and Jirsa, J.O., (1989), 'Beam-Column Connections for Composite Frames: Part 1,' *Journal of Structural Engineering*, ASCE, Vol. 115, No. 11, pp. 2858–76.

Structural Engineer (2003), 'Taipei International Financial Center Taiwan,' *Structural Engineer*, 15 April 2003 Edition, pp. 15–17.

Takanashi, K., (2003), 'SN Standards,: Rolled Steel Products for Building Structures,' *Steel Construction Today and Tomorrow*, Japan Iron and Steel Federation, No. 2, March 2003, pp. 4–6.

3
Advanced cement composites for use in civil engineering

H C WU, Wayne State University, USA

3.1 Introduction

3.1.1 Infrastructure degradation

A rising crisis facing the world today is its rapidly deteriorating infrastructure. For example, of the more than 600,000 bridges in America, 130,000 are structurally deficient, another 100,000 plus are functionally obsolete, and 143,000 are more than 50 years old and unsuitable for current or projected traffic demands (Zureick *et al.* 1995). A 1991 study in the United States estimated that transportation agencies could repair or replace only 5,000 of these nonfunctioning bridges annually; however, in the meantime, 6,000 more bridges were added to the deficiency list. The annual cost of maintaining US bridges is over $5.2 billion. Another $8.2 billion per year is needed to eliminate the backlog of bridge construction. Such rapid infrastructure degradation is not unique in the US – it is becoming a global issue. In the United Kingdom, where there are 135,000 bridges, replacement value of damaged bridges is nearly $10 billion and nearly $230 million per year is spent on maintenance (Busel and Lindsay 1997).

The primary cause of decay of America's infrastructure is the deterioration of materials used in the construction and repair of structures. One of the main causes of bridge deterioration is the corrosion of reinforcement steel in concrete decks, exacerbated by road salt used to combat winter ice and snow, especially in the northern US states with their severe weather conditions and heavy use of de-icing salt. In many types of infrastructure, such as concrete bridge decks and pavements, resistance to cracking is the key factor that limits the structure's durability and economic service life. Cracks in reinforced concrete often develop in the early years of structures from various shrinkage mechanisms such as moisture movement or thermal changes under restrained boundary conditions. Subsequently, crack openings cause durability problems in the reinforced concrete structure from corrosion of the rebars, spalling of the surface and increased permeability through the cracks. It should be noted that other durability problems also depend on the cracking and permeability of the concrete: sulfate attack, attack by other aggressive chemicals, even alkali-silica reactions, all depend on the ingress of water into the concrete.

3.1.2 Material issues

Concrete is the most commonly used material in the construction industry. Although modern Portland cement was discovered in the nineteenth century, its ancient history can be traced back several thousand years (Mindess and Young 1981). Due to their brittleness, concrete materials are designed solely to carry compressive loads; reinforcing bars are frequently incorporated in the design to carry tensile loads. However, concrete cracking and spalling often lead to exposure of steel bars. Aggressive agents such as water and chloride ions easily migrate and attack steel reinforcement, the corrosion of which causes further matrix cracking and spalling. Eventually, the integrity of the structure is lost.

Material development in response to the call for more durable infrastructures has been an active research area for decades. High strength concrete (typically with silica fume or fly ash added) with compressive strengths exceeding 100 MPa or higher has been steadily developed. However, despite the ultra high strength, these materials become even more 'brittle'. The brittleness number of high strength concrete can be several times higher than normal strength concrete (Elfgren *et al.* 1995, Bache 1995). High strength concrete is therefore potentially more vulnerable to cracking. Although high strength concrete may have very low permeability when it is intact, a cracked concrete will have similar crack related problems as normal strength concrete. It should be noted that the most efficient solution to concrete's durability problem is to reduce the extent of cracking, not just to increase the strength of concrete nor to mainly densify the concrete to improve permeability since the migration of corrosive agents is primarily achieved through concrete cracks, not by diffusion through the bulk concrete.

Parallel to the developments of high strength concrete, efforts have been made to explore fiber reinforcement (Marshall *et al.* 1985, Li *et al.* 1991a). Significant improvements on toughness have been achieved through incorporation of various types of short fibers such as glass, steel, or polymers (Shah 1991, Li *et al.* 1991b). A difference of several orders of magnitude is generally found, as is reflected in the long post-peak tails of the tensile stress/strain curves, largely attributed to fiber pull-out (Li *et al.* 1991a, Wang *et al.* 1991). In addition, advances in micromechanics modeling facilitate the creation of high ductility concrete (Aveston *et al.* 1971, Li and Leung 1992, Li and Wu 1992) and concrete with controllable crack opening (Maalej and Li 1995) when engineered fiber bridging is introduced to the composites. For fiber reinforced cementitious composites, it was also found that dry shrinkage cracking can be significantly reduced (one to two orders of magnitude less in total crack width) with the addition of only a small fiber volume fraction (typically less than 1%) (Grzybowski and Shah 1990, Balaguru and Shah 1992, Wu *et al.* 1994a).

3.2 Performance driven design with fiber reinforcement

3.2.1 Composite behavior

In the civil engineering construction industry, although fiber reinforced cementitious (FRC) composites have been experimentally exploited for almost five decades, the idea of using natural fibers, like straw, can be traced back to as early as 2500 BC. In recent years, advanced FRC composites have been developed and utilized in a wide spectrum of construction applications. These composites are typically composed of two major constituents, namely fiber and matrix. High toughness of composite is generally attributed to fibers that usually accounts for less than two volume percent of the composite, whereas a matrix phase holds fibers together and transfers loads to the high-stiffness fiber and protects them from harsh environments. Reinforcing fibers can be metal, ceramic or polymeric in forms of discontinuous or continuous filament/bundle. Matrix is cement-based with several supplementary substitutes. In this section, the function of individual constituents of the composites will be critically reviewed and discussed from a mechanics viewpoint to shed light on optimal selection of composite constituents for various kinds of infrastructure construction.

Tensile behavior

When a fiber reinforced cementitious composite is loaded beyond its tensile strength, a macroscopic crack is formed in the matrix. The composite load will then be shared by the bridging fibers. Subsequently, there are two possible scenarios. In the first scenario, the composite load will be large enough to either rupture or pull out the bridging fibers, leading to a rapidly or gradually declining post-peak behavior. This composite is still considered quasi-brittle. In the second scenario, the composite load can be sustained by the bridging fibers. These fibers then transfer the load via their interface back into the matrix. If enough load is transferred, the matrix may crack again and the process will repeat until the matrix is broken by a series of subparallel cracks (Aveston *et al.* 1971, Li and Leung 1992, Li and Wu 1992). During the process of multiple cracking, the composite load can even rise and exceed the first cracking strength of the composite (Wu and Li 1995). This composite is considered pseudo strain-hardened. The straining of the bridging fibers across the matrix cracks and within the matrix blocks gives rise to a composite strain that can be substantially higher than the matrix failure strain alone, normally one to two orders of magnitude larger than the plain brittle matrix.

As the crack expands, the bridging stress increases as the fiber/matrix interface debonds and the debonded segments of the fibers stretch. This bridging law describes the relationship between the average stress (σ) carried by the fibers bridging across a matrix crack and the opening of this crack (δ). For instance, for randomly oriented short discrete circular fibers, the bridging law has been shown as (Li 1992a):

$$\sigma(\delta) = \begin{cases} \sigma_o[2(\delta/\delta_o)^{1/2} - (\delta/\delta_o)] & \text{for } \delta \leq \delta_o \\ \sigma_o(1 - 2\delta/L_f)^2 & \text{for } \delta_o \leq \delta \leq L_f/2 \\ 0 & \text{for } L_f/2 \leq \delta \end{cases} \qquad 3.1$$

where $\delta_o = \tau L_f^2/[E_f d_f(1 + \eta)]$ is the crack opening corresponding to the maximum bridging stress.

$$\sigma_o = \frac{1}{2} g \tau V_f \frac{L_f}{d_f} \qquad 3.2$$

where g = snubbing factor; τ = bond strength; d_f = fiber diameter; L_f = fiber length; $\eta = V_f E_f / V_m E_m$; E = modulus; V = volume fraction; and subscripts f and m refer to fiber and matrix, respectively.

For multiple cracking to occur, there are some conditions that must be met. For instance, a critical fiber volume fraction, V_f^{crit}, has been defined as the minimum fiber quantity required for achieving multiple cracking (Li and Leung 1992, Li and Wu 1992, Li 1993). Similar results have also been obtained by others (see e.g. Naaman and Reinhardt 1992). The exact magnitude of this quantity depends on all relevant material parameters, including fracture toughness of matrix, fiber properties and interfacial bonds. For a short fiber randomly distributed concrete showing complete fiber pull-out when the concrete is tensioned beyond its ultimate strength, V_f^{crit} is given in eqn 3.3 (Li and Leung 1992, Li and Wu 1992, Li 1993):

$$V_f^{\text{crit}} = \frac{12 G_{\text{tip}}}{g \tau^2 L_f} E_f (d_f/L_f)^2 (1 + \eta) \qquad 3.3$$

where G_{tip} is composite toughness.

Low fracture toughness of the matrix is in favor of low V_f^{crit}. Consequently, a weak cement paste is expected to achieve multiple cracking more easily than mortar (Wu and Li 1994, Wu 2001). It would be desirable to incorporate more sand and aggregate to the mix to reduce hydration heat, shrinkage and cost. Therefore, the effect of increasing fracture toughness by adding sand or aggregate on multiple cracking needs to be properly accounted for (Li *et al.* 1995a, Wu and Arisoy 2000).

Critical fiber volume fraction (V_f^{crit}) is closely related to fiber and matrix material properties. Strain-hardening is only possible with matrices with suitable toughness for a given fiber. In other words, with the same type and same amount of fiber in different matrices, pseudo-strain hardening may or may not occur depending on the toughness of the matrix and the interfacial bond strength. Therefore V_f^{crit} can be used to guide the selection of desired material constituents.

Flexural behavior

Under flexural loading, the propagation of a tensile crack in FRC is often hindered by the bridging fibers in the wake of the crack tip, as well as by the compression zone ahead of the tip. Therefore, a much higher flexural strength than the plain concrete is usually obtained. For instance, Babut and Brandt (1978) discovered that the MOR values of steel fiber reinforced concrete (V_f = 2%) can be 2–3 times larger than plain concrete. An increase as large as fivefold was also reported for a 2% polyethylene fiber reinforced cement (Maalej and Li 1994). These significant improvements in flexural strength are also confirmed by simulations of flexural behavior using numerical techniques to solve bridged crack problems formulated as integral equations relating the crack opening profile to the bridging stress-crack opening relation $\sigma(\delta)$ (Kim *et al.* 1999, Wu and Kanda 2000). Such a numerical model can be used to predict the flexural response of FRC by incorporation of the details of the bridging law (such as eqn. 3.1). Therefore, it is possible to tailor the selection of all relevant material parameters including fiber type and fiber dimensions to improve flexural strength while keeping the fiber content low (< 2%). This upper limit is important for processing ease using conventional mixing equipment and common concrete construction practice.

Compressive behavior

Unstable propagation of a critical tensile crack accounts for the failure of brittle solids under tension. However, under compressive loading, microcracks in solids come under a local tensile field at their tips causing initiation of 'wing-cracks'. The extension of wing-cracks under such a local tension has been demonstrated to be unstable initially and then to become stable as the crack length increases (Horii and Nemat-Nasser 1986, Li 1992). However, the presence of other microcracks and the interaction between them induce instability, which results in final failure. A fracture mechanics model has been proposed to describe this chain of events (Ashby and Hallam, 1986). A critical stress is required to initiate wing crack growth, depending on initial crack length and orientation, on coefficient of friction and on stress state. These wing-cracks then grow in a stable way until they start to interact. Interaction increases the stress intensity driving the crack growth and leads to instability and final failure. Therefore, there are three distinct stages for predicting the compressive behavior of brittle solids like concrete, i.e., wing crack initiation, wing crack growth and wing crack interaction.

The presence of reinforcing fibers in FRC plays two contradicting roles for compressive strength. On the one hand, bridging fibers provide additional resistance to wing-crack sliding and crack growth, which strengthens the FRC. On the other hand, the addition of fibers produces more defects and voids from fiber entanglement and sometimes mixing difficulty, leading to a weakening effect. Li (1992b) and Li and Mishra (1992) have constructed a micromechanics model to quantify such strengthening and weakening effects on compressive strength of FRC. They found that depending on the effectiveness and amount of fiber bridging, and the degree to which fibers introduce defects to the composite, both an increase and decrease of compressive strength can be derived from increasing fiber content.

Crack opening

Desirable crack width in concrete should be less than 0.1 mm for water-retaining structures, 0.4 mm for structures in dry air conditions, and 0.8 mm for rigid pavements to permit load transfer between crack surfaces by aggregate interlocking (ACI 1972, AASHTO 1993). In most structural designs, e.g., the AASHTO pavement design guideline (AASHTO 1993), the allowable crack width is completely ensured by incorporating steel reinforcement so that it holds tightly closed any cracks that form. In such a design, only plain concrete properties, including tensile strength and shrinkage magnitude, are considered. When the additional bridging action of the reinforcing fibers is considered, the required amount of steel reinforcement may be significantly reduced if not completely eliminated. Consequently, significant savings on steel and labor cost can be achieved.

Furthermore, water flow rate is highly sensitive to crack width under constant fluid pressure (Tsukamoto 1990). In Tsukamoto's study, the flow rate was found to scale with the third power of the crack width. Water flow rate has been related to the permeability of concrete structures. Low permeability of concrete is important for long-term durability performance because permeated water through cracks can accelerate steel corrosion, which leads to further concrete cracking. Greatly reduced crack widths in concrete, therefore low permeability, can have a significant effect on the long-term behavior of concrete structures.

3.2.2 Significance of performance driven design approach

Fiber reinforced cementitious composites can be designed to provide many advantages over plain concrete, including high strength, high ductility and high toughness (Naaman and Reinhardt 1996). However, the deformation mechanisms and failure modes of FRC are very different from those of plain cement or concrete. The conditions of bridging fibers for achieving high composite strength or high ductility may counter each other, creating a tradeoff (Wu and Li 1994). Micromechanical models linking material parameters of fiber, matrix and interface properties with macroscopic behavior of the composites are critical to engineering individual components and optimizing composite performance (Li and Wu 1992, Li 1993). Therefore, desirable composite properties can be achieved with minimum fiber volume fraction, which leads to significant cost savings since fibers are usually very expensive. It is this flexibility that challenges material engineers to truly design optimum composites. Further, micromechanical models can also be used in the design of efficient fiber hybridization (mixed fiber type or mixed fiber geometry) for achieving extremely unique composite performance. Examples could be found in recent studies by Shah and his co-workers (2001, 2002).

Such a performance driven design approach has proven very effective in several studies. Some of the notable developments are highlighted as follows. Commercially available aramid short fibers with lengths of 6 mm and 12 mm were used to make conventional FRC because the fiber was commercial grade and commonly used. In contrast, an engineered fiber developed by the author using the same aramid fibers was used in making the high performance fiber reinforced cementitious composite (HPFRC). The matrix compositions were identical in both cases, as was the fixed fiber content of 2% by volume. Direct tensile test results are summarized in Fig. 3.1. For the conventional FRC reinforced with 12 mm long fibers, catastrophic failure occurred after reaching the peak load of 4 MPa. This strength level increased moderately from approximately 2–3 MPa for the plain concrete alone. The ultimate strain of 0.05% (strain at the peak load) remained nearly the same as that of plain concrete. The toughness (0.32 kJ/m^2) was the most significant gain compared to 0.01 to 0.1 kJ/m^2 in regular concrete. This was due to the long tail of the descending curve, a result of fiber pullout. Some degree of multiple cracking was observed in the 6 mm long FRC, leading to substantial improvement in ultimate strain (0.4%) and toughness (8.4 kJ/m^2). The composite strength was increased to 5 MPa. Considering the micromechanical model, these results were expected and will be briefly explained next.

The aramid fibers had a strong bond to the cement matrix, causing the majority of the 12 mm fibers to rupture near the exit point on the crack plane. Therefore, the possible pullout distance was very limited. On the other hand,

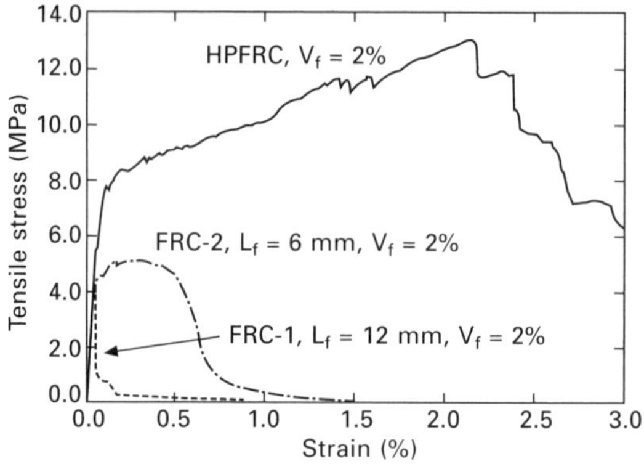

3.1 Tensile stress-strain curves of conventional FRC and HPFRC.

the 6 mm fibers endured much less fiber rupture, creating a much longer pullout distance. The long pullout distance represents stronger pullout resistance. This has a twofold implication. First, the composite is able to sustain the load directly after the occurrence of the macroscopic crack. Second, the bridging fibers are able to transfer the load back to the concrete matrix and cause several more parallel cracks. Consequently, the 6 mm fiber composite shows a much larger ductility and toughness. However, the 6 mm fiber composite did not satisfy the multiple cracking criteria discussed above; multiple cracking was not saturated and the composite showed limited ductility compared to the HPFRC.

Because the engineered fiber was designed to meet the multiple cracking criteria for the concrete compositions chosen, the HPFRC gives the most impressive results; composite strength up to 13 MPa, ultimate strain (or ductility) up to 2.2%, and toughness up to 28.7 kJ/m^2. These values represent 4–6 times, 44 times, and 2–3 orders of magnitude increase over plain concrete for strength, ductility, and toughness respectively.

3.3 Composite engineering

High performance fiber reinforced cementitious composites have been developed in recent years. Among them, short random fiber reinforcement with low fiber contents (typically less than 2% volume) appears to be the most attractive due to ease of processing and relatively low cost. The optimal selections of matrix compositions can be effectively achieved using the micromechanical model discussed above as a design tool. Such a design concept will be further elaborated in the following.

3.3.1 Matrix design (toughness control)

Concrete is a 'composite' and is the most widely used construction material.

Concrete consists of cement, sand, crushed rock or natural gravel and water. Depending on the desired material characteristics, different filling materials can be added. Typical density of ordinary concrete for structural purposes can vary between 2000 kg/m^3 (3300 lb/yd^3) and 3000 kg/m^3 (5000 lb/yd^3). Aggregate is granular material, such as sand, gravel, crushed stone or iron blast-furnace slag. Coarse aggregate usually refers to particles larger than 4.75 mm and fine aggregate refers to particles smaller than 4.75 mm but larger than 0.75 mm.

As discussed previously, parameters governing the behavior of short fiber reinforced cementitious composites can be divided into three groups. The first consists of those related to fiber, such as fiber type, geometry and strength, the second consists of matrix related parameters such as fracture toughness and elastic modulus and the third group consists of interface related parameters such as bond strength. To meet the multiple cracking criteria for pseudo-strain hardening composite, matrix compositions must be carefully selected or designed. Perhaps the most economical way to achieve this is by lowering critical fiber volume fraction (V_f^{crit}) as much as possible since fibers are the most expensive component.

The single most critical matrix parameter is fracture toughness. It governs the strength of the plain matrix and strongly affects the critical fiber volume fraction of the composite. Elastic modulus is another important performance indicator. It has a minor influence on V_f^{crit}, but a higher elastic modulus is needed for many structural applications using HPFRC. For most of the fibers investigated, the incorporation of coarse aggregate into FRC creates too high a fracture toughness resulting in very high V_f^{crit}, or never satisfying the condition (Wu and Li 1994). For this reason, coarse aggregate is not included in this study.

A systematic experimental investigation was conducted on the effect of matrix compositions on matrix fracture toughness and modulus (Wu *et al.* 1993, Li *et al.* 1995b). The investigation focused on the effects of water/cement (w/c) ratio and sand/cement (s/c) ratio on composite modulus and V_f^{crit}. The effect of s/c on matrix elastic modulus could be predicted by the Hashin model (Hashin 1962), as shown in Fig. 3.2 for two different w/c. Based on this model, the elastic modulus of a two-phase heterogeneous cement mortar could be determined by the modulus of each of the individual phases, the sand and the cement paste. Figure 3.2 can assist in selection of a particular matrix to achieve a certain composite elastic modulus.

The mix compositions of all the matrix mixes investigated are listed in Table 3.1. MR#1 is a cement paste without sand that has been included for comparison purpose. Two values of w/c ratio (0.35 and 0.45) and three values of s/c ratio (0.5, 1.0, and 2.0) have been used. The sand used is graded silica sand conforming to ASTM C 50–70. A small amount of superplasticizer (melamine) is used to adjust workability whenever needed. The two matrix

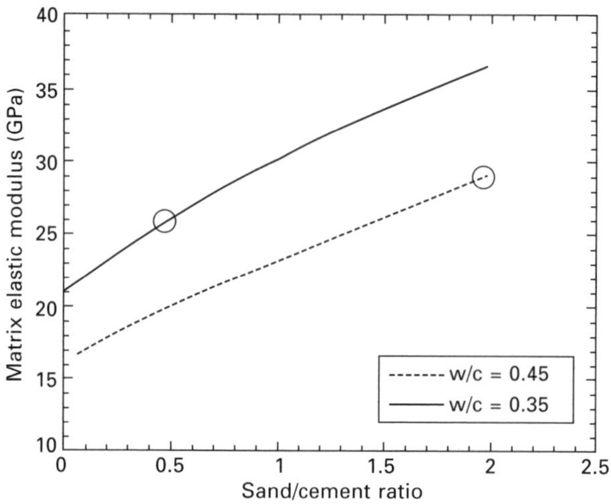

3.2 Effect of s/c and w/c on matrix elastic modulus (the points in circles indicate the matrix chosen for mixes II and III).

Table 3.1 Matrix mix proportions

Mix designation	Cement	SF[a]	Sand	w/c
MR#1	1.0	0.2	0.0	0.32
MR#2	1.0	0.2	0.5	0.45
MR#3	1.0	0.2	1.0	0.45
MR#4[b]	1.0	0.2	2.0	0.45
MR#5[c]	1.0	0.2	0.5	0.35
MR#6	1.0	0.2	1.0	0.35
MR#7	1.0	0.2	2.0	0.35

[a]SF = silica fume, [b]matrix mix used in composite mix II, [c]matrix mix used in composite mix III.

mixes (MR#4 and 5) subsequently used in composite mixes II and IIIa,b are indicated in circles in Fig. 3.2. The matrix test results are listed in Table 3.2. The average measured compressive strength, tensile strength and matrix fracture toughness of each mix are listed. The elastic modulus of the matrix has been estimated by the Hashin model.

Effect of water/cement ratio

Figure 3.3 shows the effect of w/c on matrix fracture toughness for various s/c ratios. It can be observed that a decrease in w/c from 0.45 to 0.35 results in an increase in fracture toughness at all s/c ratios, although the difference appears to diminish at high s/c ratios. For example, for the matrix with s/c = 1.0, there is an increase of 73% in fracture toughness when w/c ratio is reduced from 0.45 to 0.35. For the same w/c change, the increase in tensile

Table 3.2 Matrix test results

Mix designation	Comp. strength (MPa)	Tensile strength (MPa)	Fracture toughness (MPa $m^{1/2}$)	Elastic modulus (GPa)
MR#1	45.0	1.6	0.33	23.0
MR#2	61.3	1.9	0.31	20.1
MR#3	65.7	2.4	0.41	23.6
MR#4	82.5	3.2	0.57	30.7
MR#5	78.2	2.9	0.38	26.2
MR#6	90.6	3.2	0.71	30.3
MR#7	90.8	3.3	0.66	36.9

strength is only 29% (see Table 3.2). Such a large increase in fracture toughness makes it difficult to use this matrix and still ensure strain-hardening behavior in the composite.

Effect of sand/cement ratio

Figure 3.3 also shows the effect of s/c on matrix fracture toughness. Sand was added to increase the fracture toughness for a given w/c ratio. The fracture toughness grew by 39% for an increase in s/c from 1.0 to 2.0 at a constant w/c of 0.45. For a similar change in sand content, the tensile strength increased by 33%. These results suggest that the sand content must be carefully selected for the design of a strain-hardening composite. For a w/c of 0.35, a slight decrease in fracture toughness is observed for a higher s/c. This could be caused by the higher porosity introduced by the presence of a large amount of sand when w/c is low.

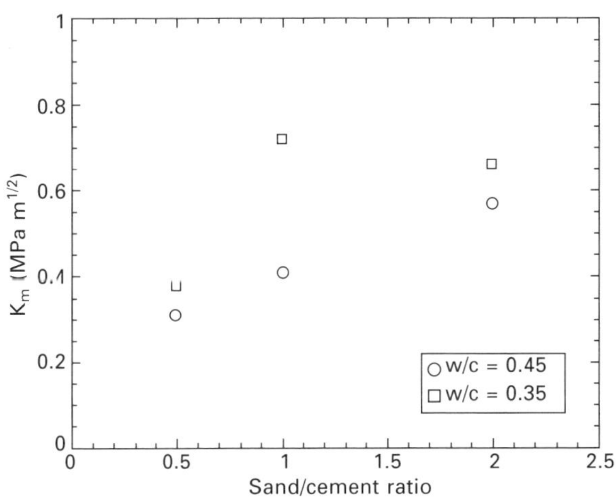

3.3 Effect of s/c and w/c on matrix fracture toughness.

Composite design

Table 3.3 shows the mix compositions of the composite mixes chosen from the matrix mix results. Mix I, without sand, is included for comparison. This mix had been previously developed and exhibited very high composite ductility but with relatively low elastic modulus (20.3 GPa) (Wu and Li 1994). The matrix MR#4 was chosen for composite mix II for its high elastic modulus (30.7 GPa). The matrix MR#5 was chosen for composite mix III for its more moderate fracture toughness and elastic modulus (26.2 GPa) compared to mixes I and II. For both MR#4 and #5, the tensile strength is close to 3 MPa. In all cases, the fiber used is a high modulus polyethylene fiber. The fiber length is 12.7 mm and diameter 38 μm. It has an elastic modulus of 117 GPa and specific gravity of 0.97. The fiber volume fraction used in each case was fixed at 2%. Two batches of tensile specimens were tested for mix III. One used fibers without plasma treatment (called mix IIIa) and the other used plasma treated fibers (called mix IIIb). Plasma treatment was found to provide significant improvement to bonding, which will be discussed in detail in the bond control section below.

The test results of the composites are summarized in Table 3.4. The interfacial bond strengths were deduced from the strength measurements. Similar to the mix I composite, mixes IIIa and IIIb composites produced substantial strain hardening behavior, whereas mix II behaved more like an ordinary quasi-brittle FRC. Figure 3.4 shows the typical stress-strain behavior of the four composites together for comparison. The graphs have been smoothed for clarity. The tensile strain capacities of mixes I and IIIb are in the same order, which is one order of magnitude higher than mix II. The much higher tensile

Table 3.3 Composite mix proportions

Mix designation	Cement	SF	Sand	SP[a]	w/c	Fiber[b]
Mix I	1.0	0.2	0.0	0.03	0.32	0.02
Mix II	1.0	0.2	2.0	0.05	0.45	0.02
Mix III	1.0	0.2	0.5	0.05	0.35	0.02

[a] SP = superplasticizer; [b] volume fraction.

Table 3.4 Composite test results

Mix designation	Comp. strength (MPa)	Tensile strength (MPa)	Strain capacity (%)	Deduced bond strength (MPa)	Elastic modulus (GPa)
Mix I	65.6	4.7	5.6	0.70	20.3
Mix II	72.2	2.9	0.2	0.43	28.3
Mix IIIa	55.7	3.5	1.7	0.52	26.0
Mix IIIb	55.7	4.8	3.8	0.72	26.0

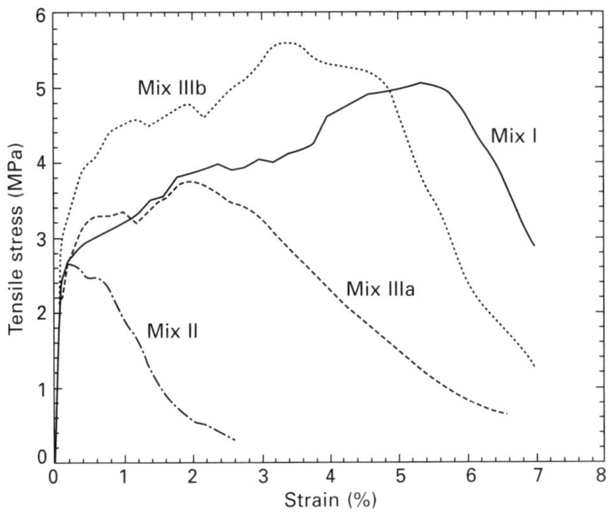

3.4 Comparison of tensile stress-strain behavior of various composites.

strength of mix IIIb over mix IIIa also demonstrated the effectiveness of plasma treatment on bond strength.

In Fig. 3.5, matrix toughness is plotted against bond strength for a fixed critical fiber volume fraction of 2%. This boundary separates the strain hardening and quasi-brittle failure modes for composites with $V_f = 2\%$. All combinations of (τ, K_m) to the left correspond to composites expected to show quasi-brittle behavior, whereas those to the right show pseudo strain

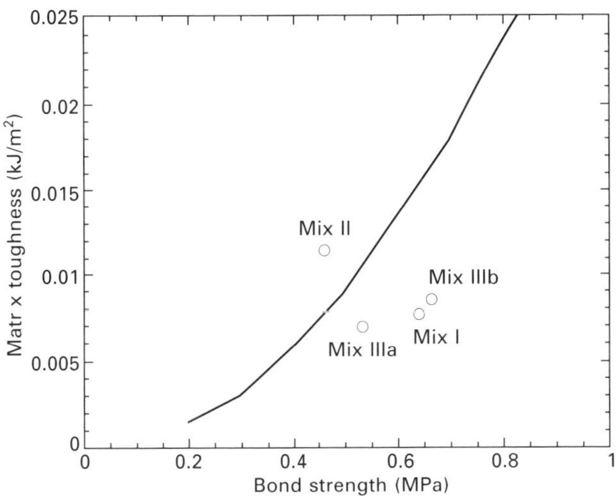

3.5 Variation of matrix fracture toughness with bond strength and critical fiber volume fraction.

hardening. Therefore, figures like Fig. 3.5 can serve as a design map for guiding selection of matrix compositions. The composite results in terms of its estimated bond strength and matrix toughness are indicated as circles. It is confirmed that mixes I, IIIa and IIIb fall in the pseudo strain hardening region, whereas mix II does not. This is consistent with the observations from the composite tensile tests. It is demonstrated that only matrices with suitable fracture toughness, as defined by the micromechanical model, retain the pseudo strain hardening property, while the composite elastic modulus is increased for all matrices with sand. In Fig. 3.5, the bond strength was deduced from the tensile strength of the composites and might not be reliable. A more suitable way of determining bond strength should involve direct measurement by single fiber pullout test.

3.3.2 Unit weight design (density control)

In this section, special attention is given to weight reduction while simultaneously maintaining good mechanical properties like flexural strength, ductility and toughness. The effect of using lightweight aggregate and foaming agents on density reduction of the composites is reviewed. This effect can also control matrix properties such as fracture toughness, as well as interface properties such as bond strength between reinforcing fibers and cement matrix. Consequently, all these factors will determine the critical fiber volume fraction, V_f^{crit}.

Concrete with a density lower than 1850 kg/m^3 is considered lightweight. Generally lightweight concrete is classified into two types; lightweight aggregate concrete and aerated concrete. Lightweight aggregate concrete is produced by using porous lightweight aggregate of specific gravity lower than 2.6. Aerated (cellular) concrete is produced by formation of voids in the concrete. Voids in the mix can be created by means of gas generated by chemical reaction (gas concrete) or by introducing foam type materials (foamed concrete). In gas concrete, voids of 0.1 to 1 mm in diameter are created by the chemical reaction of an active powder. Finely divided aluminum or zinc powder is the most commonly used material to react with calcium hydroxide or alkalis of cement hydration to produce hydrogen bubbles (FIP 1983, Neville 1995). Foamed concrete is produced either by adding preformed stable foam to the mortar or by entraining air simply by whipping the mortar. Usually hydrolyzed protein or resin soap is used as a foaming agent. This agent creates and stabilizes air bubbles during mixing at high speeds. In both gas and foamed concrete, the surface of the voids must be able to withstand mixing and compaction.

To achieve low density, high flexural strength and high ductility, one must design the composite system to meet multiple cracking criteria just as for normal weight concrete. Additionally, one must include a lightweight aggregate or foaming agent to reduce the total weight of the concrete. Arisoy (2002) has conducted a series of tests to define the effects of mix compositions including water/cement ratio, lightweight aggregate/cement ratio and the amount of foaming agent on fracture toughness of the concrete matrix and bond strength between the PVA fiber and concrete matrix. Test results have indicated that mix design and preparation are very critical to producing high strength/high ductility lightweight concrete composite with short fibers. Similar design methodology for normal weight concrete is also applicable to lightweight concrete.

Effect of sand

As shown in Fig. 3.6, fracture toughness (K_m) of the mortars tested ranges between 0.35 and 0.6 MPa-m$^{1/2}$ for # 50 sand mix compositions. Note that w/c ratio also varies to cover a wide spectrum of composition changes (see Table 3.5). The K_m is also affected by water content. For the same sand-cement ratio, the fracture toughness of the mortar increases with the reduction of the water-cement ratio (Fig. 3.7).

Effect of foaming agent

Fracture toughness (K_m) of the aerated concrete containing a MB-VR type agent increases while reducing the air entraining agent content (Fig. 3.8). With lower amounts of air entraining agent, the density of the lightweight mortar increases (Fig. 3.9).

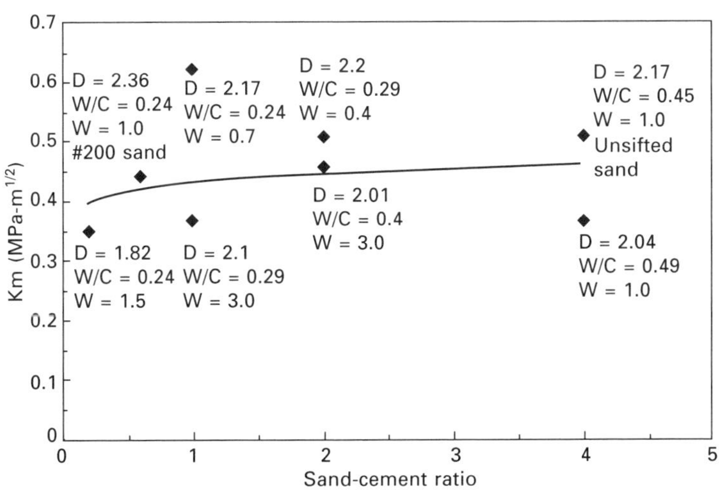

3.6 Effect of sand content on fracture toughness, K_m (D:density-g/cm^3, W/C:water-cement ratio, W:workability-cm).

Table 3.5 Mortar mix composition for fracture toughness test

Specimen No	W/C	FA/C	SF/C	S/C*	SP
Cont 1	0.24	0.20		0.20	0.02
Cont 2	0.29	0.20		1.00	0.02
Rpt Cont 2	0.24	0.20		1.00	0.02
Cont 3	0.40	0.20		2.00	0.02
Rpt Cont 3	0.29	0.20		2.00	0.02
Cont 4	0.49	0.20		4.00	0.02
Cont 5	0.45	0.20		4.00**	0.02
Cont 6	0.24		1.00	0.6***	0.02

W/C: water-cement ratio, FA/C: fly ash-cement ratio, S/C: sand-cement ratio, SP: superplasticizer, *#50 sand, **unsifted regular sand,***#200 sand

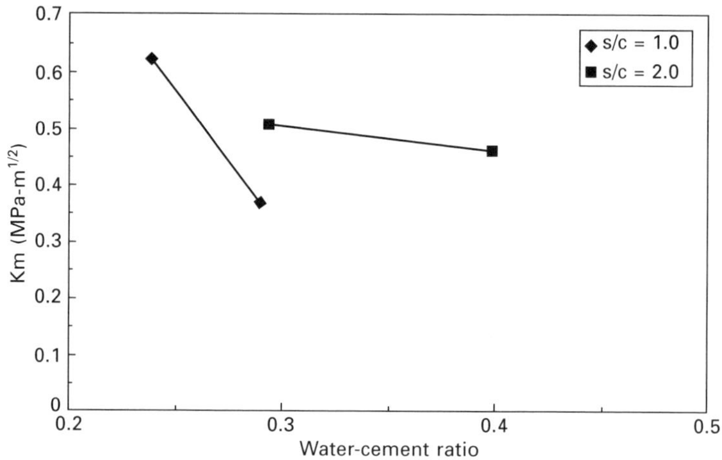

3.7 Effect of water-cement ratio content of fiber reinforced mortar on fracture toughness, K_m (#50 sand, s/c:sand-cement ratio).

Effect of lightweight aggregate

As shown in Fig. 3.10, fracture toughness (K_m) of the lightweight aggregate (LWA) concrete increased with the reduction of the LWA content. With a lower LWA content, the density of the LWA concrete increased (Fig. 3.10). Fracture toughness increased with the reduction of the water-cement ratio (Fig. 3.11). In both figures (Figs 3.10 and 3.11), the aggregate-aerated concrete (combined use of LWA and air entraining agent, represented as AA) had a low K_m, despite its low aggregate and water content. Air voids created by the air entraining agent impacted the fracture toughness and density.

Composite design: brittle-ductile boundary map

Critical fiber volume fraction (V_f^{crit}) versus frictional bond strength (τ)

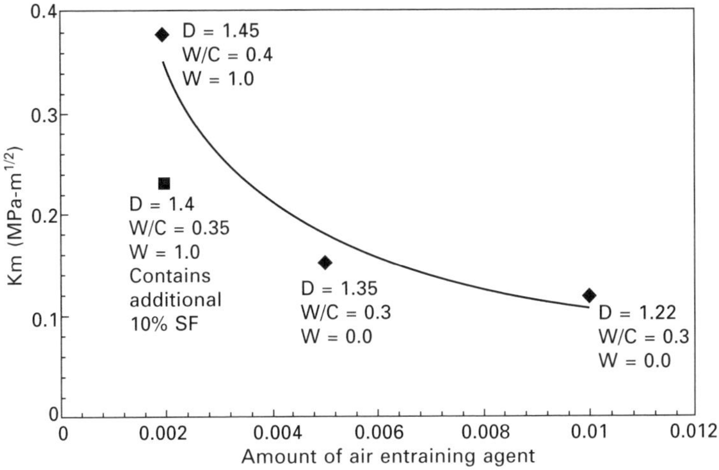

3.8 Effect of air entraining agent content on fracture toughness, K_m (D:density-g/cm^3, W/C:water-cement ratio, W:workability(cm), SF:silica fume).

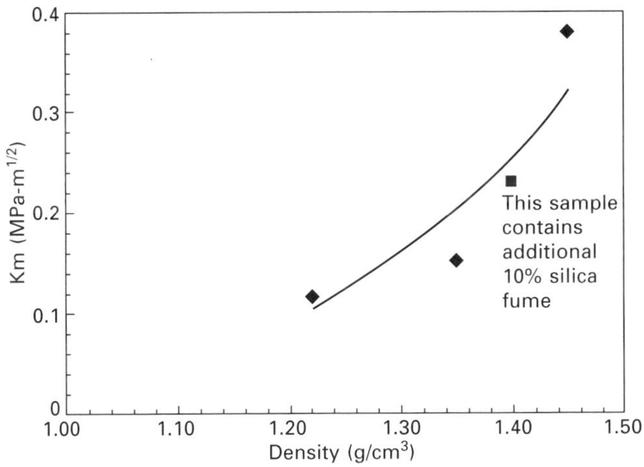

3.9 Effect of density of aerated concrete on fracture toughness, K_m.

curves for various crack tip toughness (G_{tip}) in lightweight concrete is plotted and shown in Fig. 3.12. With all other parameters being the same, the lower the matrix toughness (G_{tip}), the easier it is to achieve ductile behavior of the composite. For the lightweight compositions prepared in this study, to guarantee critical fiber volume fraction below 1.5 percent, the toughness of the matrix should be less than 0.025 kJ/m^2 for bond strength of 0.4 MPa or higher.

Figure 3.13 presents a brittle-ductile boundary plot where the $V_f^{crit} = 1.5\%$ line is labeled. The right-hand side of the 1.5% curve represents the ductile region, the left side represents the brittle region. Six mix compositions chosen

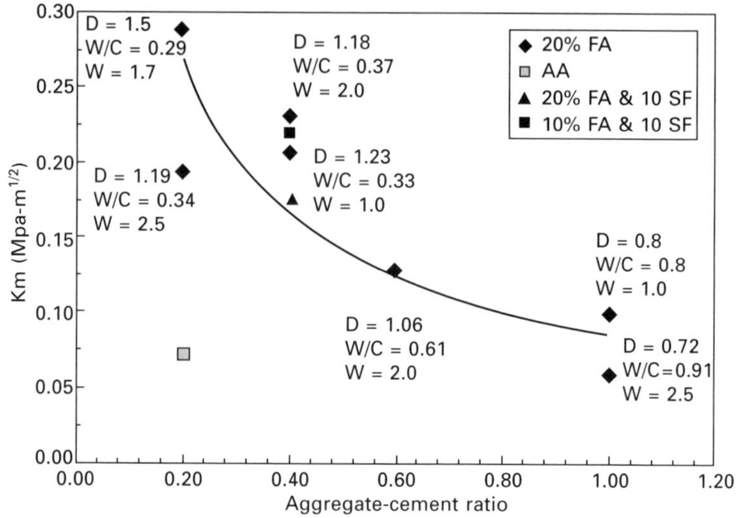

3.10 Effect of lightweight aggregate content on fracture toughness, K_m (D:density(g/cm^3), W/C:water-cement ratio, W:workability(cm), FA:fly ash, SF:silica fume, AA:aggregate-aerated concrete).

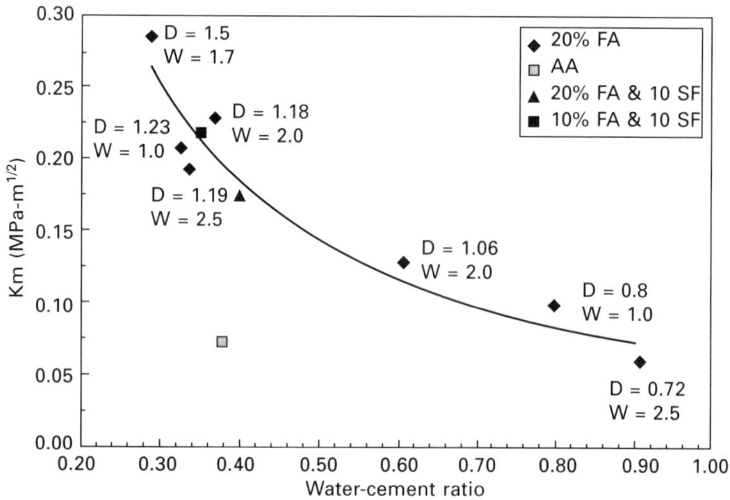

3.11 Effect of water-cement ratio on fracture toughness, K_m (D:density-g/cm^3, W:workability-cm, FA:fly ash, SF:silica fume, AA:aggregate-aerated concrete).

partly based on Fig. 3.13 are prepared for testing and listed in Table 3.6. Controls 1 and 2 fall in the ductile region as confirmed by their load-displacement curves (Fig. 3.14). On the other hand, Controls 3 and 4 fall in the brittle region since they have strong matrices and do not exhibit strain-

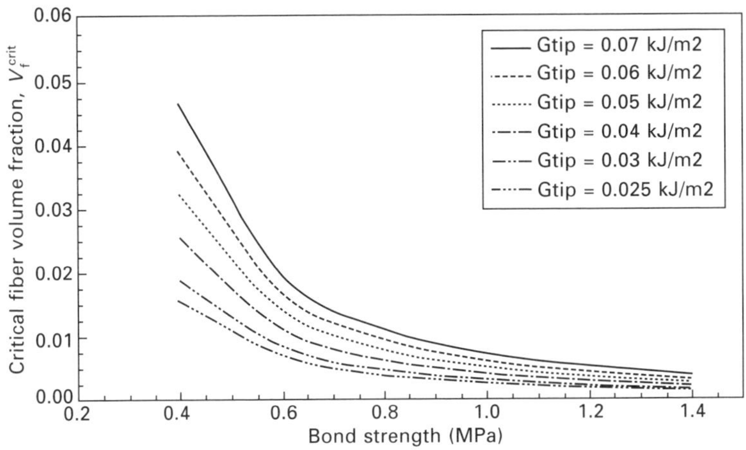

3.12 Effect of matrix fracture toughness and interfacial bond strength on critical fiber volume fraction for lightweight concrete (E_f = 40 GPa, L_f = 15 mm, d_f = 0.037 mm, g = 2, E_m = 15 GPa).

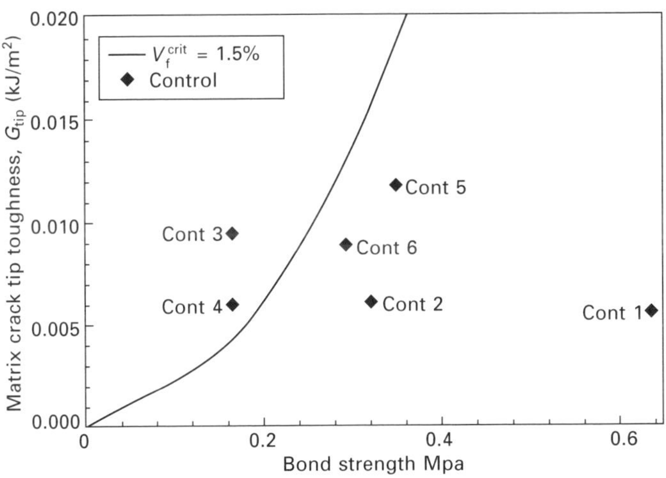

3.13 Brittle-ductile boundary for mortar with 1.5% fiber volume fraction.

hardening. The load-displacement curves for Controls 3 and 4 are also shown in Fig. 3.14. It should be noted that the V_f^{crit} curve is calculated based on the assumption that complete fiber pull-out takes place on the fracture plane. For the fiber concrete showing fiber rupture, the predictions are not valid. Therefore, much stronger matrixes (such as Controls 5 and 6) may fall in the ductile region in Fig. 3.13. However, in these two cases, the V_f^{crit} calculation is no longer valid because most of the fibers rupture and the bond strength calculated from first crack strength does not represent fiber-matrix bond strength. Their load-displacement curves show quasi-brittle qualities and are given in Fig. 3.15.

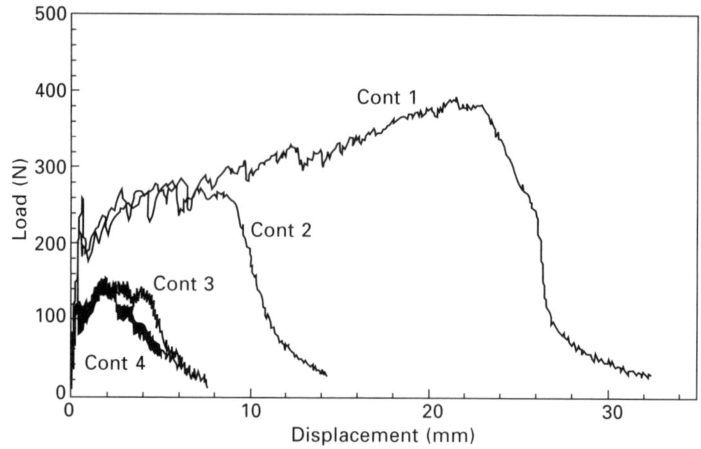

3.14 Flexural load-displacement curves of Controls 1, 2, 3 and 4 samples.

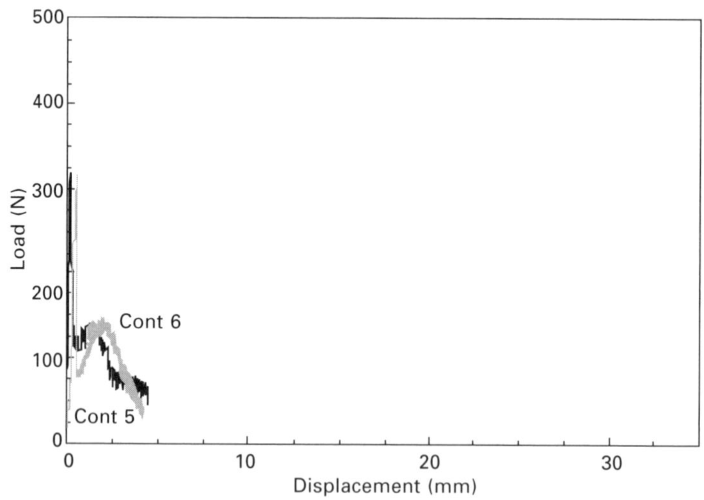

3.15 Flexural load-displacement curve of Controls 5 and 6 samples.

Table 3.6 Mortar mix composition for flexural strength test

Specimen No	Fiber content %*	W/C	FA/C	SF/C	S/C**	SP
Control 1	1.5	0.34	0.2		0.2	0.02
Control 2	1.5	0.38	0.2		1.0	0.02
Control 3	1.5	0.40	0.2		2.0	0.02
Control 4	1.5	0.49	0.2		4.0	0.02
Control 5	1.5	0.45	0.2		4.0***	0.02
Control 6	1.5	0.24	0.2	0.1	0.6	0.02

W/C: water-cement ratio, FA/C: fly ash-cement ratio, AE/C: air entraining agent-cement ratio, S/C: sand-cement ratio, SP: superplasticizer, * by volume, ** #50 sand, ***unsifted regular sand

Similar to the control mix, prediction of the ductile-brittle boundary of lightweight concrete can be prepared. The compositions of selected samples of LWA concrete are given in Table 3.7. Figure 3.16 represents the brittle-ductile boundary of the LWA concrete. The load-displacement curves of all compositions of fiber reinforced LWA concrete are shown in Fig. 3.17. As expected from Fig. 3.16, all samples exhibit ductile behavior.

In Figs 3.14 and 3.17, all samples show an ascension after the occurrence of the first crack. This gradual increase is an indication of the strong influence of the fibers. Bridging fibers could still carry loads leading to multiple cracking so there would not be a single catastrophic crack that would cause sudden failure of the specimen at the first crack strength. Very high flexural strength (maximum of 20 MPa) and very high ductility (the center displacement at peak loads approaching 20 mm) can be achieved with a fiber volume fraction of only 1.5%. Visible but very fine (less than 100 μm) multiple cracks can be observed on the tension side in some cases. In others showing low ductility one can detect only one or two macro-cracks.

Table 3.7 Selected lightweight aggregate concrete compositions for fracture toughness test

Specimen no.	W/C	A/C	FA/C	SP
1-1-MI	0.34	0.20	0.20	0.02
1-2-MI	0.37	0.40	0.20	0.02
1-9-MI	0.61	0.60	0.20	0.02
1-3-MI	0.91	1.00	0.20	0.02

W/C: water-cement ratio, FA/C: fly ash-cement ratio, A/C: lightweight aggregate-cement ratio, SP: superplasticizer

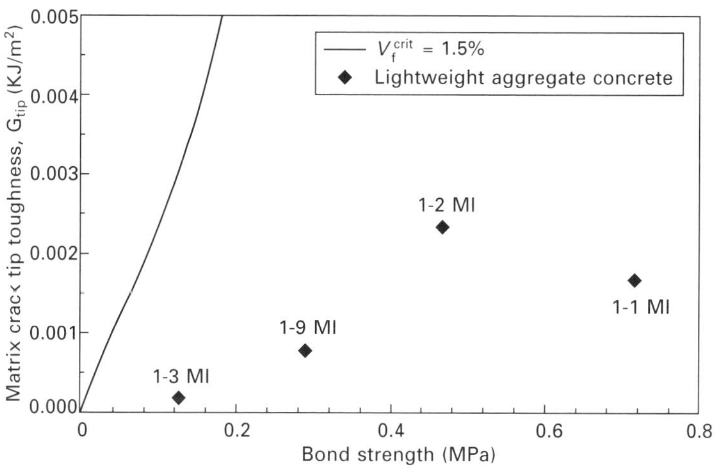

3.16 Brittle-ductile boundary for lightweight aggregate concrete with 1.5% fiber volume fraction (for mix compositions, refer to Table 3.7).

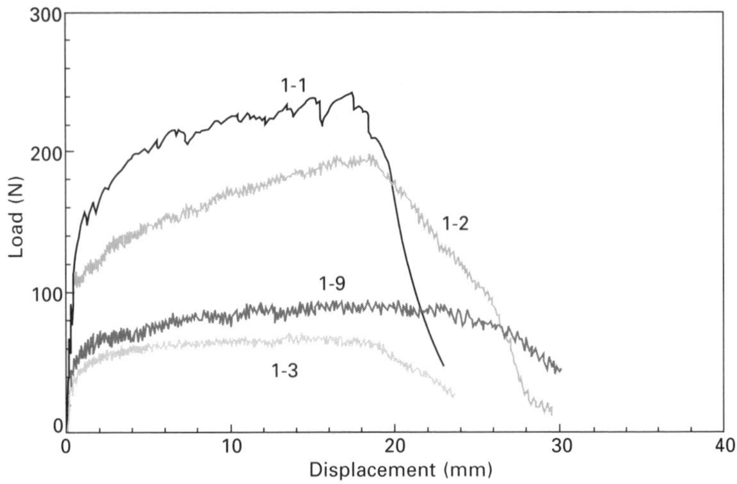

3.17 Flexural load-displacement curves of selected fiber reinforced lightweight aggregate concrete (for mix compositions, refer to Table 3.7).

Final remarks

Ultra lightweight fiber reinforced concrete with a density of 0.8 g/cm^3 (floatable in water) was achieved. However, the strength of the low-density concrete was weaker than normal weight concrete. Nevertheless, because of multiple cracking, very high ductility was still maintained. It is therefore necessary to define target performance (strength, ductility and density) prior to engaging in composite design.

3.3.3 Workability design (rheology control)

Traditionally, workability of concrete is measured by the slump test. This test consists of filling a truncated cone with concrete, removing the cone, then measuring the distance the concrete slumps (ASTM C143). The slump test can provide an indication of workability when evaluating similar mixes in an empirical manner. In contrast, it has been shown that rheological measurements are much more sensitive in characterization of the flow behavior of cement pastes that are time and strain-rate dependent (Tattersall 1973, Banfill and Tattersall 1983, Struble and Szecsy 1996, Struble *et al.* 1998). Flow parameters are measured using the shear stress (τ) and strain rate ($\dot{\gamma}$) relationship. Viscosity is defined as:

$$\eta = \frac{\tau}{\dot{\gamma}} \qquad 3.4$$

Several different types of flow behavior have been recognized (Fig. 3.18,

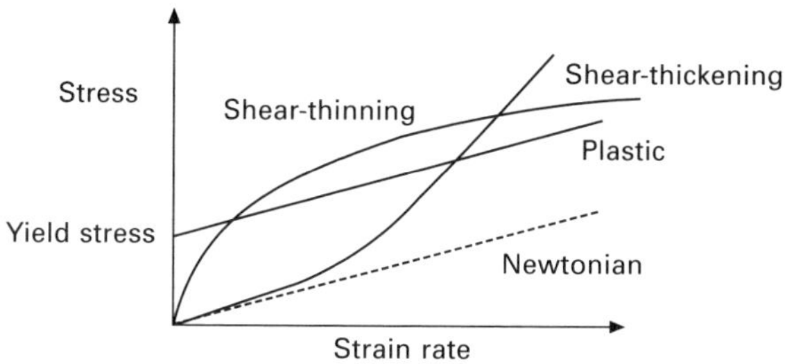

3.18 Typical flow behaviors (adapted from Struble *et al.* 1998).

adapted from Struble *et al.* 1998). Newtonian behavior displays a linear relationship between the stress and the strain rate with zero stress at the zero strain rate. This is the ideal behavior of a fluid. For a liquid containing flocculated suspensions (e.g. cement slurry), the flow behavior is characterized as plastic (also called Bingham), in which a certain level of stress (called yield stress) is required to initiate flow, and above this stress the relationship between stress and strain rate is linear. Additionally, shear thinning and shear thickening can occur in flocculated suspensions. In the former, the viscosity decreases as the strain rate increases, whereas the opposite is true for the latter.

Tattersall (1955) used a viscometer to study pastes of 0.28–0.32 w/c ratio at an age of 4.5 minutes and found that, at low rates of shear, their behavior followed the Bingham model, and the yield stress and plastic viscosity were exponential functions of water content. Tattersall (1955) also found that at higher shear rates there was evidence of structural breakdown. Therefore, the yield stress and plastic viscosity decreased under severe mixing procedure. This structural breakdown in cement pastes is not reversible and no further breakdown would occur until the previous top shear rate is exceeded. The measured rheological parameters such as yield stress and viscosity of any shear rate dependent material, such as cement paste, are very sensitive to the mixing history of the sample. Roy and Asaga (1979) studied cement pastes made by nine different mixing procedures, ranging from gentle hand-stirring, mechanical stirring and use of a colloid mill. From their published flow curves, one could find that both the yield stress and plastic viscosity decreased by about 60% when the mixing procedure was changed from the least severe to the most severe. Struble *et al.* (1998) found that the initial high shear mixing provided a uniform starting condition for rheological measurements in their study of cement paste rheology. Struble *et al.* (1998) believed it was due to the microstructural breakdown described by Tattersall and Banfill and Tattersall (1983, p. 50). In their other work, Struble and Szecsy (1996) found that the flow behavior of cement paste is controlled by the same factors that control any other suspension, the concentration of particles and the extent to which particles are flocculated.

The flow behavior of cement paste also is influenced by superplasticizer. A typical curve is shown in Fig. 3.19. Differences of several orders of magnitude in viscosity of cement pastes of the same composition can be obtained by changing the amount of superplasticizer and shear rates.

Funnel test

For practical purposes, the workability of fresh cement paste/mortar/concrete is directly correlated to the viscosity of the mix since any applied shear stress must exceed its yield stress to initiate flow. Therefore, instead of using the coaxial-cylindrical viscosity meter or the two-point workability test method (Tattersall 1991), a single point workability test method called the funnel test is usually adopted to estimate the workability of cement paste. This test consists of filling a funnel with cement paste, removing the bottom partition, then measuring the time the paste takes to vacate the funnel. This funnel time is influenced by both the viscosity and yield stress (the two most critical flow properties) of the cement paste. Cement paste flowing under its self-weight through the funnel is considered under a constant low shear rate. Because of its simplicity, this method is ideally suited for field evaluation for cement paste workability. More rigorous study of cement rheology by dilute suspension theory and viscometer can be found elsewhere (Wang 2001, Wang and Meyer 2001). Wang and Meyer determined experimentally the rheological properties of cement slurries as a function of fiber content and temperature; they also created a numerical model to predict flow behavior of cement slurry with and without short fibers through an eccentric annulus.

Effect of water-cement ratio w/c

The quantity of water in a mix is usually expressed either as liters/cubic

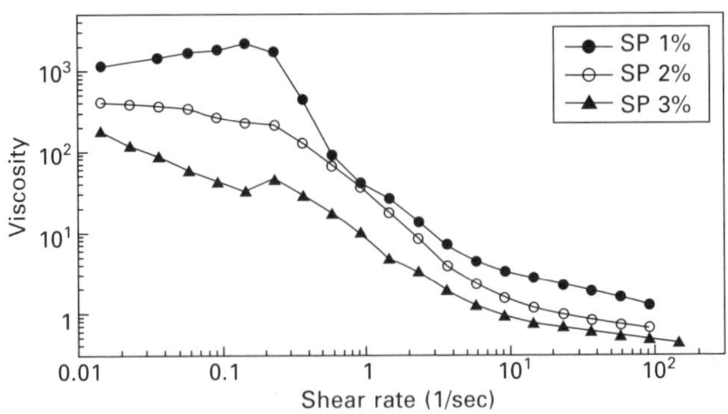

3.19 Relationship between viscosity and shear rate of cement paste of w/c = 0.3 with various amounts of superplasticizers (SP) ranging from 1.0 to 3.0% of cement by weight.

meter or as a water/ cement (w/c) ratio by weight. A portion of this water (equivalent to a w/c ratio of about 0.24) is required for complete hydration with the cement and the rest is present simply to make the mix sufficiently workable for the intended use. For any given mix, an increase in water content always results in an increase in workability so that the flow curves for a series of mixes that differ only in water content form a fan-shaped set of lines (Tattersall 1991).

A series of cement paste samples, which were different only in w/c, was made to perform a funnel test. The experimental results show a gradual reduction in funnel test time with increases in the water contents (Table 3.8). For comparison, the funnel test time (FTT) for pure water is 2.0 seconds. When the w/c ratio was less than 0.43, the funnel test failed because the cement paste was not able to flow through the funnel. It seems that the shear stress produced by the self weight of the cement paste inside the funnel was less than the yield stress. It should be noted that some degree of cement segregation occurred when the w/c was high.

Effect of superplasticizer

Superplasticizer (SP) plays an important role in modern concrete technology. It can be used for the following three individual purposes or a combination of these (Collepardi 1993):

1 to increase the workability of cement slurry without changing the mix composition;
2 to reduce the mixing water and the water/cement (w/c) ratio in order to increase strength and improve durability;
3 to reduce both water and cement in order to decrease creep, shrinkage and thermal strains resulting from cement hydration.

For a water/cement ratio of 0.30–0.40, Faroug *et al.* (1999) suggest a superplasticizer percentage of 2% to 5% relative to the cement weight. Within this range, the cement paste shows low yield stress and low plastic viscosity. Although their experimental results are obtained from fresh concrete, similar

Table 3.8 The effect of w/c (pure cement) on slurry workability

Mix no.	w/c	Funnel time (seconds)	Observation
1	0.40	–	Very thick, will not flow
2	0.43	4.95	No segregation
3	0.47	3.72	No segregation
4	0.55	2.38	Minor segregation
5	0.80	2.06	Severe segregation
6	Water	2.00	Pure water

results are expected to apply for fresh cement paste. It is possible that this range may vary for different types of SP.

As shown in Fig. 3.20, the addition of SP (DARACEM-100) into cement paste causes a substantial change in its workability. In the previous test with plain cement, the cement paste with w/c = 0.40 showed very poor workability. It could not flow at all when poured into the funnel, indicating a very high yield stress. By the addition of 2% superplasticizer, the funnel time dropped to five seconds. When the amount of SP was further increased, the funnel time remained the same (Fig. 3.20). Too much superplasticizer however retards the setting, causing possible segregation and high costs. A 10% metakaolin also is used in the mix, which will be discussed in the next section. Such results confirm that superplasticizer decreases not only yield stress but also plastic viscosity. A similar trend was reported by Faroug *et al.* (1999).

Effect of metakaolin

Metakaolin is a manufactured pozzolanic mineral derived from purified kaolin clay. It is a white, amorphous, alumino-silicate that reacts aggressively with calcium hydroxide (free lime) to form cementitious compounds. It is milled and classified to exact particle size distribution specifications. The material is 99.9% finer than 16 μm and has a mean particle size of 3 μm. Metakaolin has similar chemical properties as silica fume (SF), an extremely fine powder containing particles with an average diameter of about 0.10 μm. This extremely fine particle size can sometimes lead to handling difficulties.

Silica fume exerts significant influence on rheological characteristics of cement paste (Ivanov and Roshavelov 1990). When the SF content in the

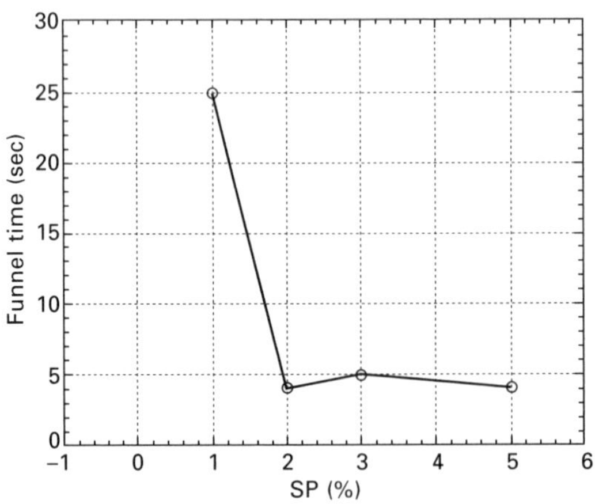

3.20 Effect of superplasticizer on slurry workability (w/c = 0.4, meta = 10%).

mix increases to 7.5%, both the yield stress and plastic viscosity of the mix decrease. Moreover, when the SF content exceeds 15%, a rapid increase in both yield stress and plastic viscosity is observed. This threshold value, depending on cement content and other factors, is in the range of 7.5 to 15% SF. This suggests an optimum dosage of SF to the workability of cement paste. The observed minimum of the viscosity can be ascribed to a state of suspension, in which the most compact packing of the cement grains and SF is reached due to their different particle sizes. It is believed that broader ranges of particle size result in maximum particle packing because the finer particles fit into the gaps between the coarser particles (Nehdi *et al.* 1998). What is known as the ball bearing effect is produced by the very fine spherical particles, which increases the workability of cement paste. However, too many fine particles will increase the total specific surface area and absorb more water, which then causes increased viscosity. Similar to the microfiller effect of SF, the addition of metakaolin can improve particle packing and bonding at the cement-aggregate interface. It is expected that the workability of cement paste would improve significantly when adding metakaolin to reach a 7.5 and 15% ratio of metakaolin to cement weight.

The effect of metakaolin on the workability of cement paste is shown in Fig. 3.21. When the water-binder (cement plus metakaolin) ratio of the cement paste is 0.35 and machine mixing is used, experimental results show a small linear increase in funnel time until the metakaolin content reaches 15%. After that point, the change accelerates. Unlike as expected from the bearing effect discussed above, the funnel time should first show a decrease then an increase with respect to the amounts of metakaolin. It was determined that W/(C + M) = 0.35 already provided very good workability, therefore, the effect of metakaolin was not profound. Consequently, the amount of water was reduced to W/(C + M) = 0.3 and a slow hand mixing was used. In the case of the lower water content and hand mixing, the funnel time curve was very consistent with the prediction explained by the bearing effect. High-speed machine mixing (hence high shear rate) appears to greatly improve workability, also shown in Fig. 3.21.

Effect of HPMC: hydroxypropyl methylcellulose

It has been shown that superplasticizer can greatly enhance the workability of cement paste, but too much superplasticizer can cause cement segregation. Therefore, various polymers have been used as viscosity agents to suppress possible segregation of cement, aggregate or short fiber (Lewis *et al.* 2000, Kong *et al.* 2003). Specifically, the effect of Hydroxypropyl methylcellulose (HPMC) is most noticeable. HPMC is a water-soluble polymer derived from cellulose, the most abundant polymer in nature. According to Kong *et al.* (2003, 2003a), due to the high rate of built-up flocculated structure, the

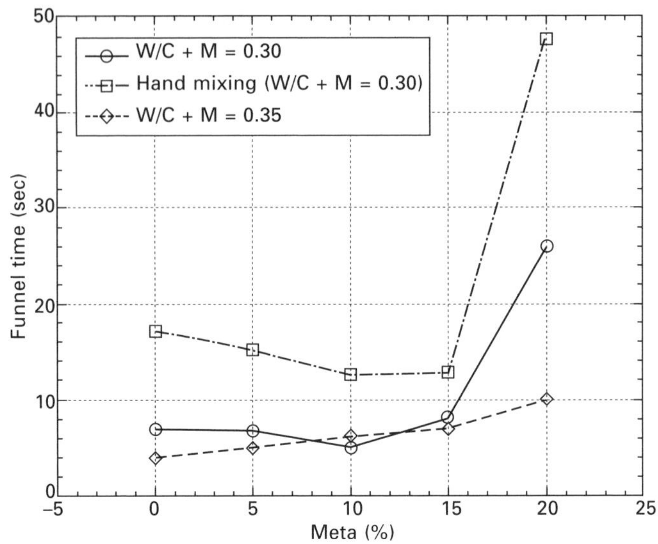

3.21 Effect of metakaolin content on slurry workability (w/(c + M) = 0.3, SP = 3%).

viscosity of fresh mix increases with time. The cementitious suspension without HPMC or with high HPMC (> 0.025%) concentration induces the strongly flocculated structure and shows a monotonous increase in viscosity with time. In the suspension system, the fresh mix dispersed by 0.013% of HPMC shows a decrease in initial viscosity followed by a slow increase with time. The initial decrease in viscosity indicates a weakly flocculated structure. On the other hand, the rate of viscosity growth with time is higher at HPMC concentration above 0.025%.

The effects of HPMC on workability at various time intervals are shown in Fig. 3.22. For HPMC = 0%, the curve represents the effect of SP alone on workability over time. In all cases, workability appears to remain more or less constant during the first 120 minutes. The somewhat reduced funnel time does not necessarily indicate improved workability. It was observed that some minor segregation occurred over time leading to higher water content at the top portion of the container. All funnel tests are carried out using cement paste from this top portion. In addition, the repeatability and accuracy of time measurement might play some role. A more profound change was observed for HPMC = 0.026%. The reason for this is yet to be discovered. It is clearly shown in Fig. 3.22 that workability decreases (longer funnel time) with higher HPMC content. It is particularly interesting to note that a small dose of HPMC = 0.007% actually helped improve workability compared to the control. A similar conclusion was reached by Kong *et al.* (2003), attributing to the electrosteric stabilization imparted by SP and HPMC. They further explained that SP electrostatically disperses flocculated cement particles,

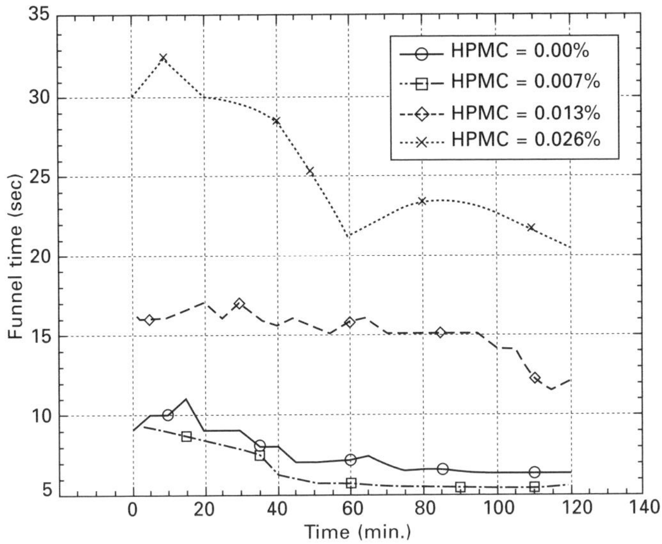

3.22 Effect of HPMC content and time on slurry workability (w/c = 0.35, SP = 3%, Meta = 10%).

and HPMC sterically prevents re-flocculation enhanced by van der Waals attraction (Kong *et al.* 2003).

Figure 3.23 shows the effect of HPMC concentration on the funnel time (workability) of fresh cement paste. Water-cement ratios of 0.35 and 0.4

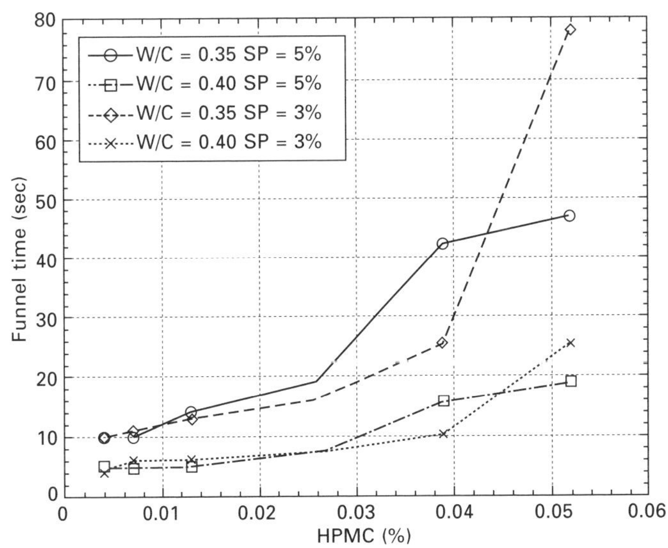

3.23 Effect of HPMC and superplasticizer on slurry workability (w/c = 0.35 and 0.4, SP = 3% and 5%, Meta = 10%).

were investigated. In these tests, the HPMC concentration varied from 0.004% to 0.052% of cement weight. The SP dosage was 3% and 5%. This plot shows that the SP dosage does not play an important role in paste workability when the contents of HPMC are low. This is consistent with our previous results that the SP dosage above 2% does not change the workability of the cement paste. The results also suggest a critical HPMC content. Exceeding this critical content of 0.026%, workability of the cement paste was greatly reduced (see Fig. 3.23). However, this critical content changed for pastes with different water-cement ratios. As shown in Fig. 3.24, the critical content became 0.013% for w/c = 0.30.

3.3.4 Interface design (bond control)

Interfacial bond strength governs many important composite properties. These include composite strength, fracture energy and ductility. Most polymer fibers bond poorly to cementitious materials. Such poor bonding characteristics are a severe limitation to the effective use of polymeric fibers in making high performance cementitious composites. Consequently, developing techniques to enhance interfacial bond strength has been a major research focus. The need for enhancing interface bond properties is especially important for high modulus/high strength polymeric fibers, which are being introduced commercially to the construction industry at increasingly attractive prices. For these fibers, better interfacial bond strength is necessary to translate the

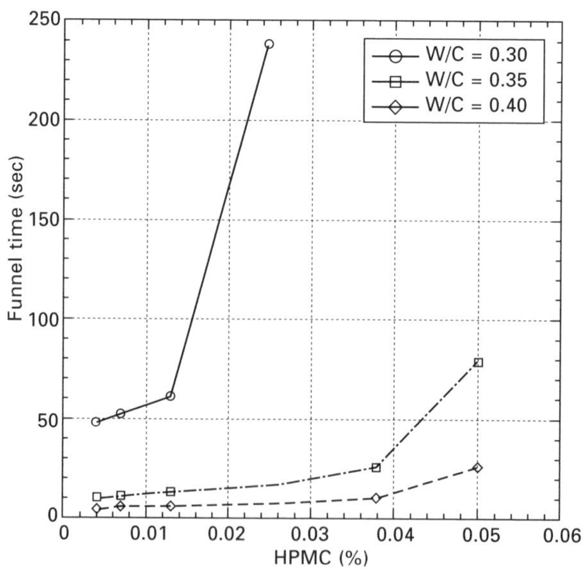

3.24 Effect of HPMC and w/c on slurry workability (SP = 3%, Meta = 10%).

improved fiber property into high performance of the composite. Therefore, interfacial bond dictates composite performance, including strengths, fracture energy and ductility. Therefore, developing interface (between fiber and cement matrix) modification strategies in order to optimize composite performance and cost efficiency on the basis of sound scientific understanding is of great importance.

Cold plasma treatment

Plasma, taking into consideration the energy of the particles constituting it, is generically the fourth state of the matter, in addition to solid, liquid, and gas. Plasma generally consists of electrons, positive ions, molecules (or atoms) and molecular fragments (free radical). The charge density of ions, which is approximately equal to the density of the ions, is equal to the density of electrons (and is called plasma density), making the plasma electrically neutral. Energy from an applied electric field is mainly received by the free electrons in the plasma due to their extremely light weight. They are accelerated and absorb large amount of energy. The electrons then transfer energy by colliding with gas molecules causing their ionization and dissociation. Therefore, the mechanism for the surface modification of polymer fibers in gas plasma is the removal of hydrogen atoms from the polymer backbone followed by their replacement with polar groups. The presence of polar or functional chemical groups on the fiber surface enhances reactivity with the matrix, thus promoting excellent adhesion (Kaplan *et al.* 1988). The selection of reaction gases and process conditions such as generator power and reactor pressure provides opportunities for tailoring fiber surface chemistry and reactivity most adequate for a given fiber/matrix.

Various processing conditions of plasma treatment were conducted on a high strength/high modulus polyethylene fiber (Li and Wu 1998, Wu and Li 1999). To evaluate interfacial modification of the fiber by plasma treatment on bonding property, continuous polyethylene fibers were treated and tested on the single fiber level by single fiber pull-out test. Four types of gases, i.e., argon, air, ammonia, and oxygen, were employed. Various power levels and treatment times (see Table 3.9) were used along with a fixed flow rate of 58 ml/min, and an initial pressure of 100 mtorr. After each treatment, the plasma

Table 3.9 Plasma treatment conditions

Gas	Power (Watt)	Time (min)
Air	100, 300	1, 5, 10
Ar	100, 300	1, 5, 10
NH_3	100, 300	1, 5, 10
O_2	100, 300	1, 5, 10

chamber was flushed with the same type of gas for 30 seconds before the chamber was devacuumed to atmospheric pressure. This was to ensure that no contamination arose from the purging air.

Single fiber direct pull-out tests were conducted to evaluate interfacial bond properties. Cement paste with a w/c ratio of 0.3 was used throughout. Other compositions of the cement matrix included 10% SF and 1% superplasticizer of cement weight. A high strength/high modulus polyethylene fiber (trade name Spectra 900 by Allied Signal, diameter of 38 μm) was employed.

Single fiber pull-out tests

The single fiber pull-out test and specimen configuration are shown in Fig. 3.25. A computer data acquisition system was employed to collect data during the tests, including the applied load P. The displacement of the fiber protruded end u was obtained by subtracting the elastic stretch of the fiber free length between the matrix base and the fiber grip from the measured cross-head displacement. The elastic stretch of the fiber free length at any given applied load, in turn, was calculated based on the initial fiber free length, fiber cross-sectional area and fiber elastic modulus. The interfacial bond properties were then determined from these P-u curves.

A standard fiber with a free length of 5 mm, a fiber embedment length of 6 mm and pull-out speed of 0.02 mm/sec was used throughout the study. The effect of a longer free length (50 mm) and various pull-out speeds (from

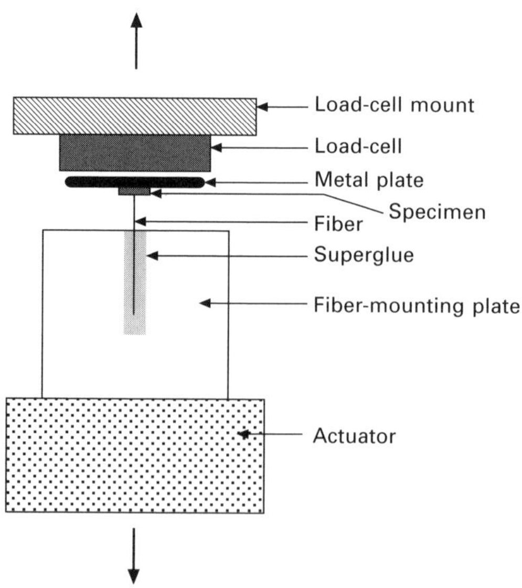

3.25 Single fiber pull-out test setup.

0.002 to 0.2 mm/sec) on interfacial properties were also investigated. In addition to monotonic loading to complete fiber pull-out, ten times precycling between 0 to 80% of the bend-over-point load followed immediately by ramp loading to complete pull-out was also employed to examine the cycling effect on bond properties. In the precycling tests, a longer embedment length of 12 mm was adopted to promote the slip-hardening phenomenon.

Figure 3.26 shows a typical pull-out curve. Generally, the pull-out curve includes a near-linear portion, corresponding to the debonding process, and a non-linear portion which covers most of the pull-out curve, representing pull-out process. Initial frictional bonds can be calculated from the load at full debonding which, in this case, refers to the onset of the nonlinear branch (bend-over point, coincided with the maximum peak in Fig. 3.26) of the pull-out curve divided by the initial fiber/matrix contact area, $\pi d_f l_f$, where l_f and d_f are the fiber embedment length and fiber diameter, respectively. The total toughness is determined by the total area under the load vs. displacement (P-u) curves.

Effect of free length

Apparent frictional bonds (τ) for two different free length (5 and 50 mm separately) specimens, together with the data from three testing speeds, are shown in Fig. 3.27. The independence of τ on free length seems to suggest that frictional debonding is governed by the strength-based criterion rather than the fracture-based criterion (Kim *et al.* 1992, Zhou *et al.* 1992, Li and Chan 1994). The fracture-based criterion would involve an elastic energy contribution from the fiber free length portion, suggesting an influence of the magnitude of the free length on the apparent bond strength. However, at

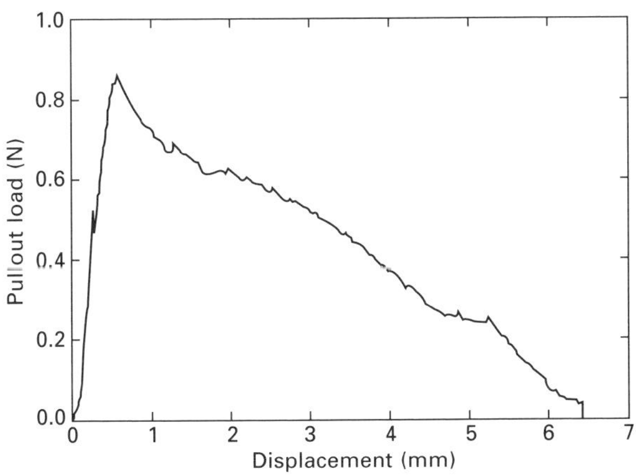

3.26 Typical pull-out (P-u) curve, l_f = 6 mm.

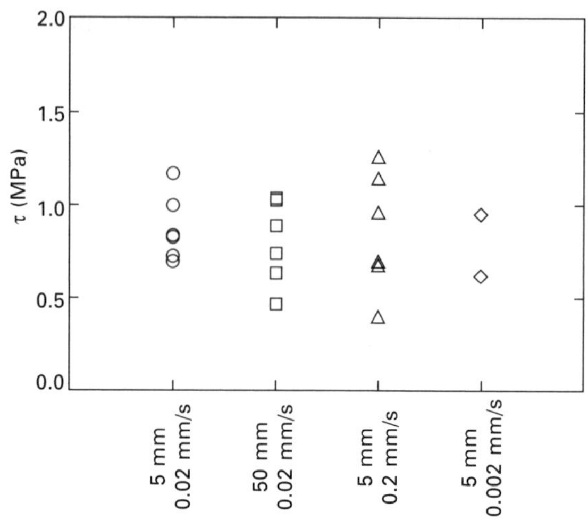

3.27 Effect of fiber free length and pull-out speed on frictional bonds.

present, there are no model predictions available that take this free length effect into account, making unequivocal comparisons difficult.

Effect of pull-out speed

As shown in Fig. 3.27, pull-out speed has little or no effect on bond strength.

Effect of precycling

The effect of ten times precycling between 0 and 80% of the bend-over-point (BOP) load is shown in Table 3.10. The τ_{max} was calculated based on the maximum peak load, which was the second peak after the BOP and exceeded the BOP value due to severe slip-hardening by the long embedment length specimens (l_f = 12 mm). Such profound slip-hardening was not observed for the shorter embedment length specimens (l_f = 6 mm). Unlike the steel fiber case for which bond degradation after precycling was reported (Li and Matsumoto 1998), this study shows that precycling is actually beneficial to soft polymeric fibers such as polyethylene. Polymeric fibers are more

Table 3.10 Comparison of interfacial properties obtained from monotonical and precycling tests

	τ (MPa)	τ_{max} (MPa)	T (N-mm)
Monotonic	0.64	0.72	7.1
Precycling	0.73 (+14%)	0.88 (+22%)	8.6 (+21%)

Values represent the average of 4 tests, () indicates % of changes.

susceptible to surface abrasion, whereas steel fibers tend to cause damage on the matrix side during load cycling (Geng and Leung 1995).

Effect of plasma treatment

Since details of the plasma treatment have been published elsewhere (Wu and Li 1999), only a summary is given below. For the fibers treated with Ar plasma, the concave-downward shape of the descending nonlinear branch indicates slip-hardening behavior during fiber pull-out, resulting from fiber surface abrasion. The average frictional bond increased with pull-out distance. Profound abrasion damage might give rise to a second peak that is higher than the first peak in magnitude, as shown in Fig. 3.28. A significant increase in friction bond and energy dissipation was achieved. In the case of oxygen and air plasmas, a distinct interfacial failure mode was noticed. There was an immediate load drop after the peak followed by a descending branch (see Fig. 3.29). This is a typical characteristic of chemical bond (or elastic bond) failure. It was the first time that chemical bonding was achieved for a polyethylene/cement system. In fact, the bond strength was so strong that the embedment length of the fibers in the cement matrix had to be shortened from 6 mm to 3 mm in order to avoid fiber rupture. The total toughness of the ten-minute oxygen specimen was, however, significantly lower than that of other treatments, although it should be noted that the embedment length was half of other specimens. Such a reduction in energy dissipation capability was expected when a significant load drop occurred in the pull-out behavior.

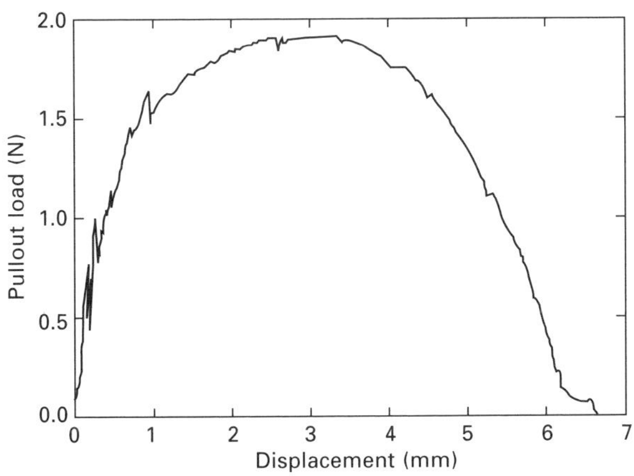

3.28 Typical pull-out curves of Ar plasma treated fibers with 300 watt power showing slip-hardening portion after the onset of complete debonding, l_f = 6 mm.

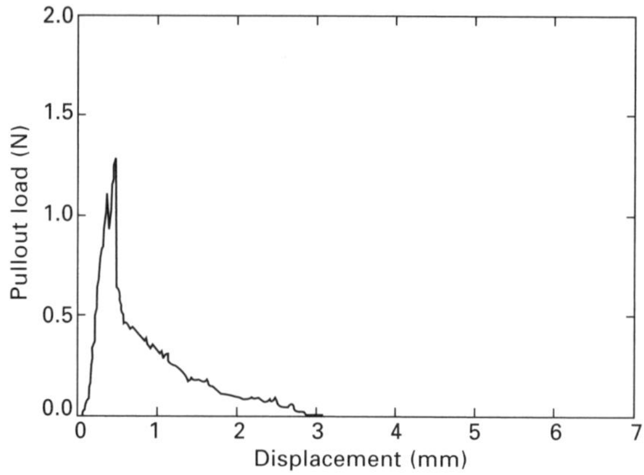

3.29 Typical pull-out curves of oxygen plasma treated fibers with 300 watt power, showing strong chemical debonding; treatment time ten minutes, $l_f = 3$ mm.

A comparison of maximum improvements on interfacial property (frictional bond and total toughness) by various plasma treatment conditions can be found in Table 3.11, together with the optimum conditions. It is clearly shown that high power plasma is more effective in modifying fiber surface chemistry, leading to improved bond strength, with the exception of ammonia plasma. Very high chemical bonds can be achieved by using more aggressive gases containing oxygen, but at the expense of total toughness. Under this category, high power air plasma might be more desirable than oxygen plasma because it is easy to process and is inexpensive. Energy dissipation can be another important consideration for composite design since very high composite toughness might outweigh strength requirement for certain applications. In this regard, high power argon plasma exhibits the highest improvement in total toughness as well as friction bond with a five minute treatment. When

Table 3.11 Maximum improvement of interfacial property by various plasma treatments

Gas type	Power level (Watt)	Bond, τ/τ_0	Toughness, T/T_0	Optimum time (min)
Argon	100	3.5	4.9	10
	300	4.5	6.9	5
Ammonia	100	3.7	3.6	1
	300	2.8	3.8	1
Oxygen	100	4.0	5.4	10
	300	6.3	1.3	10
Air	100	4.0	3.8	10
	300	6.0	0.7	10

τ: frictional bond, T: toughness, τ_0 and T_0: refers to the control.

the treatment time is increased to ten minutes a transition from frictional bond to chemical bond takes place, resulting in lower energy dissipation. Prolonged treatment with ammonia plasma gives adverse results for both power levels used in this study. This phenomenon is not observed for other plasmas, and the reason is yet to be discovered.

With the optimum plasma treatment, a sixfold increase in bond strength or sevenfold increase in total toughness could be achieved separately. Furthermore, a distinct interfacial chemical bond, as opposed to a common frictional bond, is observed for the first time in the polyethylene/cement composite. This finding represents a significant breakthrough for interface tailoring. This is because one can now truly design interfacial properties such as bond magnitude, post peak pull-out behavior or slip hardening through fiber surface tailoring. This flexibility allows us a greater freedom in satisfying various composite performance requirements. Fiber interface can be modified to show high bond strength for high composite strength or high energy dissipation for high composite toughness. The choice is obviously dependent on individual applications.

3.4 Advanced cementitious composites

In the following section, some newly developed high performance fiber reinforced cementitious (HPFRC) composites will be reviewed and discussed. Attention will be given to their extraordinary features such as high strength, high ductility and crack resistance. These HPFRC developments are guided and facilitated by the above-described performance driven design concept.

3.4.1 Short fiber composites

Tensile behavior

Short fibers can be conveniently mixed with other concrete constituents when fiber content is low ($< 2\%$). This upper limit is important for processing ease when using conventional mixing equipment and common concrete construction practice. Reinforcing fibers could be metallic, polymeric, ceramic or natural. Fiber types and fiber volume fraction are important factors. For a given fiber type, energy-absorbing capacity or toughness of composites generally increases when fiber volume and length of fiber increase. Uniform fiber distribution is necessary to ensure isotropic properties (Balaguru and Shah 1992). Several analyses (e.g. Li and Wu 1992, Li and Leung, 1992) have shown that first crack strength of composites depends strongly on matrix parameters for fiber contents of up to 2% by volume. With higher fiber contents, first crack strength would be more dependent on fiber content and fiber property (Bentur and Mindess 1990, Li and Wu 1992, Li and Leung 1992).

As shown in Fig. 3.30, the HPFRC shows a direct tensile strength of 5.8 MPa and tensile strain of 7.2%. These values represent increases of 1.32 and 240 times that of the unreinforced composite. It should be noted that this HPFRC is reinforced with 2% polyethylene fibers by volume. Visible but very fine multiple cracks (crack widths less than 120 μm) can be seen on the surfaces of the HPFRC specimens after the tension test.

Flexural behavior

It has been well recognized that flexural strength of concrete is particularly important for many current applications such as highway pavements, airfield runways, roofing tiles, sewage pipes and wall panels. It should be noted that under three-point flexural loading, a beam specimen is subject to non-uniform tensile stresses on the tension face. Tensile stress reaches its maximum directly under the loading pin but linearly decays towards the support pins near the edges. At the same time, flaw distribution may not concentrate in the highly stressed area near the loading pin. Therefore, it is likely that several other cracks may develop away from the first crack that occurred at the center of the beam. This multiple crack phenomenon sometimes is mistaken to be pseudo strain hardening. The number of cracks under such flexural loading is usually limited and the crack spacing is irregular. These are vastly distinct from the multiple cracking as described in the previous section. In a true pseudo strain hardening example, multiple cracks must develop until they saturate throughout the entire specimen leading to more-or-less constant spacing. Such crack spacing can be predicted from statistical considerations (Wu and Li 1992).

As shown in Fig. 3.31, a high performance fiber reinforced cementitious composite has been designed to show pseudo strain hardening. Very high

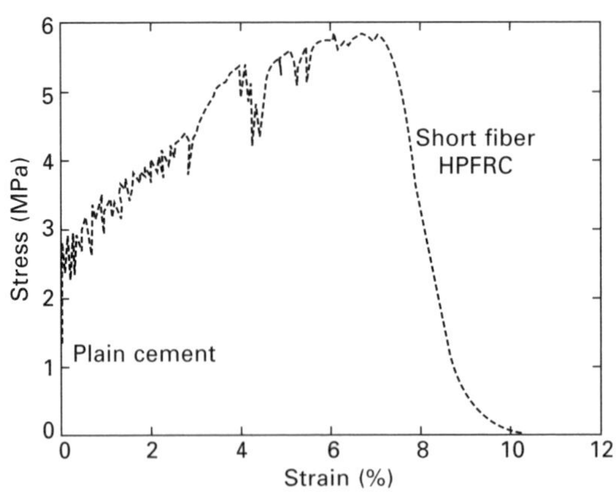

3.30 Tensile stress-strain curves of plain concrete and HPFRC (2% polyethylene fibers).

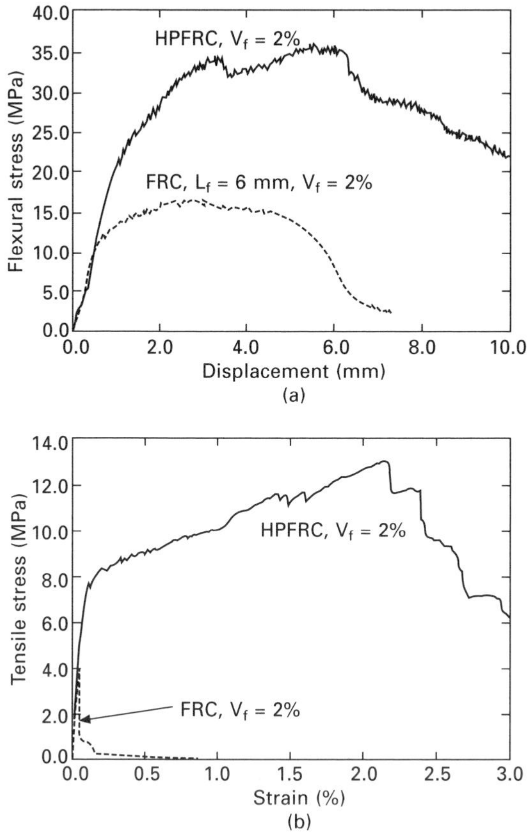

3.31 (a) Under flexural loading, (b) under direct tension.

ductility due to the saturation of multiple cracks is achieved for both specimens loaded by flexure and direct tension. On the other hand, a conventional FRC can exhibit some limited ductility under flexure but not under tension.

Crack resistance

When FRC cracks, bridging fibers can provide stress transfer and restrain crack opening. A series of tensile tests were performed to evaluate crack openings at various load levels. Coupon specimens of 12.7 mm by 76.2 mm by 304.8 mm (0.5 in. by 3 in. by 12 in.) long were gripped at both ends, and tensioned to failure. During testing, a video microscope was used to continuously monitor selected cracks (Wu and Li 1995).

For a conventional FRC such as Nylon fiber composite, its tensile load vs. tensile strain curve was depicted in Fig. 3.32(a) where a second peak was observed after the concrete cracked. The second arising branch corresponds to increasing pull-out resistance of the bridging nylon fibers across the crack.

However, the ultimate bridging resistance is not sufficient to exceed the first peak load corresponding to the tensile strength of this FRC, therefore there are no additional cracks developed before the load curve goes into its descending branch. Consequently, there is only one crack as with the plain concrete; nevertheless, the toughness is significantly improved by the fiber bridging and pull-out process. As shown in Fig. 3.32(b), the crack width enlarges monotonically as the applied load increases.

When pseudo strain hardening is achieved, such as in a high performance polyethylene fiber reinforced concrete ($V_f = 2\%$), the individual crack openings of the multiple cracks are initially delayed and then gradually increase until reaching the ultimate strength, as shown in Fig. 3.33. Complete crack closure could not occur upon unloading due to a reversed frictional stress which prevents fibers from slipping back into the matrix when the specimen is

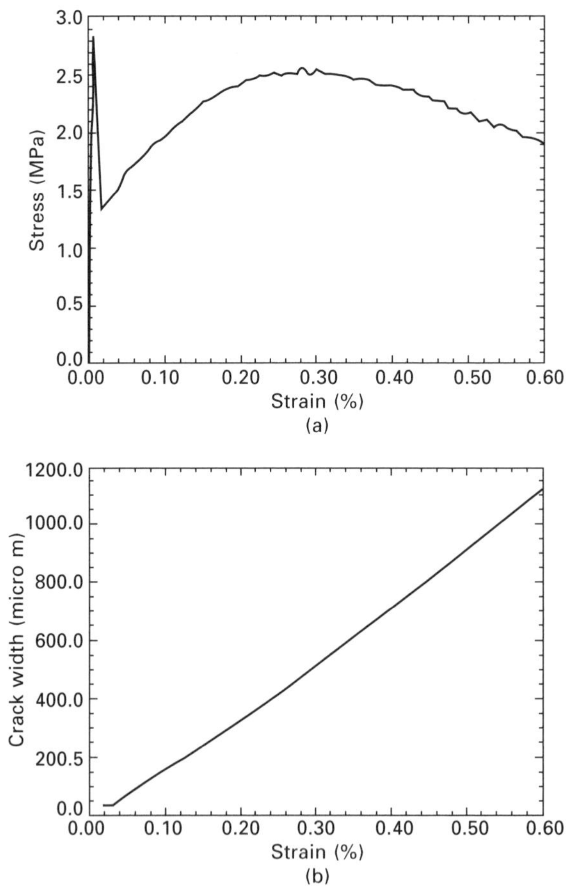

3.32 Nylon fiber reinforced concrete, V_f = 2% (a) tensile stress-strain curve (b) crack width.

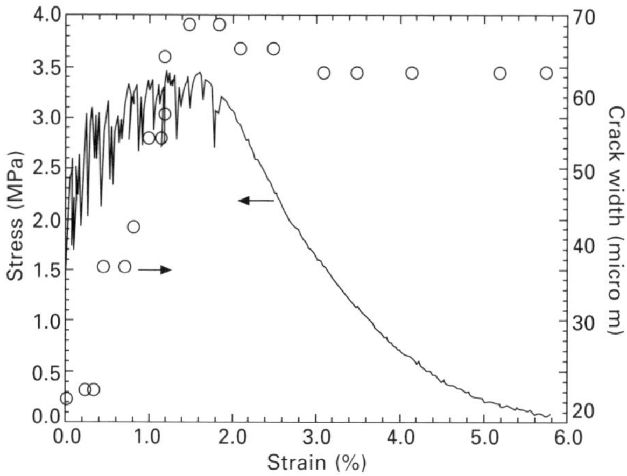

3.33 Crack opening history of high performance FRC with 2% PE fiber.

unloaded (Wu *et al.* 1994b). The reduction in crack width is typically less than 20% after complete unloading from peak load.

In another study, free shrinkage and dry shrinkage cracking behavior of HPFRC under restrained condition were experimentally measured (Wu *et al.* 1996). In the restrained cracking tests, ring specimens were adopted. A microscope with 100 times magnification was used to monitor crack width on the side surface of the ring specimens over a period of 56 days. Three points (1.27 cm, 7.62 cm and 13.97 cm from the top of the specimen) were measured along each individual crack and the average values were reported. Hooked end steel fiber (30 mm long and 0.5 mm diameter) and mesh type polypropylene fiber (19 mm long and 50 × 200 μm cross section) were both employed as reinforcement. These two types of fibers are commonly used for controlling plastic shrinkage cracking in concrete at typically low fiber volume fractions (<0.5 %). The fiber content used by Wu *et al.* (1996) was 1% in volume.

Free shrinkage of the steel FRC was approximately 7% lower than the plain concrete. On the other hand, free shrinkage of the polypropylene FRC was 50% higher than in the plain concrete (Wu *et al.* 1996). It should be noted that although w/c (0.45) was the same for both plain concrete and FRCs, superplasticizer was not used in the plain concrete and steel FRC. The restrained shrinkage cracking behavior of various composites is illustrated in Fig. 3.34. For the plain concrete, there was only one crack. As expected, several cracks developed in the two types of composites and each crack had only a limited opening. The total crack widths (openings) were significantly reduced in the FRCs, especially the steel FRC.

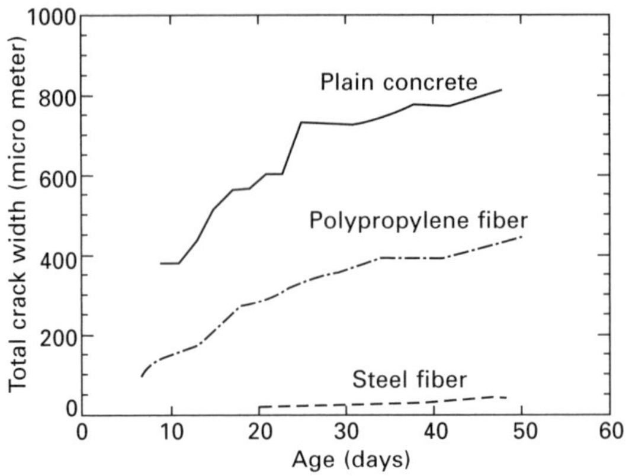

3.34 Total crack width vs. ages.

Compressive behavior

Typical compressive load-displacement curves of HPFRC samples are shown in Fig. 3.35. Although it was observed that the compressive strengths of the LWA concrete with and without PVA fibers were close to each other, their failure modes were distinctly different. While the samples without fibers exhibited explosive failure, the samples with fibers did not and sustained their integrity after testing.

It is generally recognized that the influence of fibers on compressive strength is not significant at low volume fraction. For the same material

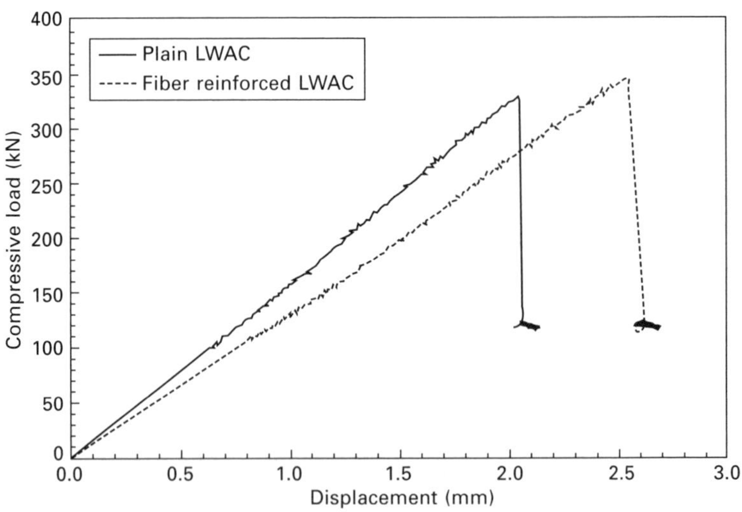

3.35 Typical compressive load-displacement curve of fiber reinforced lightweight aggregate concrete (LWAC: composition contains 40% lightweight aggregate, 10% fly ash, 10% silica fume proportional to cement and 1.5% fiber by volume).

system, compressive strength may first increase then decrease when the fiber volume fraction is increased (Li *et al.* 1995a). This trend is in general agreement with the model predictions by Li and Mishra as discussed under 'Compressive behavior' (p. 67) When special processing techniques are employed, presumably leading to lessened matrix defects, a substantial increase in compressive strength could be achieved (e.g. Rossi *et al.* 1995).

3.4.2 Continuous fiber composites

Short fiber FRC is attractive because it is easy to process and is inexpensive. The latter attribute allows the minimum use of fibers that are expensive and causes no change in concrete mixing and placement practice. Although ductility and toughness of short fiber FRC can amazingly approach that of ductile metals, tensile strength is several times higher than that of concrete but still significantly lower than that of other structural materials such as steel and wood. It is desirable to incorporate more fibers into the composite to enhance tensile strength allowing for the use of HPFRC in tensile strength demanding areas as well. Motivated by the success of fiber reinforced polymer composites (FRP) in the aerospace and automotive industries, continuous fiber reinforcement offers a unique opportunity to achieve this target goal.

Functions of fiber and matrix in composites

A typical advanced composite is primarily made of continuous fibers and matrix. The advantages of fiber reinforced polymer composites (FRP) as opposed to more conventional materials are often related to the high ratios of stiffness and strength to weight. A typical FRP is about four times lighter than steel of an equal strength. The strength/stiffness of FRP composite is almost entirely attributed to fibers (Bogner 1990, Swanson 1997), since a typical polymeric matrix has negligible strength/stiffness compared to engineering fibers. The matrix serves three important functions: (i) it holds the fiber in place, (ii) it transfers loads to the high-stiffness fiber, and (iii) it protects the fiber. Instead of using polymeric resins, cement-based materials can also be used to make high performance composites. Cement-based materials have many advantages over polymeric resins, including low cost, higher fire and vandalism rcsistancc, and thcy arc uscr fricndly to thc construction industry.

In developing continuous fiber reinforced cement-based composites, it is of ultimate importance to retain processing simplicity of concrete materials and eliminate the need for post curing that is typical for FRP, including heat, pressure or vacuum. This is desired to keep the process simple and the cost down. Significant progress has been made recently (Cox 1994, Peled and Bentur 2003, Kruger *et al.* 2003, Reinhardt *et al.* 2003, Curbach and Hegger

2003). For instance, Wu and his co-workers (Wu and Teng 2002, Wu 2004) have created very high performance continuous fiber reinforced cementitious composites. In their studies, the same kind of fiber reinforcement (unidirectional tapes or woven fabrics) as are employed in regular FRP composite was used, and their attention was given to developing a process with minimum modifications to the usual process for plain concrete.

The manufacturing procedure is analogous to regular FRP. Unidirectional fiber tapes and woven fabrics can be used, including various fiber types such as carbon, PVA and glass. It is recognized that glass fiber suffers from degradation when in contact with a cement pore solution that is highly alkaline. The glass fiber was included in their study mainly for comparison purposes. Nevertheless, alkali-resistant glass (AR-glass) under development may have the potential to overcome this deficiency. Three-dimensional fabric, consisting of two biaxial woven fabrics connected by vertical fibers can be manufactured to have various thickness (or height), ranging from 3 mm and above.

Dry fiber sheet, unidirectional tape or woven fabric, is first precut to specified dimensions and then impregnated with cementitious slurry to form composite. Infiltration time is certainly dependent on viscosity of the cementitious slurry. Full penetration has been achieved within 10–20 seconds when appropriate cementitious slurries are used. Fresh cementitious composite is immediately cast against the mold and left for curing in air. It may be necessary to add more layers of fiber sheets to build up enough thickness or to increase the distance between adjacent fiber layers. Apparently, either way affects the final fiber count of the composite and should be determined based on performance requirements for each application.

Ultra high performance

Because of high fiber contents in continuous fiber FRC (fiber volume fractions range typically between 5% to 20%, and can be as high as 50%), continuous fiber composites usually have very high strength and stiffness. As shown in Fig. 3.36 a very high tensile strength of 200 MPa was achieved with continuous carbon fibers whereas polymer fiber composite was, as usual, not as strong and much more flexible. In addition, just like its short fiber counterpart, the ductility and toughness were significantly better than that of plain concrete and approached regular FRP. Such high strength and high ductility are not the norm of cementitious materials, but it should not be too surprising since the high strength comes from the fibers and the high ductility from the multiple cracking phenomenon of the brittle cementitious matrix (Aveston *et al.* 1971, Li and Wu 1992).

Under flexural loading, similar results were obtained (see Fig. 3.37). Again, the carbon composite showed higher flexural strength/stiffness than the glass composite. However, the final failures of both composites were

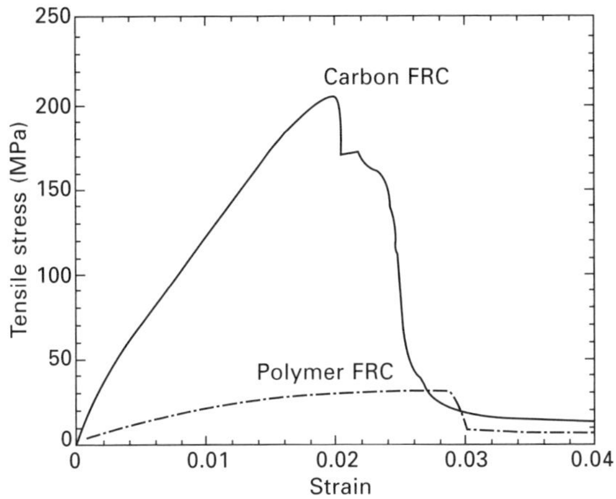

3.36 Tensile stress-strain curves of continuous fiber reinforced cementitious composites (fiber volume fractions are approximately 20% and 5% in the carbon composite and polymer fiber composite respectively).

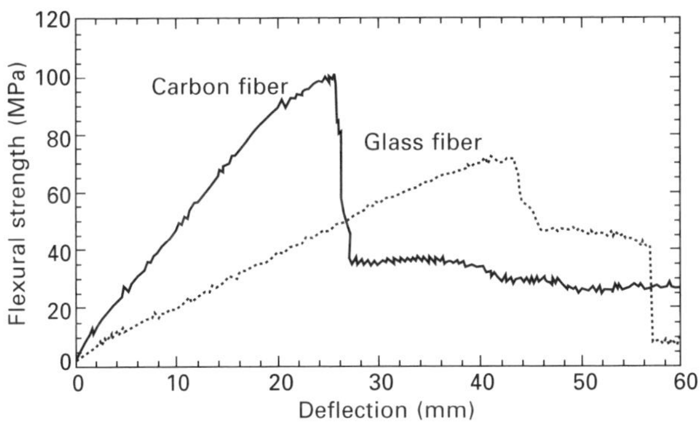

3.37 Three-point flexural stress-deflection curves of continuous fiber reinforced cement composites (fiber volume fractions are approximately 15% in both cases).

dominated by concrete crushing on the compressive side of the specimens. The flexural strength can be further improved when the cementitious compositions are modified to improve their crushing resistance. Potential applications for ultra high performance continuous-fiber FRC include retrofit, architectural panels/facades and tunnel lining. Structural retrofit is further described in section 3.5.

3.4.3 Durability

Although a large number of studies on fiber reinforced cementitious composites have been reported, little is known about the long term durability of these composites. It is expected that changing environments such as temperature, humidity and presence of chemicals or CO_2 could influence fiber, matrix and interfacial properties of the composite. Consequently, some or all of the mechanical properties (strength, ductility, or toughness) may change over time. For cellulose fiber cement composites, Akers and Studinka (1989) carried out a series of accelerated tests to examine the aging effect in natural weathering conditions. They discovered that carbonation cycles could simulate accurately the effects of long-term natural weathering on cellulose composites. They also found that the cement composite became slightly stiffer, stronger but more brittle after aging. After aging, cellulose fibers petrify and fail in a brittle manner.

Most recently, Li *et al.* (2004) have investigated the durability of a high performance engineered cementitious composite (ECC) reinforced with polyvinyl alcohol (PVA) fiber. This ECC-PVA has a tensile strength of 6 MPa and a tensile strain of 4.7% at the age of 28 days. They adopted a hot water immersion test to simulate a long-term hot and humid environment, and concluded that after 26 weeks immersion in 60 °C water, the composite showed minimum change in strength but a 40% reduction in strain (from 4.7% to 2.7%). Nevertheless, the tensile strain capacity of 2.7% still remained over 200 times higher than that of normal concrete. It is estimated that 26 weeks of hot water immersion is equivalent to 70 years of natural weathering in Japan (Building Material Industry Society of Japan 2001).

3.5 Engineering applications

3.5.1 Structural retrofit for compressive strength

The US has an estimated $20 trillion investment in civil infrastructure systems. Because of aging, overuse, exposure, misuse and neglect, many of these systems are deteriorating and becoming more vulnerable to catastrophic failure when earthquakes or other natural hazards strike. It would be prohibitively costly and disruptive to replace these vast networks. They must instead be renewed in an intelligent manner. It is generally recognized that fiber reinforced polymer (FRP) composite is one of the most vital materials for repair, strengthening and rehabilitation of existing structures. Applications typically involve externally bonded composite fabrics or jackets on beams, columns and bridge decks (Priestley *et al.* 1996).

The typical density of common engineering fibers is around 1.0 g/cm^3 for most polymeric fibers, 1.7–2.0 g/cm^3 for carbon and 2.5–2.7 g/cm^3 for glass. For polymeric matrix, epoxy and polyester have a density between 1.2 and 1.4 g/cm^3, giving a lightweight composite density between 1.5 and 2.2 g/cm^3 (Ashby and Jones 1986). It is clear from the discussions in the continuous fiber composites section that polymer matrix provides a negligible contribution to composite strength/stiffness that is needed for effective retrofit of concrete structures. However, polymers have many other problems, including the lack of fire resistance and degradation under UV light.

Inspired by the many successful studies using thin FRP sheets for structural retrofit (Nanni *et al.* 1992, Karbhari *et al.* 1993, Saadatmanesh *et al.* 1994, Labossiere *et al.* 1995), a preliminary study on structural retrofit with high performance continuous fiber reinforced cementitious (FRC) composite jackets has recently been carried out at Wayne State University (Wu and Teng 2002). The purpose of these tests is to develop an innovative thin FRC composite that could be employed in structural retrofit. The target FRC composite under development is expected to provide similar or improved retrofit efficiencies with a lower cost and easier installation in comparison to FRP composite.

Advantages of cement-based matrix

The following comparisons are made between the conventional FRP composite and the new cement-based composite being developed at Wayne State University.

1. Concrete surface preparation for better bonding. It is generally required to apply a labor-intensive sequence of concrete surface preparation to achieve good bonding between the FRP sheet and the concrete. General procedure includes: (i) concrete grinding, (ii) priming, (iii) putty application, (iv) resin application, (v) fiber sheet application, (vi) for multiple plies repeat steps iv and v, (vii) resin application (cover and protective coating), (viii) finishing; this procedure generates noise and dust, and sometimes odors are generated from the chemical reactions of the resins. These problems are particularly acute in a closed environment. Therefore, continuous operation of the structures may not be possible during the installation of the FRP retrofit, especially for buildings. Much less or no concrete surface preparation will be needed when the polymer resin is replaced by ordinary cement due to its natural affinity to concrete. No unpleasant odors will be generated.
2. Cost benefit. Typical cost of polymer resin (e.g. epoxy or polyester) is about $1.0 to $2.0 per pound, whereas cement costs about $0.1 per pound. Additional cost savings can be attained by eliminating stringent surface preparation procedures for concrete.
3. Improved stiffness. Typical stiffness of polymer matrix (e.g. epoxy or polyester) is less than 5 GPa, whereas cement/mortar is about 20 GPa. FRC composites can possess much higher stiffness than their FRP counterparts. Improved composite stiffness can further enhance retrofit efficiency.

4. Fire resistance. One of the major problems/obstacles of using FRP sheets for structural retrofit is that the epoxy resin begins to soften at 180 °F (or 82 °C). This limits a widespread use of FRP composites particularly in buildings due to fire concerns. To improve fire resistance of FRP retrofitted concrete, additional fire resistant coating or fireproof panels are installed on top of the FRP sheets. This procedure is labor intensive and very costly. Cement matrix composite is fire resistant at a high temperature (1000 °F or 538 °C). The development of a fire resistant matrix material to replace epoxy resin would be an important technical breakthrough in the field. Certainly, the fibers chosen for fire resistant applications must be temperature-resistant as well.

It should be noted that typical density of cementitious materials ranges from 0.8 to 2.2 g/cm^3, depending on their compositions. Therefore HPFRC can also be made lightweight. It should also be emphasized that *in-situ* applicability of cementitious matrix can be achieved through control of the rheological properties of cementitious materials that can range from water-like to dough-like.

Test results

A cement based matrix developed at Wayne State University and conventional epoxy resin were used separately to make thin CFRC sheets and CFRP sheets. Both CFRC and CFRP composites contained two layers of unidirectional carbon fiber tapes. The average thickness of the CFRC sheets was 3.0 mm, whereas the CFRP was between 2–3 mm. These thin composite sheets were then used to wrap 101.6 mm by 203.2 mm (4 in. by 8 in.) concrete cylinders. The bond length of the CFRC samples was 76.2 mm (3 in.) and 50.8 mm (2 in.) for the CFRP. A 38.1 mm (1.5 in.) gap existed between the top of the cylinder and the top of the composite sheet at both ends. These concrete cylinders, unconfined and confined with CFRC or CFRP composites were tested using a high-stiffness, high-capacity MTS testing machine following ASTM C39-96, *Standard Test Method for Compressive Strength of Cylindrical Concrete Specimens*. Per ASTM Practice C1231-93, steel restraining rings and rubber pads were used without other capping during testing.

The axial compressive stress-versus-axial displacement relationships of the unconfined and confined concrete are shown in Fig. 3.38. The initial portions of the stress-strain responses of the confined specimens essentially follow the curves of the unconfined concrete. The compressive strength of the unconfined concrete is 54 MPa. The CFRC group shows compressive strength two times higher than the control, an increase from 54 to 100 MPa. In addition, the ductility increased by 2.5 times, from the maximum axial displacement of 2.7 mm to 6.7 mm. The CFRP concrete had the highest

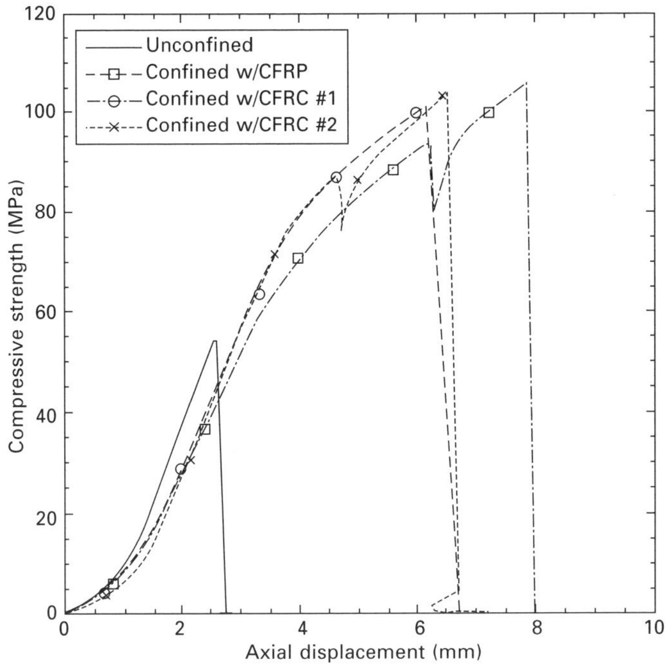

3.38 Compressive behavior of unconfined and confined concrete.

compressive strength (105 MPa) and ductility. Nevertheless, the differences between the CFRC sample and CFRP sample are insignificant.

The CFRP concrete showed explosive failure at the onset of final rupture. This was triggered by complete rupture of the CFRP jacket. The CFRC cylinder also showed fiber rupture failure similar to the CFRP cylinder. The CFRC concrete, however, had a much less violent global failure than the CFRP.

3.6 Conclusions

The above discussion has dealt with the unique properties of several advanced fiber reinforced cement based composites. As mentioned earlier, the intent was to introduce the concept of performance driven design, describe some of the salient composite properties, and to exploit potential applications; it was not intended to be an exhaustive account of all of the fiber reinforced concrete that can be considered to be 'high performance'. Much other useful information can be found in the collection of papers edited by Naaman and Reinhardt (1992, 1996, 1999, 2003). In addition, those books by Bentur and Mindess (1990), Shah and Ahmad (1992), and Shah *et al.* (1995) provide the most complete and broad overview of all aspects of fiber reinforced concrete.

3.7 Acknowledgments

Special thanks are offered to Professor Christian Meyer of Columbia University and Professor Sidney Mindess of University of British Columbia, for their reviews and valuable input.

3.8 References

AASHTO (1993). *AASHTO Guide for Design of Pavement Structures*, American Association of State Highway and Transportation Officials.

ACI Committee 224 (1972). 'Control of Cracking in Concrete Structures,' *Journal of the American Concrete Institute*, Vol. 69, No. 12, pp. 717–753.

Akers, S.A.S. and Studinka, J.B. (1989). 'Aging Behavior of Cellulose Fiber Cement Composites in Natural Weathering and Accelerated Tests,' *Inter. J. Cement Composites and Light Weight Concrete*, Vol. 11, No. 2, pp. 93–97.

Arisoy, B. (2002). *Development and Fracture Evaluation of Lightweight Fiber Reinforced Concrete*, Ph.D. Dissertation, Department of Civil and Environmental Engineering, Wayne State University, Detroit, MI.

Ashby, M.F. and Hallam, S.D. (1986). 'The Failure of Brittle Solids Containing Small Cracks under Compressive Stress States,' *Acta Metall.* Vol. 34, No. 3, pp. 497–510.

Ashby, M.F. and Jones, D.R.H. (1986). *Engineering Materials*, Pergamon Press, Oxford.

Aveston, J., Cooper, G.A. and Kelly, A. (1971). 'Single and Multiple Fracture', in *The Properties of Fiber Composites, Conf. Proc.*, IPC Sci. & Technology Press Ltd, pp. 15–24.

Babut, R. and Brandt, A.M. (1978). 'The Method of Testing and Analyzing of Steel Fiber Reinforced Concrete Elements in Flexure,' *Proc. RILEM Symp., Testing and Test Methods of Fiber Cement Composites*, R.N. Swamy ed., The Construction Press, pp. 479–486.

Bache, H.H. (1995). 'Concrete and Concrete Technology in a Broad Perspective', *Nordic Symp. On Modern Design of Concrete Structures*, ed. K. Aakjar, Aalborg University.

Balaguru, P.N. and Shah, S.P. (1992). *Fiber-Reinforced Cement Composites*, McGraw-Hill Inc.

Banfill, P.F.G. and Tattersall, G.H. (1983). *The Rheology of Fresh Concrete*, Pitman Publishing, London.

Bentur, A. and Mindess, S. (1990). *Fiber Reinforced Cementitious Composites*, Elsevier Applied Science, London.

Bogner, B.R. (1990). 'Isopolyester Pultrusion Resin Study,' *Proc. SPI Composite Institute 45th Conf.,* New York.

Building Material Industry Society of Japan (2001). *Research on the Evaluation Method of the Durability of Fiber Reinforced Concrete.*

Busel, J.P. and Lindsay, K. (1997). *Special Report, Composite Design and Application*, January/February, pp. 14–23.

Collepardi, M. (1993). 'Superplasticizers and Air Entraining Agents: State of the Art and Future Needs,' *Concrete Technology Past, Present, and Future*. ACI, Detroit SP-144.

Cox, B. N. (1994). 'Delamination and Buckling in 3D Composites,' *J. Composite Materials*, Vol. 28, No. 12, pp. 1114–1126.

Curbach, M. and Hegger, J. (2003). *2nd Colloquium on Textile Reinforced Structures*, ed. Dresden, Germany.

Elfgren, L., Noghabai, K., Ohlsson, U. and Olofsson, T. (1995). 'Applications of Fracture Mechanics to Anchors and Bond', *Proc. of Workshop on Fracture Mechanics and Design of Concrete Structures*, FRAMCOS II, Zürich.

Faroug, F., Szwabowski, J. and Wild, S. (1999). 'Influence of Superplasticizers on Workability of Concrete,' *Journal of Materials in Civil Engineering*, May, pp. 151–157.

FIP (1983). *FIP Manual of Lightweight Aggregate Concrete.*

Geng, Y. and Leung, C.K.Y. (1995). 'Damage Evolution of Fiber/Mortar Interface during Pull-out,' in *Proc. Symp. Vol. 370, Materials Research Society*, ed. S. Diamond, *et al.*, pp. 519–528.

Grzybowski, M. and Shah, S.P. (1990). 'Shrinkage Cracking of Fiber Reinforced Concrete,' *ACI Materials Journal*, Vol. 87, No. 2, pp. 138–148.

Hashin, Z. (1962). 'The Elastic Moduli of Heterogeneous Material,' *J. Appl. Mech. Trans. ASME* 29, pp. 143–150.

Horii, H. and Nemat-Nasser, S. (1986). 'Brittle Failure in Compression: Splitting, Faulting, and Brittle-Ductile Transition,' *Phil. Trans. Royal Soc.* London, Vol. 319, pp. 337–374.

Ivanov, Y.P. and Roshavelov, T.T. (1990). 'The Effect of Condensed Silica Fume on the Rheological Behavior of Cement Pastes,' *Rheology of Fresh Cement and Concrete*, edited by P.F.G. Banfill, E&FN Spon, London.

Kaplan, S.L., Rose, P.W., Nguyen, H.X. and Chang, H.W. (1988). 'Gas Plasma Treatment of Spectra Fiber,' *SAMPE Quarterly*, Vol. 19, No. 4, pp. 55–59.

Karbhari, V.M., Eckel, D.A. and Tunis, G.C. (1993). 'Strengthening of Concrete Column Stubs Through Resin Infused Composite Wraps,' *J. of Thermoplastic Composite Materials*, Vol. 6, pp. 92–107.

Kim, J.K., Baillie, C. and Mai, Y.W. (1992). 'Interfacial Debonding and Fiber Pull-out Stresses Part I: Critical Comparison of Existing Theories with Experiments,' *J. Mater. Sci.*, Vol. 27, pp. 3143–3154.

Kim, P., Wu, H.C., Lin, Z., Li, V.C., de Lhoneux, B. and Akers, S. (1999). 'Micromechanics Based Durability Study of Cellulose Cement in Flexure', *Journal of Cement and Concrete Research*, Vol. 29, No. 2, pp. 201–208.

Kong, H.J., Bike, S.G. and Li, V.C. (2003). 'Development of a Self-Consolidating Engineered Cementitious Composite Employing Electrosteric Dispersion/Stabilization,' *Cement and Concrete Composites*, Vol. 25, pp. 301–309.

Kong, H.J., Bike, S.G. and Li, V.C. (2003a). 'Constitutive Rheological Control to Develop a Self-Consolidating Engineered Cementitious Composite Reinforced with Hydrophilic Poly(vinyl alcohol) Fibers,' *Cement and Concrete Composites*, Vol. 25, pp. 333–341.

Kruger, M., Reinhardt, H.W. and Yong, X. (2003). 'Sulphoaluminate Cement Matrices Used for Textile and Glass Fiber Reinforced Concrete Elements,' in *Proc. High Performance Fiber Reinforced Cement Composites*, 4 (HPFRCC 4), Naaman, A. and Reinhardt, H.W. (eds), Cachan, France.

Labossiere, P., Neale, K.W., Demers, M. and Picher, F. (1995). 'Repair of Reinforced Concrete Columns with Advanced Composite Materials Confinement,' in *Repair and Rehabilitation of the Infrastructure of the Americas*, H.T. Toutanji (ed.), University of Puerto Rico, pp. 153–165.

Lewis, J.A., Matsuyama, H., Kirby, G., Morissette, S. and Young, J.F. (2000). 'Polyelectrolyte Effects on the Rheological Properties of Concentrated Cement Suspensions,' *J. Am. Ceram. Soc.*, Vol. 83, No. 98, pp. 1905–1913.

Li, V.C. (1992a). 'Postcrack Scaling Relations for Fiber Reinforced Cementitious Composites,' *J. Materials in Civil Engineering*, Vol. 4, No. 1, pp. 41–57.

Li, V.C. (1992b). 'A Simplified Micromechanical Model of Compressive Strength of Fiber Reinforced Cementitious Composites,' *J. of Cement and Concrete Composites*, Vol. 14, pp. 131–141.

Li, V.C. (1993). 'From Micromechanics to Structural Engineering – The Design of Cementitious Composites for Civil Engineering Applications,' *J. Struct. Mech. Earthquake Eng.*, Vol. 47, pp. 1–24.

Li, V.C. and Chan, Y.W. (1994). 'Determination of Interfacial Debond Mode for Fiber-Reinforced Cementitious Composites,' *ASCE, J. Engineering Mechanics*, Vol. 120, No. 4, pp. 707–719.

Li, V.C. and Leung, C.K.Y. (1992). 'Theory of Steady State and Multiple Cracking of Random Discontinuous Fiber Reinforced Brittle Matrix Composites,' *ASCE J. of Engng. Mechanics*, Vol. 118, No. 11, pp. 2246–64.

Li, V.C. and Matsumoto, T. (1998). 'Fatigue Crack Growth Analysis of Fiber Reinforced Concrete with Effect of Interfacial Bond Degradation,' *J. Cement and Concrete Composites*, Vol. 20, No. 5, pp. 339–351.

Li, V.C. and Mishra, D.K. (1992). 'Micromechanics of Fiber Effect on the Uniaxial Compressive Strength of Cementitious Composites,' in *RILEM 4th Inter. Symp. on Fiber Reinforced Concrete*, R.N. Swamy (ed.), E&FNSpon, London.

Li, V.C. and Wu, H.C. (1992). 'Conditions for Pseudo Strain-Hardening in Fiber Reinforced Brittle Matrix Composites,' *Applied Mechanics Review*, Vol. 45, pp. 390–398.

Li, V.C. and Wu, H.C. (1998). US patent 5,788,760.

Li, V.C., Wang, Y. and Backer, S. (1991a). 'Fracture Energy Optimization in Synthetic Fiber Reinforced Cementitious Composites,' *Mat. Res. Soc. Symp. Proc.* Vol. 211, pp. 63–69.

Li, V.C., Wang, Y. and Backer, S. (1991b). 'A Micromechanical Model of Tension-Softening and Bridging Toughening of Short Random Fiber Reinforced Brittle Matrix Composites,' *J. Mech. Phys. Solids*, Vol. 39, No. 5, pp. 607–625.

Li, V.C., Mishra, D.K. and Wu, H.C. (1995a). 'Matrix Design for Pseudo-Strain-Hardening Fiber Reinforced Cementitious Composites,' *Materials and Structures*, Vol. 28, pp. 586–595.

Li, V.C., Mihashi, H., Wu, H.C., Alwan, J., Brincker, R., Horii, H., Leung, C.K.Y., Maalej, M. and Stang, H. (1995b). 'Micromechanical Models of Mechanical Response of HPFRCC,' in *High Performance Fiber Reinforced Cement Composites*, Naaman, A.E. and Reinhardt, H.W. (eds), E&FN Spon.

Li, V.C., Horikoshi, T., Ogawa, A., Torigoe, S. and Saito, T. (2004). 'Micromechanics-based Durability Study of Polyvinyl Alcohol-Engineered Cementitious Composite,' *ACI Materials J.*, Vol. 101, No. 3, pp. 242–248.

Maalej, M. and Li, V.C. (1994). 'Flexural/Tensile Strength Ratio in Engineered Cementitious Composites,' *ASCE J. of Materials in Civil Engineering*, Vol. 6, No. 4, pp. 513–528.

Maalej, M. and Li, V.C. (1995). 'Introduction of Strain Hardening Engineered Cementitious Composites in the Design of Reinforced Concrete Flexural Members for Improved Durability,' *American Concrete Institute Structural J.*, Vol. 92, No. 2, pp. 167–176.

Marshall, D.B., Cox, B.N. and Evans, A.G. (1985). 'The Mechanics of Matrix Cracking in Brittle-Matrix Fiber Composites,' *Acta Metall.* Vol. 33, No. 11, pp. 2013–2021.

Mindess, S. and Young, J.F. (1981). *Concrete*, Prentice-Hall Inc., Englewood Cliffs, N.J.

Naaman, A.E. and Reinhardt, H.W. (1992). 'High Performance Fiber Reinforced Cement Composites, 1 (HPFRCC1)', *RILEM Proceedings 15*, E&FN Spon, London.

Naaman, A.E. and Reinhardt, H.W. (1996). 'High Performance Fiber Reinforced Cement Composites, 2 (HPFRCC2)', *RILEM Proceedings 31*, E&FN Spon, London.
Naaman, A.E. and Reinhardt, H.W. (1999). 'High Performance Fiber Reinforced Cement Composites, 3 (HPFRCC 3),' *RILEM Proceedings, PRO 6*, RILEM Publications S.A.R.L., Cachan, France.
Naaman A.E. and Reinhardt H.W. (2003). 'High Performance Fiber Reinforced Cement Composites, 4 (HPFRCC 4)', *RILEM Proceedings, PRO 30*, RILEM Publications S.A.R.L., Cachan, France.
Nanni, A., Norris, M.S. and Bradford, N.M. (1992). 'Lateral Confinement of Concrete Using FRP Reinforcement,' *ACI SP 138, Fiber Reinforced Plastic Reinforcement for Concrete Structures*, pp. 193–209.
Nehdi, M., Mindess, S. and Aitcin, P.-C. (1998). 'Rheology of High-Performance Concrete: Effect of Ultra Fine Particles,' *Cement and Concrete Research* Vol. 28, No. 5, pp. 687-3-697.
Neville, A.M. (1995). *Properties of Concrete*. Longman Group Limited, Essex, England.
Peled, A. and Bentur, A. (2003). 'Cement Impregnated Fabrics for Repair and Retrofit of Structural Concrete,' in *Proc. High Performance Fiber Reinforced Cement Composites, 4* (HPFRCC 4), eds Naaman, A. and Reinhardt, H.W., Cachan, France.
Priestley, M.J.N., Seible, F. and Calvi, G.M. (1996). *Seismic Design and Retrofit of Bridges,* John Wiley & Sons, New York.
Reinhardt, H.W., Kruger, M. and Grosse, C.U. (2003). 'Concrete Prestressed with Textile Fabric,' *Journal of Advanced Concrete Technology*, Vol. 1, No. 3, pp. 231–239.
Rossi, P., Sedran, T., Renwez, S. and Belloc, A. (1995). 'Ultra High Strength Steel Fiber Reinforced Concretes: Mix Design and Mechanical Characterization,' in *Fiber Reinforced Concrete – Modern Developments*, Banthia, N. and Mindess, S. (eds), UBC Press, Canada, pp. 181–186.
Roy, D.M. and Asaga, K. (1979). 'Rheological Properties of Cement Mixes: III The Effects of Mixing Procedures on Viscometric Properties of Mixes Containing Superplasticisers,' *Cement and Concrete Research*, pp. 731–739.
Saadatmanesh, H., Ehsani, M.R. and Li, M.W. (1994). 'Strength and Ductility of Concrete Columns Externally Reinforced with Fiber Composite Straps,' *ACI Structural Journal*, Vol. 91, No. 4, pp. 434–447.
Shah, S.P. (1991). 'Toughening of Cement-based Materials with Fiber Reinforcement', *Mat. Res. Soc. Sym. Proc.* Vol. 211, pp. 3–13.
Shah, S.P. and Ahmad, S.H. (1992). *Fiber Reinforced Cement Composites*. McGraw Hill.
Shah, S.P., Swartz, S.E. and Ouyang, C. (1995). *Fracture Mechanics of Concrete*. John Wiley & Sons.
Shah, S.P., Cyr, M.F. and Peled, A. (2001). 'Extruded Hybrid Fiber Reinforced Cementitious Composites,' in *FERRO-7: Seventh Inter. Symp. on Ferrocement and Thin Reinforced Cement Composites*, Singapore.
Shah, S.P., Lawler, J.S. and Zampini, D. (2002). 'Permeability of Cracked Hybrid Fiber-Reinforced Mortar under Load,' *ACI Materials Journal*, July–August, Vol. 99, No. 4, pp. 379–386.
Struble, L. and Szecsy, R. (1996). 'Rheology of Fresh Concrete,' in *Proc. 4th Materials Engineering Conf*, pp. 1121–1128, ed. K. Chong, ASCE.
Struble, L., Szecsy, R., Lei, W.G. and Sun, G.K. (1998). 'Rheology of Cement Paste and Concrete,' *Cement, Concrete, and Aggregate, CCAGDP*, Vol. 20, No. 2, pp. 269–277.
Swanson, S.R. (1997). *Advanced Composite Materials*, Prentice Hall, New Jersey.

Tattersall, G.H. (1955). Structural Breakdown of Cement Paste at Constant Rate of Shear,' *Nature*, Vol. 175 (4447), pp. 166.

Tattersall, G.H. (1973). 'The Rationale of a Two-Point Workability Test,' *Mag. of Concrete Res.*, Vol. 25, pp. 169–172.

Tattersall, G.H. (1991). *Workability and Quality Control of Concrete*, E &FN Spon.

Tsukamoto, T. (1990). 'Tightness of fiber concrete,' in *Darmstadt Concrete*, Vol. 5, pp. 215–225.

Wang, Y. (2001). *Flow of Fiber-Reinforced Cement Slurries at Elevated Temperatures*, PhD Dissertation, Columbia University, New York, NY.

Wang, Y. and Meyer, C. (2001). 'Flow of Fiber-Reinforced Cement Slurries at Elevated Temperatures,' in *Proc. 1st MIT Conference on Computational and Fluid Mechanics*, MIT Cambridge, MA.

Wang, Y., Li, V.C. and Backer, S. (1991). 'Tensile Failure Mechanisms in Synthetic Fiber-reinforced Mortar,' *J. Mater. Sci.* Vol. 26, pp. 6565–6575.

Wu, H.C. (2001). 'Discussion on Mechanical Properties of Steel Microfiber Reinforced Cement Pastes and Mortars,' *ASCE Journal of Materials*, Vol. 13, No. 3, pp. 240–241.

Wu, H.C. (2004). 'Design Flexibility of Composites for Construction,' in *Proc. Inter. Conf. on Fiber Composites, High Performance Concretes and Smart Materials*, ed. V.S. Parameswaran, Chennai, India, pp. 421–432.

Wu, H.C. and Arisoy, B. (2000). 'Lightweight High Strength/Ductility Cementitious Composites,' in *Proc. 6th Inter. Symp. on Brittle Matrix Composites,* eds. A.M. Brandt, V.C. Li and I. Marshall, Warsaw, Poland, pp. 118–126.

Wu, H.C. and Kanda, T. (2000). 'FRP Wrap Interaction with Concrete Cracks through Bridging Stress,' in *Proc. 3rd International Conference on Advanced Composite Materials in Bridges and Structures*, eds J. Humar and A.G. Razaqpur, CSCE, pp. 275–282.

Wu, H.C. and Li, V.C. (1992). 'Snubbing and Bundling Effects on Multiple Crack Spacing of Discontinuous Random Fiber Reinforced Brittle Matrix Composites,' *Journal of the American Ceramic Society*, Vol. 75, No. 12, pp. 3487–89.

Wu, H.C. and Li, V.C. (1994). 'Trade-off between Strength and Ductility of Random Discontinuous Fiber Reinforced Cementitious Composites,' *Cement and Concrete Composites*, Vol. 16, pp. 23–29.

Wu, H.C. and Li, V.C. (1995). 'Stochastic Process of Multiple Cracking in Discontinuous Random Fiber Reinforced Brittle Matrix Composites,' *Inter. J. Damage Mechanics*, Vol. 4, pp. 83–102.

Wu, H.C. and Li, V.C. (1999). 'Fiber/Cement Interface Tailoring with Plasma Treatment,' *Journal of Cement and Concrete Composites*, Vol. 21, pp. 205–212.

Wu, H.C. and Teng, J. (2002). 'Innovative Cement Based Thin Sheet Composites for Retrofit,' in *CD Proc. 3rd Inter. Conf. on Composites in Infrastructures*, San Francisco, CA.

Wu, H.C., Mishra, D.K. and Li, V.C. (1993). 'Influence of Matrix Fracture Toughness on Tensile Behavior of Fiber Reinforced Cementitious Composites,' in *Proc. 3rd Beijing Int'l Symposium on Cement and Concrete*, Beijing, China, Vol. 3, pp. 222–237.

Wu, H.C., Lim, Y.M. and Li, V.C. (1994a). 'Shrinkage Behavior of Cementitious Composites with Recycled Fibers', in *Proc. 2nd Annual Great Lakes Geotechnical/ Geoenvironmental Conference on Waste Materials and Their Geotechnical/ Geoenvironmental Applications*.

Wu, H.C., Matsumoto, T. and Li, V.C. (1994b). 'Buckling of Bridging Fibers in Composites,' *J. Mater. Sci. Lett.*, Vol. 13, pp. 1800–1803.

Wu, H.C., Lim, Y.M., Li, V.C. and Foremsky, D.J. (1996). 'Utilization of Recycled Fibers in Concrete,' in *ASCE Proc. Materials Engineering Conference*, ed. K. Chong, pp. 799–808.

Zhou, L.M., Kim, J.K. and Mai, Y.W. (1992). 'Interfacial Debonding and Fiber Pull-out Stresses Part II: A New Model Based on the Fracture Mechanics Approach,' *J. Mater. Sci.*, Vol. 27, pp. 3155–3166.

Zureick A.H., Shih B. and Munley E. (1995). 'Fiber Reinforced Polymeric Bridge Decks,' *Structural Engineering Review*, Vol. 7, No. 3, pp. 257–266.

4

Advanced fiber-reinforced polymer (FRP) structural composites for use in civil engineering

J F DAVALOS, West Virginia University, USA and
P QIAO and L SHAN, The University of Akron, USA

4.1 Introduction

Polymeric composites are advanced engineering materials with the combination of high-strength, high-stiffness fibers (e.g., E-glass, carbon, and aramid) and low-cost, light-weight, environmentally resistant matrices (e.g., polyester, vinylester, and epoxy resins). The use of fiber-reinforced polymer or plastic (FRP) composite materials can be traced back to the 1940s in the military and defense industry, particularly in aerospace and naval applications. Because of their excellent properties (e.g., lightweight, noncorrosive, nonmagnetic, and nonconductive), composites can meet the high performance requirements of space exploration and air travel, and for this reason, composites were broadly used in the aerospace industry during the 1960s and 1970s (Bakis *et al.* 2002). From the 1950s, composites have been increasingly used in civil engineering for semi-permanent structures and rehabilitation of old buildings. A concise review on FRP composites for construction applications in civil engineering is given by Bakis *et al.* (2002).

Structures made of FRP composites have been shown to provide efficient and economical applications in bridges and piers, retaining walls, airport facilities, storage structures exposed to salts and chemicals, and others. In addition to light-weight, noncorrosive, nonmagnetic, and nonconductive properties, FRP composites exhibit excellent energy absorption characteristics – suitable for seismic response – high strength, fatigue life, and durability; competitive costs based on load capacity per unit weight; and ease of handling, transportation, and installation. FRP materials offer the inherent ability to alleviate or eliminate the following four construction related problems adversely contributing to transportation deterioration worldwide (Head 1996): corrosion of steel, high labor costs, energy consumption and environmental pollution, and devastating effects of natural hazards such as earthquakes. A great need exists for new materials and methods to repair and/or replace deteriorated structures at reasonable costs.

With the increasing demand for infrastructure renewal and the decreasing of cost for composite manufacturing, FRP materials began to be extensively used in civil infrastructure from the 1980s and continue to expand in recent years. Attention has been focused on FRP shapes as alternative bridge deck materials, because of their high specific stiffness and strength, corrosion resistance, light weight, and potential modular fabrication and installation that can lead to decreased field assembly time and traffic routing costs. In 1986, the first highway bridge using composites-reinforcing tendons in the world was built in Germany. The first all-composites pedestrian bridge was installed in 1992 in Aberfeldy, Scotland. The first FRP reinforced concrete bridge deck in the US was built in 1996 at McKinleyville, WV, followed by the first all-composite vehicular bridge as a sandwich deck built in Russell, Kansas in 1997.

Most currently available commercial bridge decks are constructed using assemblies of adhesively bonded pultruded FRP shapes. Such shapes can be economically produced in continuous lengths by numerous manufacturers using well-established processing methods (see details in section 4.2). Secondary bonding operations of cellular section are best accomplished at the manufacturing plant for maximum quality control. Design flexibility in this type of deck is obtained by changing the constituents of the shapes (such as fiber fabrics and fiber orientations) and, to a lesser extent, by changing the cross-section of the shapes. Due to the potentially high cost of pultrusion dies, however, variations in the cross-section of shapes are feasible only if sufficiently high production warrants the tooling investment.

There are numerous theoretical and experimental investigations which have been conducted to study stiffness, strength, and stability characteristics of FRP composite bridge structures. Henry (1985) and Ahmad and Plecnik (1989) examined the performance of several glass-reinforced polymer bridge deck configurations using finite element analysis, and indicated that the design was always controlled by deflection limit state rather than strength limit states. Later, FRP decks were fabricated using a combination of filament winding and hand lay-up processes, and the static and fatigue behaviors of decks were determined experimentally by Plecnik and Azar (1991); from the experimental observation, it was evident that damage under fatigue loading consisted primarily of interfacial delamination initiation and local buckling of thin delaminated layers under compressive service loads. Bakeri and Sunder (1990) used balanced symmetrical laminate to study the feasibility of two FRP bridge deck systems, one made entirely of glass-reinforced polymer and the other by a hybrid combination of materials including glass-reinforced polymer, carbon-reinforced polymer and light-weight concrete; they concluded that the hybrid system concept achieved less deflection and had a promising future in infrastructure applications. With the objective of developing an FRP bridge deck system, Mongi (1991) tested three full-scale floor systems which differ in their size, joint type, and loading conditions. GangaRao and

Sotiropoulos (1991) assembled and tested two FRP bridge superstructure systems consisting of bridge decks and stringers, they used a simplified finite-element model with equivalent plates representing the stringers and the cellular deck, and results were correlated with experimental data.

Burnside *et al.* (1993) presented a design optimization of an all-composite bridge deck; cellular-box and stiffened-box geometries were optimized with consideration of deflection and buckling. In 1997, the first all-composite bridge deck in the US, consisting of a honeycomb sandwich, was built in Russel, Kansas (Plunkett 1997). Later, two highway bridges were constructed with a modular FRP composite deck in West Virginia (Lopez-Anido *et al.* 1998a); one was built as an all-composite short-span deck/stringer system and the other as a modular FRP deck supported by steel stingers. Fatigue and failure characteristics of a modular deck were investigated by Lopez-Anido *et al.* (1998a;b) and satisfactory performance was observed. An overview on research and applications of fiber-reinforced polymeric bridge decks was presented by Zureick (1997) and Bakis *et al.* (2002).

A critical obstacle to the widespread use and application of FRP structures in construction is the lack of simplified and practical design guidelines. Unlike standard materials (e.g., steel and concrete), FRP composites are typically orthotropic or anisotropic, and their analyses are much more complex. For example, while changes in the geometry of FRP shapes can be easily related to changes in stiffness, changes in the material constituents do not lead to such obvious results. In addition, shear deformations in FRP composite materials are usually significant, and therefore, the modeling of FRP structural components should account for shear effects. For applications to pedestrian and vehicular FRP bridges, there is a need to develop simplified design equations and procedures, which should provide relatively accurate predictions of bridge behavior and be easily implemented by practicing engineers.

Closed-form, mechanics-based methods for designing sectional stiffness properties of composite shapes were detailed by Barbero *et al.* (1993) and Davalos and Qiao (1999). These mechanics concepts combined with elastic equivalent analysis can be translated into approximate methods for estimating the equivalent orthotropic plate behavior of decks. In this way, deck designs can be defined as assemblies of structurally efficient and easy-to-manufacture pultruded composite sections. A systematic analysis and design approach for single-span FRP deck-and-stringer bridges was presented by Qiao *et al.* (2000); while systematic methods for optimizing pultruded shapes have been developed (Davalos *et al.* 1996a; Qiao *et al.* 1998), and optimized deck designs are largely derived by trial and error.

A systemic design approach and procedure for FRP composite structures is presented in this chapter. First, the material properties involving constituent materials, ply properties, and laminated panel engineering properties are analyzed and given in the format of carpet plots for design convenience. Then the mechanics of laminated beam (MLB) theory is used to study the member stiffness, and several critical mechanical behaviors of FRP shapes including bending, buckling and material failure are presented, followed by the equivalent analysis of FRP cellular panels and macro-flexibility analysis of deck-and-stringer bridge systems. Finally, design guidelines and examples for FRP beams and bridge systems are given in details to illustrate the application and design procedures.

4.2 Manufacturing process by pultrusion

There are a number of available processes for composites manufacturers to produce cost-efficient products, such as pultrusion, resin transfer molding (RTM), vacuum assisted resin injection molding (VARIM), hand lay-up, compression molding, and filament winding. Each fabrication process has its own characteristics that define the type of products to be produced. For civil infrastructure related applications, the pultrusion process is the most prevalent for fabricating FRP shapes, due to its continuous process and mass production capabilities.

Pultrusion is a continuous manufacturing process capable of delivering one to five feet per minute of prismatic thin-walled members. Pultrusion composites exhibit all of the characteristics produced by other composite processes, such as high strength-to-weight ratio, corrosion resistance, electrical insulation, and dimensional stability. In addition, complex thin-walled profiles including hollow sections can be manufactured by pultrusion, and also relatively large pultruded shapes can be made since the equipment can be built to accommodate large sections.

There are two types of pultrusion products: (i) solid rods, bars and plates, and (ii) structural profiles. The first category is usually produced from axial fiberglass reinforcements and polyester resins, and primarily used to make items like electrical insulation bars and fishing rods which require high tensile strength. Recent applications have shown that FRP rebars can be efficiently used to reinforce concrete structures to minimize deterioration due to steel corrosion. FRP profiles are manufactured by combining the unidirectional roving fibers, various continuous strand mats, and stitched (braided, woven and knitted) fiber fabrics (see Fig. 4.1(a)) together with thermosetting resins. The fiber systems are continuously impregnated and pulled through a heated die where they are shaped and cured (see Fig. 4.1(b)). The die can be designed to meet the special requirement of different FRP shapes, such as I-, C-, and box-sections.

The process is normally continuous and highly automated (CP Design Manual 2000). A schematic diagram of a simple Pultrusion line is shown in Fig. 4.2. Reinforcement materials (e.g., roving, mat or fabrics) are positioned

(a) (b)

4.1 Pultrusion process (a) fiber guiding and impregnating; (b) forming structural profiles.

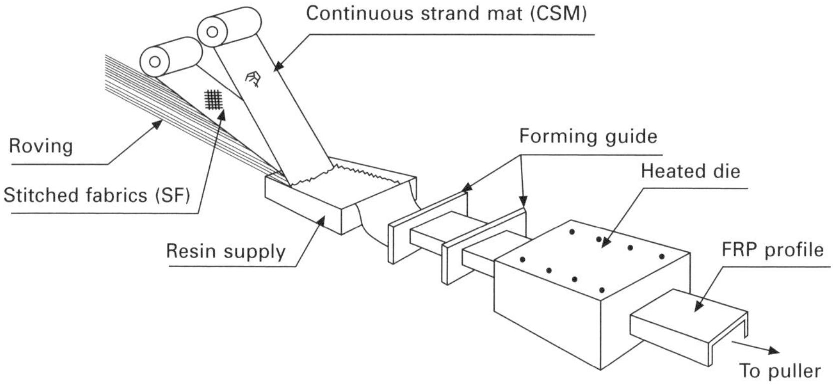

4.2 Schematic diagram of pultrusion process.

in a specific location using preformed shapers or guides to form the profile. It is easier to form an open section than a closed one, which is usually fabricated by using a mandrel cantilevered behind the entrance to the die. A rotating winder can also be used to apply reinforcements at an angle (usually symmetric $\pm\theta°$) around the product, which is commonly used to manufacture pipes. There are other additions from fiber pre-heaters to radio-frequency heaters for thick parts or using a thermoplastic matrix. The reinforcements are drawn through a resin bath or wet-out where the material is thoroughly coated or impregnated with a liquid thermosetting resin. The resin-saturated reinforcements enter a heated metal pultrusion die. The dimensions and shape of the die define the finished part being fabricated. Inside the metal die, heat is transferred, initiated by precise temperature control to the reinforcements and liquid resin. The heat energy initiates an exothermic reaction in the thermosetting resin and activates the curing or polymerization with changing its state from a liquid to a solid. The process is driven by a system of pullers, and the solid laminate emerges from the pultrusion die to the exact shape of the die cavity. The laminate solidifies when cooled and it is continuously pulled through the pultrusion machine and cut to the desired length.

Any fiber types and various resins including thermoplastics can be used in the pultrusion process, but E-glass fibers and polyester or vinylester resins dominate the market because of their relatively low cost. The main fiber types (Fig. 4.3) include roving, continuous strand mat (CSM), and stitched

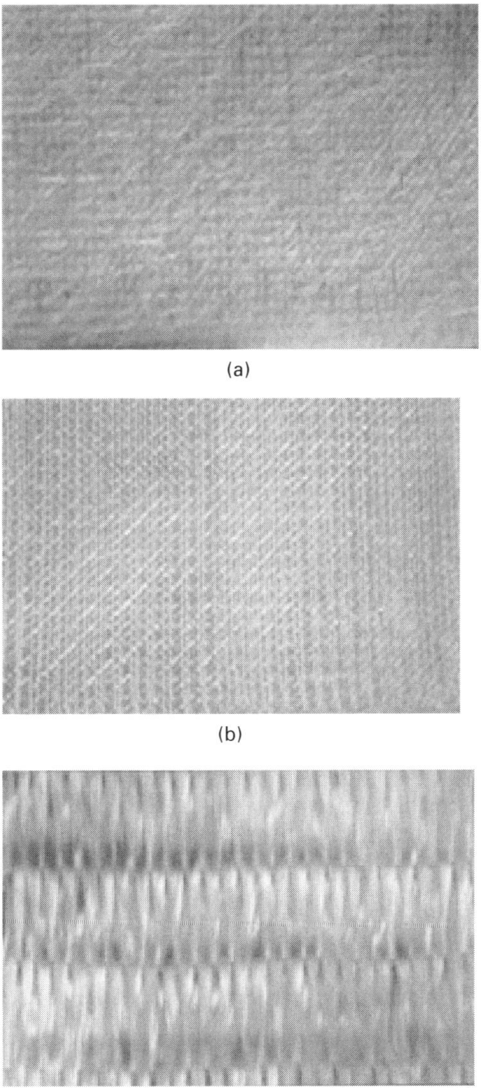

(a) (b) (c)

4.3 Typical fiber layers used in pultrusion process (a) continuous strand mat (CSM); (b) ± 45° stitched fabrics (SF); (c) roving bondles.

fabrics (SF). Pultrusion has some constraints regarding fiber volume fraction and thickness. Due to the need to pull the products during the process, fiber volume fraction is seldom over 45%, and 30%~35% is a typical value being used. The thickness of the pultruded panels is commonly limited to about 12.7 mm (1/2 in) when using standard conduction heaters because of their limitations and potential for interlaminar cracking. The pultrusion rates of the products are different, based on the type of machine being used and the cross-section being pultruded. On average, a standard beam cross-section can be produced at about 2 m per minute, while a bridge deck panel section can be manufactured at about 20 m^2 per minute in a single production line.

4.3 Material properties and systematic analysis and design

In this section, a systematic approach for analysis and design of all FRP composite structures is presented. This approach (Fig. 4.4) is based on analyses at micro-level (material), macro-level (structural component) and system level (structure) to design all FRP deck-and-stringer bridge systems. First, based on manufacturer's information and material lay-up, ply properties are predicted by micromechanics. Once the ply stiffnesses are obtained, macromechanics is applied to compute the panel mechanical properties which are then presented in the form of carpet plots for design convenience. Beam or stringer stiffness properties are then evaluated from mechanics of thin-walled laminated beams (MLB). The elastic deformation including shear effects and maximum stress of FRP beams are correspondingly predicted. Then, based on the stability analyses, formulas for global and local buckling loads of FRP shapes (columns and beams) are presented. From the panel strength data and carpet plots, the material failure strengths of FRP beams are evaluated. Using elastic equivalence, apparent stiffnesses for composite cellular decks are formulated in terms of panel and single-cell beam stiffness properties, and their equivalent orthotropic material properties are further obtained. For design analysis of FRP deck-and-stringer bridge systems, an approximate series solution for first-order shear deformation orthotropic plate theory is applied to develop simplified design equations, which account for load distribution factors for various load cases. As illustrated in Fig. 4.4, the present systematic approach, which accounts for the microstructure of composite materials and geometric orthotropy of a composite structural system, can be used to design and optimize efficient FRP decks and deck-and-stringer systems.

4.3.1 Constituent materials and ply properties

FRP shapes are not laminated structures in a rigorous sense. However, they

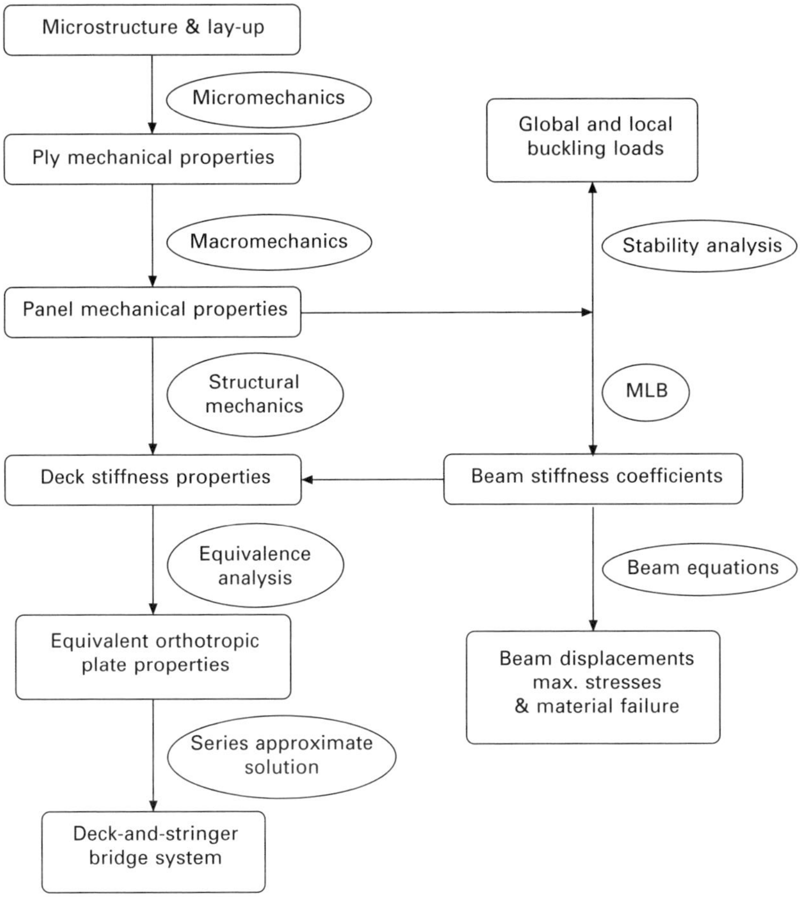

4.4 Systematic analysis protocol for all FRP composite structures.

are produced with material architectures that can be simulated as laminated configurations (Davalos *et al.* 1996b). For a typical pultruded FRP section shown in Fig. 4.5, the panels in the thin-walled structure can be simulated as laminates consisting of combinations of the following four types of layers: (i) A thin layer of randomly oriented chopped fibers (Nexus) placed on the surface of the composite, which is a resin-rich layer primarily used as a protective coating, and its contribution to the laminate response can be neglected; (ii) continuous or chopped strand mats (CSM) of different weights consisting of either continuous or chopped randomly-oriented fibers; (iii) stitched fabrics (SF) with different fiber orientation; and (iv) roving layers that contain continuous unidirectional fiber bundles. In this section we present constituent material properties, modeling of materials based on best evaluations of volume fractions of constituents, and computation of ply stiffness by micromechanics.

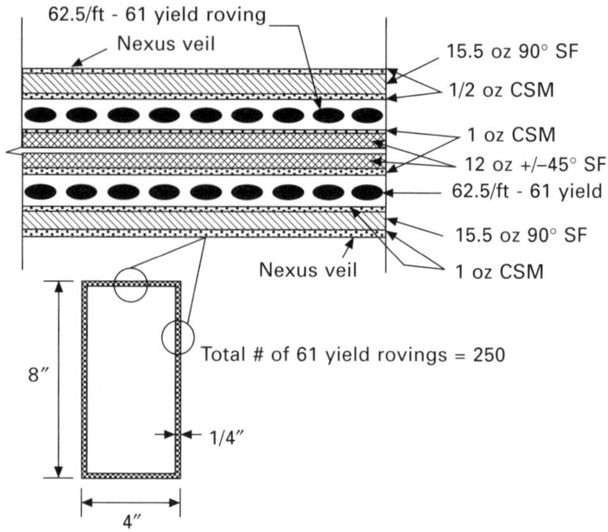

4.5 Lay-up of a typical FRP box-section (4″ × 8″ × 1/4″).

Constituent materials

Ply stiffnesses of pultruded panels cannot be readily evaluated experimentally, since the material is not produced by lamination lay-up, but they can be computed from micromechanics formulas for roving, CSM, and SF layers (Qiao 1997). The ply stiffnesses can then be used in classical lamination theory (CLT) (Jones 1999) to predict the laminate stiffnesses. Several textbooks on composite materials provide information on constituent material properties, and as a illustration, we present in Table 4.1 properties for E-glass fiber and vinylester resins. As for isotropic materials, the Lamé constants for fiber and matrix are expressed in terms of modulus E and Poisson's ratio ν, as

$$\lambda = \frac{E\nu}{(1+\nu)(1-2\nu)} \qquad 4.1$$

$$\mu = G = \frac{E}{2(1+\nu)}$$

Table 4.1 Material properties of the constituents

Material	E (psi)	G (psi)	ν	ρ (lb/in^3)
E-glass fiber	10.5×10^6	4.1833×10^6	0.255	0.092
Vinylester resin[(a)]	7.336×10^5	2.3714×10^5	0.30	0.041

(a) obtained experimentally (Tomblin 1994).

Fiber volume fraction

The properties of a pultruded composite section are controlled by the relative volumes of fiber and matrix used. Each layer is modeled as a homogeneous, linearly elastic and generally orthotropic material (Davalos *et al.* 1996b), and to evaluate its properties, the information provided by the material producer and pultrusion manufacturer are used to compute the fiber volume fraction (V_f). The fiber volume fraction (V_f) is defined as the ratio of the volume of fibers present to the total volume of the layer. The relative volumes of fiber and matrix are first determined in order to evaluate the ply stiffnesses. Similarly, the fiber volume fraction (V_f) of the whole section can be defined. For the CSM and SF layers, which are, respectively, specified in oz/ft^2 and oz/yd^2, V_f can be determined as follows:

$$(V_f)_{\text{CSM/SF}} = \frac{w}{\rho t_f} \qquad 4.2$$

where, w is the weight per unit area in lb/in^2 (i.e., convert the units of oz/ft^2 and oz/yd^2 to the one of lb/in^2), ρ is the unconsolidated density of the CSM or SF fibers in lb/in^3, and t_f is the 'as manufactured' thickness of the fabrics (inches) as provided by the material producer. For the roving layers, the fiber volume fraction (V_f) is defined as:

$$(V_f)_r = \frac{n_r A_r}{t_r} \qquad 4.3$$

where, n_r is the number of rovings per unit width (in^{-1}) provided by the manufacturer, t_r is the thickness of the composite layer, which can be evaluated by subtracting the CSM and SF layer thickness from the total panel thickness. A_r is the area of one roving computed from:

$$A_r = \frac{1}{Y\rho_r} \qquad 4.4$$

where, Y is the yield specified in yard/lb and converted to in/lb, and ρ_r is the density of the fibers given by the producer. Once the V_f for all the typical layers is computed, the ply stiffnesses are predicted using selected micromechanics formulas.

Prediction of ply stiffness by periodic microstructure model (PMM)

There are several micromechanics models available to predict the effective elastic properties of composite materials (Chamis 1984). Because of its accuracy, it is recommended to compute the ply stiffnesses for the roving and SF layers using preferentially the micromechanics model for composites with periodic microstructure by Luciano and Barbero (1994). Detailed expressions for the computations of the elastic constants E_1, E_2, G_{12}, and ν_{12} are given in the original paper along with experimental correlations, and only the elastic constants for unidirectional composites are included in this section, as follows:

$$\begin{aligned} E_1 &= C_{11}^*(s) \\ E_2 &= \frac{3}{4}C_{22}^*(s) + \frac{1}{4}C_{23}^*(s) + \frac{1}{2}C_{44}^*E(s) \\ G_{12} &= C_{66}^*(s) \\ \nu_{12} &= \frac{C_{12}^*(s)}{C_{22}^*(s) + C_{23}^*(s)} \end{aligned} \tag{4.5}$$

where

$$\begin{aligned} C_{11}^*(s) &= \lambda_0 + 2\mu_0 - V_f \\ &\times \left[\frac{S_3^2}{\mu_0^2} - \frac{2S_6S_3}{\mu_0^2 g} - \frac{aS_3}{\mu_0 c} + \frac{S_6^2 - S_7^2}{\mu_0^2 g^2} + \frac{aS_6 + bS_7}{\mu_0 gc} + \frac{a^2 - b^2}{4c^2}\right] / D \\ C_{12}^*(s) &= \lambda_0 + V_f b\left[\frac{S_3}{2c\mu_0} - \frac{S_6 - S_7}{2c\mu_0 g} - \frac{a+b}{4c^2}\right] / D \\ C_{23}^*(s) &= \lambda_0 + V_f\left[\frac{aS_7}{2\mu_0 gc} - \frac{ba + b^2}{4c^2}\right] / D \\ C_{22}^*(s) &= \lambda_0 + 2\mu_0 - V_f\left[-\frac{aS_3}{2\mu_0 c} + \frac{aS_6}{2\mu_0 gc} + \frac{a^2 - b^2}{4c^2}\right] / D \\ C_{44}^*(s) &= \mu_0 - V_f\left[-\frac{2S_3}{\mu_0} + (\mu_0 - \mu_1)^{-1} + \frac{4S_7}{\mu_0(2 - 2\nu_0)}\right]^{-1} \\ C_{66}^*(s) &= \mu_0 - V_f\left[-\frac{S_3}{\mu_0} + (\mu_0 - \mu_1)^{-1}\right]^{-1} \end{aligned} \tag{4.6}$$

and, V_f is the fiber volume fraction, and the subscripts 0 and 1 stand for the matrix and fiber, respectively; the coefficients of a, b, c, g and D are given by

$$a = \mu_1 - \mu_0 - 2\mu_1\nu_0 + 2\mu_0\nu_1$$

$$b = \nu_1\mu_1 - \mu_0\nu_0 - 2\mu_1\nu_0\nu_1 + 2\mu_0\nu_0\nu_1$$

$$\begin{aligned} c = (\mu_0 - \mu_1)\,(&\mu_1 - \mu_0 - 2\mu_1\nu_0 - \mu_0\nu_0 + \mu_1\nu_1 + 2\mu_0\nu_1 + \mu_1\nu_1 \\ &+ 2\mu_0\nu_0\nu_1 - 2\mu_1\nu_0\nu_1) \end{aligned}$$

$$g = 2 - 2\nu_0$$

$$D = \frac{aS_3^2}{2\mu_0^2 c} - \frac{aS_6 S_3}{\mu_0^2 g_c} + \frac{a(S_6^2 - S_7^2)}{2\mu_0^2 g^2 c} + \frac{S_3(b^2 - a^2)}{2\mu_0 c^2}$$
$$+ \frac{S_6(a^2 - b^2) + S_7(ab + b^2)}{2\mu_0 g c^2} + \frac{a^3 - 2b^3 - 3ab^2}{8c^3}$$

The series S_3, S_6, and S_7 accounting for the geometries of the fibers are obtained from Nemat-Nasser and Taya (1981) and are expressed with the parabolic fittings (Luciano and Barbero 1994) as

$$S_3 = 0.49247 - 0.47603V_f - 0.02748\, V_f^2$$
$$S_6 = 0.36844 - 0.14944V_f - 0.27152\, V_f^2 \qquad 4.7$$
$$S_7 = 0.12346 - 0.32035V_f + 0.23517\, V_f^2$$

The SF and roving layers are usually modeled as unidirectional composites in two orthogonal directions. The CSM layer, however, is assumed to be isotropic in the plane and the properties can be obtained from a set of approximate relations proposed by Hull (1981) for randomly oriented composites. A model for determining the elastic stiffness of CSM layer was presented by Harris and Barbero (1998), where the Young's modulus (E_{CSM}), shear modulus (G_{CSM}) and Poisson's ratio (ν_{CSM}) of CSM layer are obtained by adopting Hull's averaging procedure (1981) in the plane expressed in terms of the elastic constants of unidirectional composites (eqn 4.5) as

$$G_{CSM} = \frac{1}{8}\frac{E_1}{\Delta} - \frac{1}{4}\frac{\nu_{12}E_2}{\Delta} + \frac{1}{8}\frac{E_2}{\Delta} + \frac{1}{2}G_{12}$$

$$\nu_{CSM} = \frac{\frac{1}{8}\frac{E_1}{\Delta} + \frac{3}{4}\frac{\nu_{12}E_2}{\Delta} + \frac{1}{8}\frac{E_2}{\Delta} - \frac{1}{2}G_{12}}{\frac{3}{8}\frac{E_1}{\Delta} + \frac{1}{4}\frac{\nu_{12}E_2}{\Delta} + \frac{3}{8}\frac{E_2}{\Delta} + \frac{1}{2}G_{12}} \qquad 4.8$$

$$E_{CSM} = 2G_{CSM}(\nu_{CSM} + 1)$$

$$\Delta = 1 - \nu_{12}\nu_{21} = 1 - \nu_{12}^2\frac{E_2}{E_1}$$

4.3.2 Laminated panel engineering properties and carpet plots

Panel stiffnesses by macromechanics

Once the ply stiffnesses for each flat panel or wall section of a pultruded shape are computed, the stiffness of a panel can be computed from classical lamination theory (CLT) (Jones 1999). For a laminated panel, the general constitutive relation between the resultant forces (N_x, N_y, N_{xy}) and moments (M_x, M_y, M_{xy}) and the midsurface strains (ε_x^0, ε_y^0, γ_{xy}^0) and curvatures (κ_x, κ_y, κ_{xy}) is defined by CLT (Jones 1999) as

$$\begin{Bmatrix} N_x \\ N_y \\ N_{xy} \\ M_x \\ M_y \\ M_{xy} \end{Bmatrix} = \begin{bmatrix} A_{11} & A_{12} & A_{16} & B_{11} & B_{12} & B_{16} \\ A_{12} & A_{22} & A_{26} & B_{12} & B_{22} & B_{26} \\ A_{16} & A_{26} & A_{66} & B_{16} & B_{26} & B_{66} \\ B_{11} & B_{12} & B_{16} & D_{11} & D_{12} & D_{16} \\ B_{12} & B_{22} & B_{26} & D_{12} & D_{22} & D_{26} \\ B_{16} & B_{26} & B_{66} & D_{16} & D_{26} & D_{66} \end{bmatrix} \begin{Bmatrix} \varepsilon_x^0 \\ \varepsilon_y^0 \\ \gamma_{xy}^0 \\ \kappa_x \\ \kappa_y \\ \kappa_{xy} \end{Bmatrix} \tag{4.9}$$

where $[A]$, $[B]$ and $[D]$ are the panel stiffness submatrices. By full inversion of the panel stiffness matrix, we can express the midsurface strains and curvatures in terms of the compliance coefficients and panel resultant forces as:

$$\begin{Bmatrix} \varepsilon_x^0 \\ \varepsilon_y^0 \\ \gamma_{xy}^0 \\ \kappa_x \\ \kappa_y \\ \kappa_{xy} \end{Bmatrix} = \begin{bmatrix} \alpha_{11} & \alpha_{12} & \alpha_{16} & \beta_{11} & \beta_{12} & \beta_{16} \\ \alpha_{12} & \alpha_{22} & \alpha_{26} & \beta_{12} & \beta_{22} & \beta_{26} \\ \alpha_{16} & \alpha_{26} & \alpha_{66} & \beta_{16} & \beta_{26} & \beta_{66} \\ \beta_{11} & \beta_{12} & \beta_{16} & \delta_{11} & \delta_{12} & \delta_{16} \\ \beta_{12} & \beta_{22} & \beta_{26} & \delta_{12} & \delta_{22} & \delta_{26} \\ \beta_{16} & \beta_{26} & \beta_{66} & \delta_{16} & \delta_{26} & \delta_{66} \end{bmatrix} \begin{Bmatrix} N_x \\ N_y \\ N_{xy} \\ M_x \\ M_y \\ M_{xy} \end{Bmatrix} \tag{4.10}$$

where $[\alpha]$, $[\beta]$ and $[\delta]$ are the panel compliance submatrices. In particular, the panel compliance matrix $[\alpha]$ is used to compute the equivalent in-plane elastic properties of the panels as

$$E_x = 1/(t\alpha_{11}),\ E_y = 1/(t\alpha_{22}),\quad \nu_{xy} = -\alpha_{12}/\alpha_{11},\quad G_{xy} = 1/(t\alpha_{66}) \tag{4.11}$$

where t is the thickness of the panel. The panel moduli given in eqn 4.11 are valid for in-plane loads only. The laminate stiffnesses of the panels as indicated in eqn. 4.11 can be directly obtained from tests of coupon samples, which are cut from the FRP sections and tested in tension and shear (Iosipescu) (see Table 4.2). As an application, the panel engineering properties obtained by micro/macromechanics for 23 customized sections produced by Creative Pultrusions, Inc. are shown in Table 4.3.

Strength of pultruded FRP panels

In design of FRP shapes, strength of individual panels is a critical design factor which determines the ultimate material failure load of the component.

Table 4.2 Panel stiffness properties of FRP shapes

FRP Shapes	E_{xx} ($\times 10^6$ psi)		G_{xy} ($\times 10^6$ psi)	
	Tension test	Micro/macro-mechanics	Iosipescu test	Micro/macro-mechanics
WF-Beam 6″ × 6″ × 3/8″ (**WF6×6**)	4.155 (COV = 5.28%)	4.206	0.686 (COV = 8.39%)	0.682
I-Beam 4″ × 8″ × 3/8″ (**I4 × 8**)	5.037 (COV = 2.24%)	4.902	0.745 (COV = 9.79%)	0.794
WF-Beam 4″ × 4″ × 1/4″ (**WF4 × 4**)	4.391 (COV = 5.55%)	4.167	0.778 (COV = 11.28%)	0.676
Square Tube 4″ × 4″ × 1/4″ (**Box 4 × 4**)	4.295 (COV = 10.70%)	3.604	0.548 (COV = 8.39%)	0.550

In this section, simplified formulas for compressive strength of pultruded FRP panels are presented, and the conventional methods for compressive and shear strength predictions based on first-ply-failure (FPF) and fiber failure (FF) (Barbero 1999) are introduced. Since the composites are relatively weak in compression and shear loadings, either of these two strength parameters is likely to control the design in case of material failure.

Panel strength prediction by compressive failure

As aforementioned, most pultruded FRP panels consist of three types of layer: CSM, SF and rovings. Experimental data (Barbero *et al.* 1999) indicated that the compressive strengths of CSM and SF layers are relatively lower than that of a roving layer. This is further confirmed by performing a progressive failure analysis of full-sized samples (Kim *et al.* 1996). All the CSM and SF layers failed before the rovings reached their ultimate capacity, and therefore the roving layers sustain the applied load up to ultimate failure. Thus, the compressive strength of a panel laminate is proportional to the fiber volume fraction of the roving layers and the percentage of roving layers in the laminate, and it is therefore possible to derive a simplified formula for prediction of compressive strength of FRP panels based on the assumption that the roving layers are the last to fail at ultimate load.

For pultruded FRP panels, the expression for the ultimate compressive load can be written as

$$P = F_{1c}^{*}\, n_r\, A_r \tag{4.12}$$

where, F_{1c}^{*} is defined as the specific roving compressive strength and can be

Table 4.3 Panel engineering properties of 23 FRP shapes

Section	E_{xx} ($\times 10^6$ psi)	E_{yy} ($\times 10^6$ psi)	G_{xy} ($\times 10^6$ psi)	v_{xy}	v_{yx}
Channel 6″×1-5/8″ ×1/4″ **(C6×2)**	3.728	1.843	0.651	0.359	0.177
Channel 3″×7/8″ ×1/4″ **(C3×1)**	2.941	1.646	0.575	0.372	0.208
Channel 4″×1-1/8″ ×1/4″ **(C4×1)**	2.857	1.633	0.568	0.373	0.213
Channel 8″×2-3/16″ ×3/8″ **(C8×2)**	3.292	1.616	0.570	0.360	0.177
Channel 10″×2-3/4″ ×1/2″ **(C10×3)**	3.704	1.709	0.606	0.355	0.164
WF-Beam 4″×4″ ×1/4″ **(WF4×4)**	4.167	1.471	0.676	0.359	0.126
WF-Beam 6″×6″ ×3/8″ **(WF6×6)**	4.206	1.482	0.682	0.353	0.127
I-Beam 4″×8″ ×3/8″ **(I4×8)**	4.902	1.754	0.794	0.346	0.124
I-Beam 3″×6″ ×3/8″ **(I3×6)**	5.012	1.814	0.821	0.342	0.124
I-Beam 2″×4″ ×1/4″ **(I2×4)**	4.479	1.575	0.717	0.353	0.124
I-Beam 6″×12″ ×1/2″ **(I6×12)**	4.292	1.600	0.758	0.367	0.137
Rectangular Tube 4″×6″×1/4″ **(Box4×6)**	3.448	1.688	0.595	0.360	0.176
Rectangular Tube 4″×7″×1/4″ **(Box4×7)**	3.077	1.600	0.561	0.364	0.189
Square Tube 1-3/4″ ×1-3/4″×1/4″ **(Box1×1)**	3.773	1.587	0.571	0.342	0.144
Square Tube 2″×2″ ×1/8″ **(Box2×2T8)**	3.738	1.587	0.571	0.343	0.146
Square Tube 2″×2″ ×1/4″ **(Box2×2T4)**	3.703	1.562	0.563	0.343	0.145
Square Tube 3″×3″ ×1/4″ **(Box3×3)**	3.604	1.538	0.552	0.346	0.148
Square Tube 4″×4″ ×1/4″ **(Box4×4)**	3.604	1.533	0.550	0.345	0.147
WF 8″×8″×3/8″ **(WF8×8)**	4.645	1.646	0.749	0.348	0.123
WF Beam 3″×3″ ×1/4″ **(WF3×3)**	3.571	1.444	0.522	0.339	0.137
WF Beam 12″×12″ ×1/2″ **(WF12×12A)**	3.773	1.905	0.913	0.398	0.201
WF Beam 12″×12″ ×1/2″ **(WF 12×12C)**	4.049	2.114	0.760	0.302	0.157
WF Beam 12″×12″ ×1/2″ **(WF12×12T)**	4.264	1.587	0.749	0.369	0.137

obtained from the roving layer compressive strength (F_{1c}) divided by the fiber volume fraction of roving layer $(V_f)_r$ as

$$F_{1c}^* = \frac{F_{1c}}{(V_f)_r} \tag{4.13}$$

and n_r is the number of rovings per unit width (in^{-1}), and A_r is the area of one bundle of roving and given in eqn 4.4. Combining eqns 4.12 and 4.13, we can obtain the panel compressive strength as

$$F_{xc} = \frac{P}{1.0'' \times t} = \frac{F_{1c}^* n_r A_r}{t} = \frac{F_{1c}^* n_r}{\rho_r Y t} \tag{4.14}$$

where t is the thickness of the panel in inch. Hence, we can rewrite eqn 4.14 as

$$F_{xc} = \frac{F_{1c}^* n_r A}{t} = \frac{F_{1c}^* t_r (V_f)_r}{t} \tag{4.15}$$

where t_r is the thickness of the roving layers.

By defining the percentage of roving layers in the panel as $\alpha = \frac{t_r}{t}$, we can simplify eqn 4.15 as

$$F_{xc} = F_{1c}^* \alpha (V_f)_r \tag{4.16}$$

For typical rovings used for pultruded FRP shapes, the specific roving compressive strengths are given in Table 4.4 (Barbero *et al.* 1999). Based on eqn 4.16, the compressive strength of the FRP panels can be predicted once the percentage of the roving layer in the panel and corresponding roving layer fiber volume fraction are known.

Table 4.4 Specific compressive strength for most common pultruded materials (Barbero *et al.* 1999)

Resin type	Roving type (yield)	Fiber volume fraction	F_{1c}^* (ksi)
Vinyl ester	56	40.2	174
Vinyl ester	113	43	176
Vinyl ester	250	43.9	178
Polyester	56	40.2	172
Polyester	113	43	156
Polyester	250	43.9	161
Brittle polyester	56	40.2	190
Brittle polyester	113	43	184
Brittle polyester	250	43.9	163

Panel strength design by failure criteria

The classical approach for strength design of laminates includes the predictions of first-ply-failure (FPF) and fiber failure (FF) loads (Barbero 1999). The FPF load can be obtained when failure first occurs in any layer of the laminate, and the FPF criterion can be based on any failure criteria commonly used (e.g., Tsai-Wu). Due to the weakness of transverse strength of polymer matrix composites, FPF is usually associated with matrix cracking. Following the FPF, the material is degraded and the load is increased until a fiber failure occurs (called the FF load).

In FF, a single degradation factor (f_d) is commonly used to reduce all the stiffness values of a degraded layer, except for the stiffness in the fiber direction which is assumed to be unaffected by FPF. The stiffnesses of a degraded layer are expressed as

$$\begin{aligned} E_1 &= E_1^o & E_2 &= f_d E_2^o \\ G_{12} &= f_d G_{12}^o & G_{23} &= f_d G_{23}^o \\ \nu_{12} &= f_d \nu_{12}^o & f_{12} &= f_{12} = f_d f_{12}^o \end{aligned} \qquad 4.17$$

where the superscript o indicates the original, intact, or undegraded properties. In this chapter, both the compressive and shear strength values for common pultruded FRP panels based on FPF and FF load concepts are obtained (see carpet plots in Figs 4.18–4.29) and further used as baseline data for failure design of FRP shapes.

Carpet plots for simplified design of FRP panel properties

Based on the analyses presented above for the panel stiffness and strength properties, carpet plots are developed and used for simplified design of FRP panels. In carpet plots, diagrams of apparent moduli and strength for various laminate configurations can be produced beforehand and effectively applied to design of FRP structures. Most pultruded FRP shapes consist of typical three layers, and their lay-ups are usually balanced-symmetric and can be simply defined as

$$[0_{\alpha/2}/\mathrm{CSM}_{\beta/2}/(\pm 45°)_{\gamma/2}]_s \qquad 4.18$$

where α, β and γ represent the percentages of roving, CSM, and SF (± 45°) layers in the laminated panel, and they should satisfy the following relationship:

$$\alpha + \beta + \gamma = 1 \qquad 4.19$$

Subsequently, the carpet plots for panel moduli and strengths are produced using the micro/macromechanics and strength data introduced above.

Carpet plots for panel stiffnesses

Based on eqn 4.11, the carpet plots for panel stiffness properties (E_x, E_y, G_{xy},

ν_{xy}) are produced for a E-glass/polyester composite panel with $V_f = 20\%$ and $V_f = 50\%$, respectively. Figures 4.6 to 4.9 show the carpet plots for $V_f = 20\%$; whereas Figs 4.10 to 4.13 provide those for $V_f = 50\%$. As indicated in Figs 4.6 to 4.13, similar trends for corresponding stiffness properties are observed for $V_f = 20\%$ and $V_f = 50\%$. Further, a statistical study is performed (Li 2000) to compare the corresponding stiffness coefficients with same fiber lay-up and fiber percentage combinations but different fiber volume fractions. The ratios for the stiffness coefficients at $V_f = 50\%$ over the ones at $V_f = 20\%$ almost remain constant for every pair of the comparisons and are given as follows:

$$(E_x)_{50}/(E_x)_{20} = 1.91 \qquad (\text{COV} = 0.02823)$$
$$(E_y)_{50}/(E_y)_{20} = 2.0309 \qquad (\text{COV} = 0.03260)$$
$$(G_{xy})_{50}/(G_{xy})_{20} = 1.9921 \qquad (\text{COV} = 0.03042) \qquad 4.20$$
$$(\nu_{xy})_{50}/(\nu_{xy})_{20} = 1.0254 \qquad (\text{COV} = 0.02165)$$

where, $(E_x)_{50}$, $(E_y)_{50}$, $(G_{xy})_{50}$, and $(\nu_{xy})_{50}$ are the elastic coefficients for $V_f = 50\%$, and $(E_x)_{20}$, $(E_y)_{20}$, $(G_{xy})_{20}$, and $(\nu_{xy})_{20}$ are those for $V_f = 20\%$.

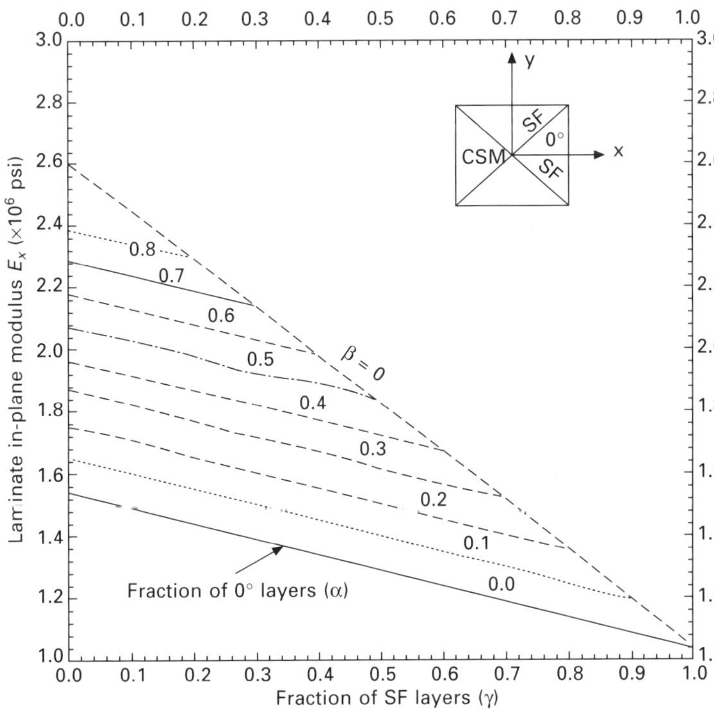

4.6 Carpet plot for laminate in-plane modulus E_x ($V_f = 20\%$).

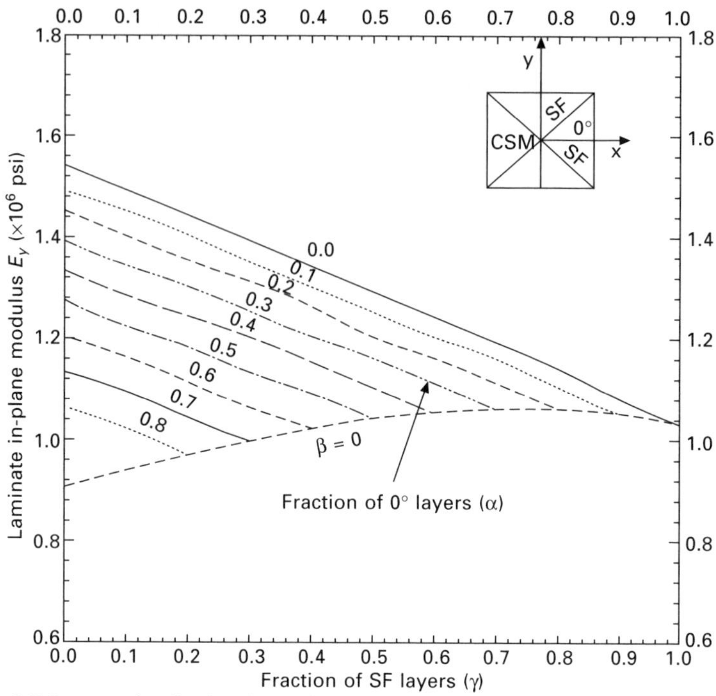

4.7 Carpet plot for laminate in-plane modulus E_y (V_f = 20%).

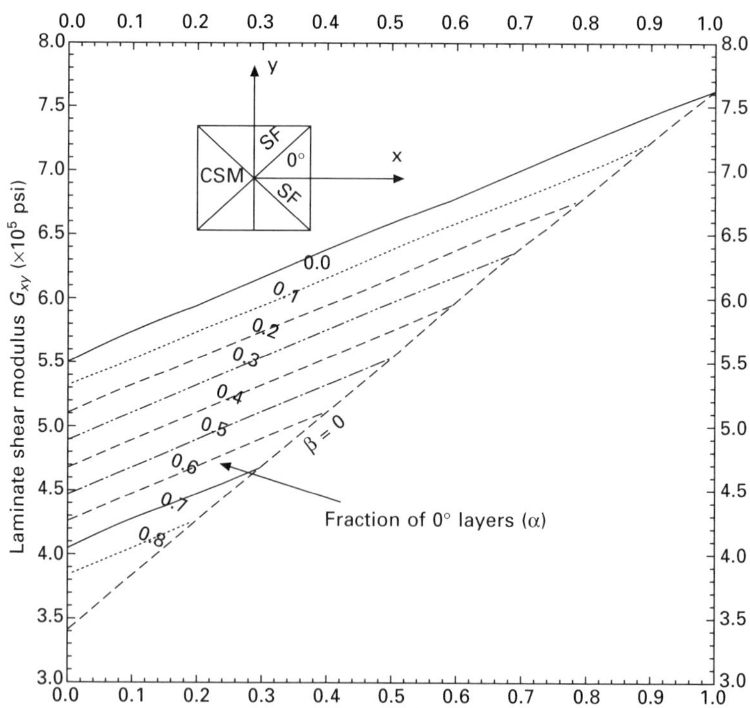

4.8 Carpet plot for laminate shear modulus G_{xy} (V_f = 20%).

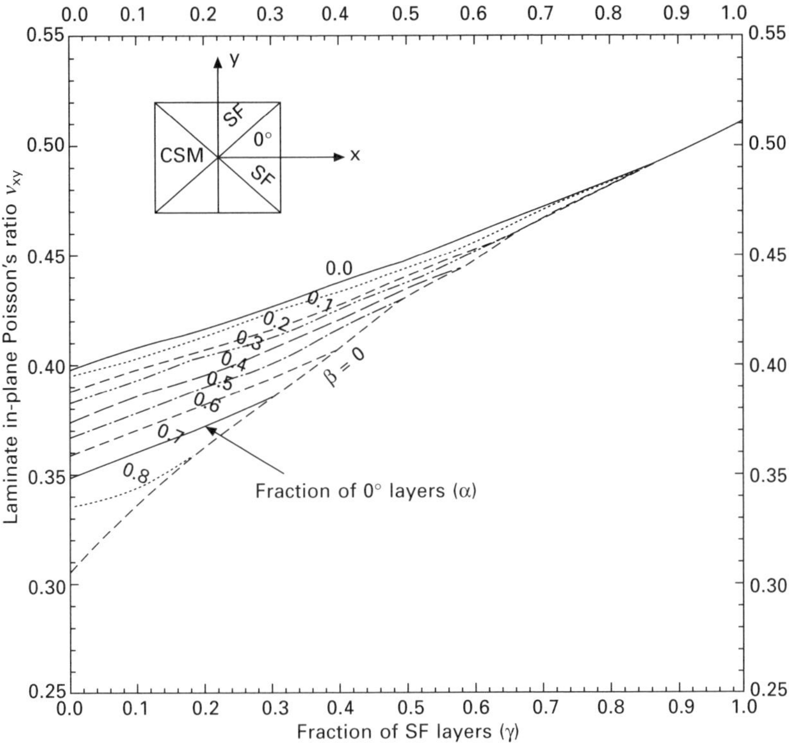

4.9 Carpet plot for laminate in-plane Poisson's ratio v_{xy} (V_f = 20%).

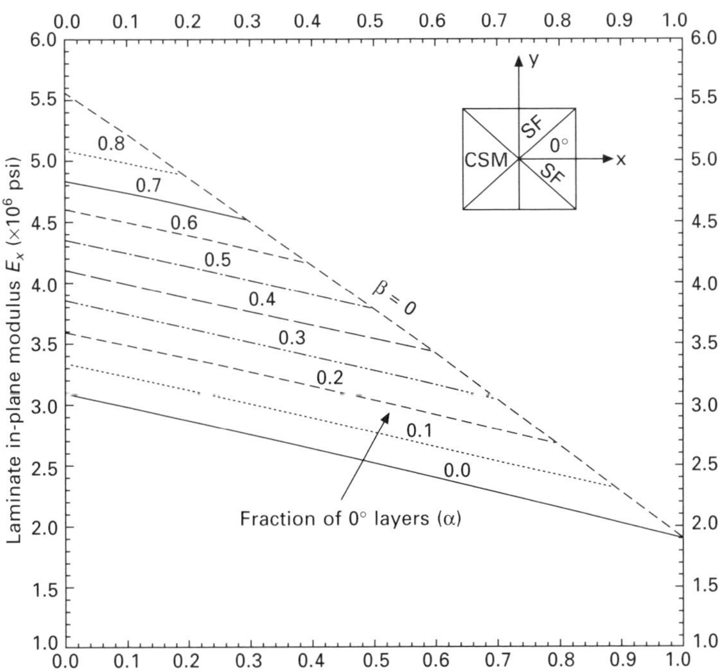

4.10 Carpet plot for laminate in-plane modulus E_x (V_f = 50%).

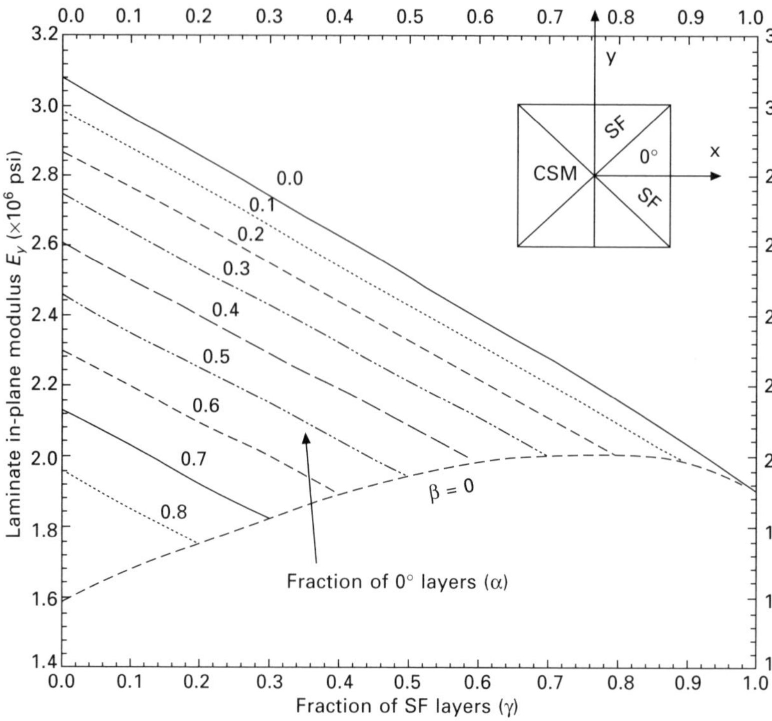

4.11 Carpet plot for laminate in-plane modulus E_y (V_f = 50%).

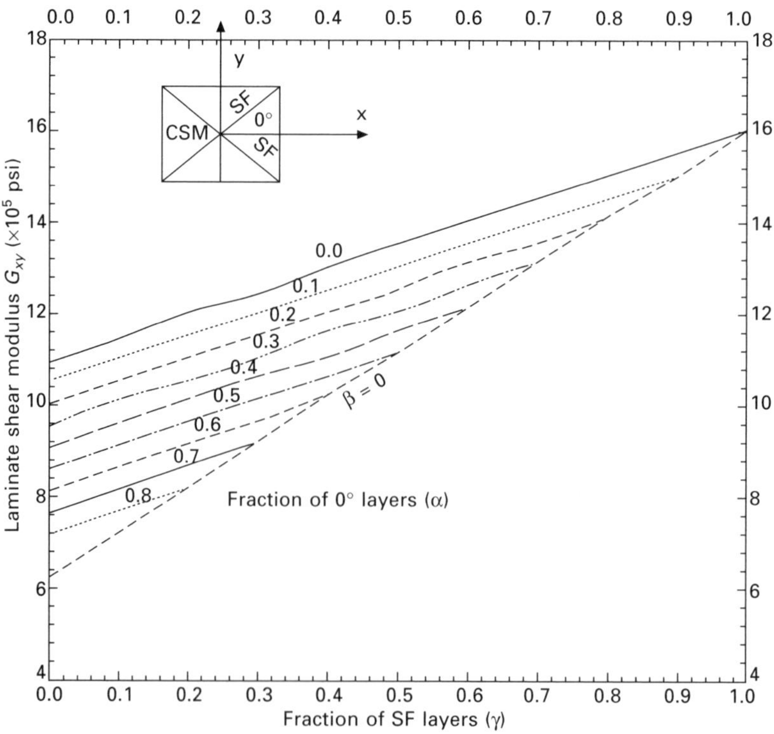

4.12 Carpet plot for laminate shear modulus G_{xy} (V_f = 50%).

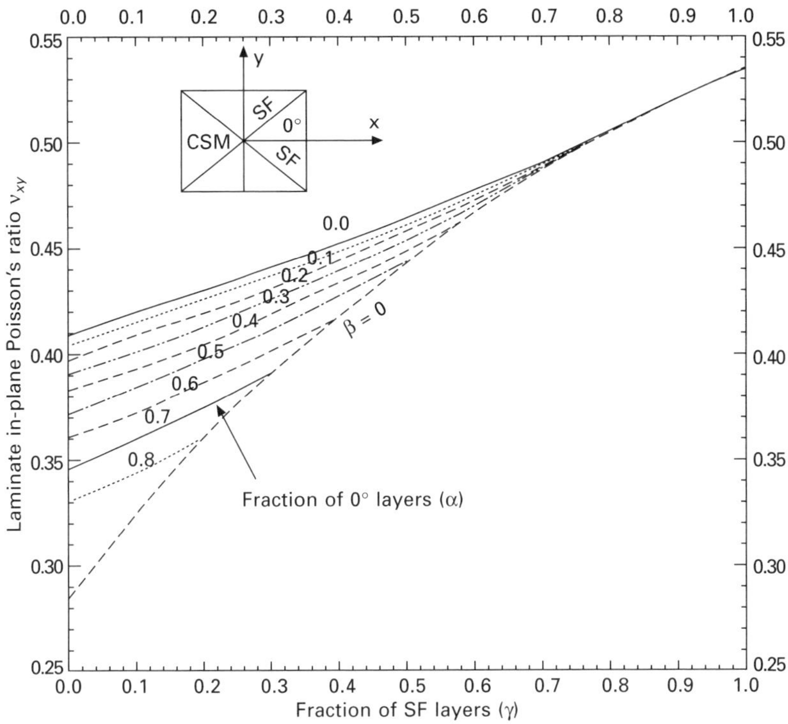

4.13 Carpet plot for laminate in-plane Poisson's ratio v_{xy} (V_f = 50%).

Due to the similar trends among carpet plots for panel stiffness coefficients of different fiber volume fractions, it can be sufficient to produce only one set of master plots (e.g., with V_f = 50% of Figs 4.10–4.13) and the corresponding ratio plots (Figs 4.14 to 4.17), which can be used effectively to estimate stiffness properties for panels of various fiber volume fractions. For example, with respect to the one for V_f = 50%, we can determine the equivalent panel moduli for different fiber volume fractions by multiplying the corresponding ratios by the equivalent panel moduli for V_f = 50%.

Using the concept described above, we define the carpet plots for V_f = 50% (Figs 4.10 to 4.13) as master plots, and the corresponding ratio plots for E_x, E_y, G_{xy}, and v_{xy} with various fiber volume fractions are given in Figs 4.14 to 4.17. As an example, we illustrate how to obtain the panel stiffness coefficients for a pultruded panel with V_f = 36% and α = 60%, β = 20% and γ = 20%. First, we obtain the panel stiffness properties for V_f = 50% using carpet plots in Figs 4.10 to 4.13 with α = 60%, β = 20% and γ = 20% as:

$$(E_x)_{50} = 4.37 \times 10^6 \text{ psi}, \quad (E_y)_{50} = 2.09 \times 10^6 \text{ psi}$$

$$(G_{xy})_{50} = 0.913 \times 10^6 \text{ psi}, \quad (v_{xy})_{50} = 0.386 \qquad 4.21$$

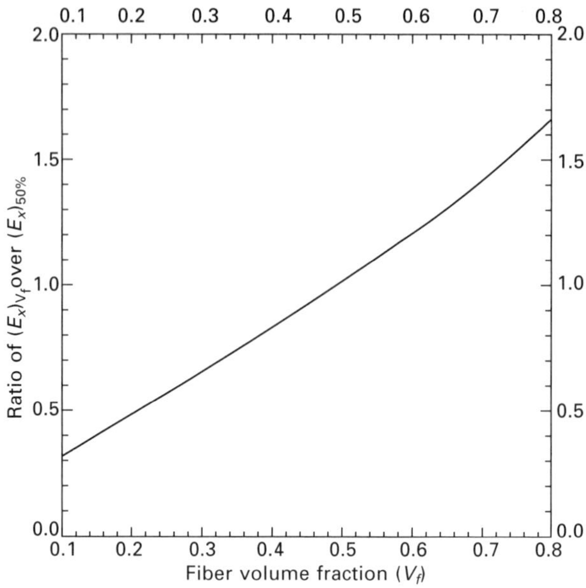

4.14 Ratio plot for E_x.

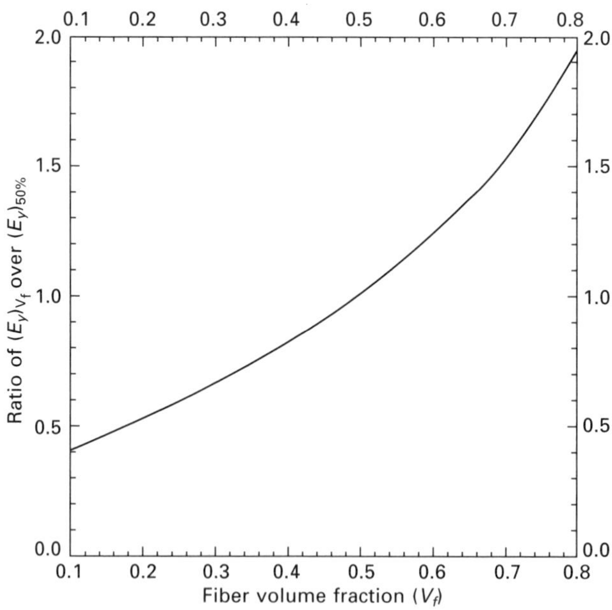

4.15 Ratio plot for E_y.

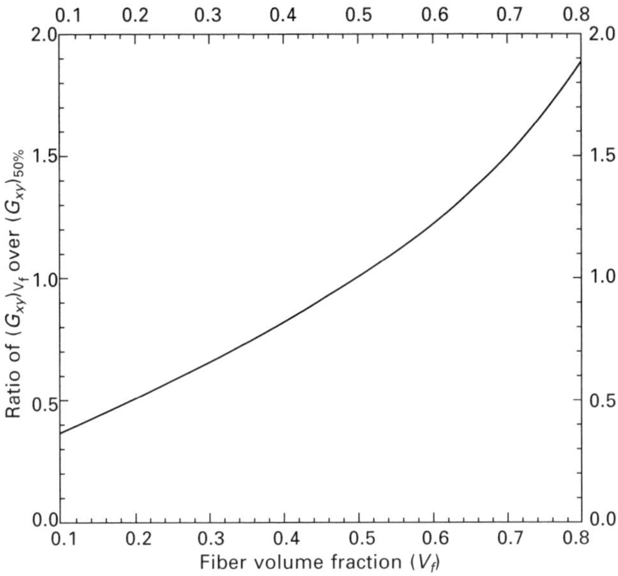

4.16 Ratio plot for G_{xy}.

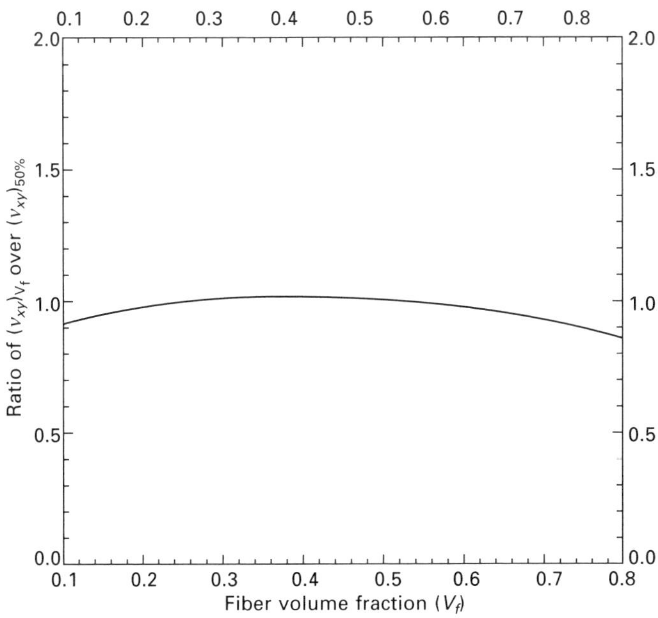

4.17 Ratio plot for v_{xy}.

Then, based on the ratio plots in Figs 4.14 to 4.17, we intercept the coefficient ratios as:

$$(E_x)_{36}/(E_x)_{50} = 0.75, \quad (E_y)_{36}/(E_y)_{50} = 0.75$$

$$(G_{xy})_{36}/(G_{xy})_{50} = 0.75, \ (v_{xy})_{36}/(v_{xy})_{50} = 1.1 \qquad 4.22$$

The above ratios can lead to the panel stiffness coefficients for $V_f = 36\%$ with $\alpha = 60\%$, $\beta = 20\%$ and $\gamma = 20\%$ as:

$$(E_x)_{36} = 3.278 \times 10^6 \text{ psi}, \quad (E_y)_{36} = 1.568 \times 10^6 \text{ psi}, \qquad 4.23$$

$$(G_{xy})_{36} = 0.685 \times 10^6 \text{ psi}, \quad (v_{xy})_{36} = 0.425$$

Carpet plots for panel strength

Compressive strength

As mentioned before, the roving compressive strength dominates the compressive strength of the whole laminated panel. A graphical representation of panel compressive strength values for various fiber volume fractions is very useful for preliminary design. In Figs 4.18 and 4.19, the carpet plot for compressive strength of FRP panels is produced based on eqn 4.16, and the

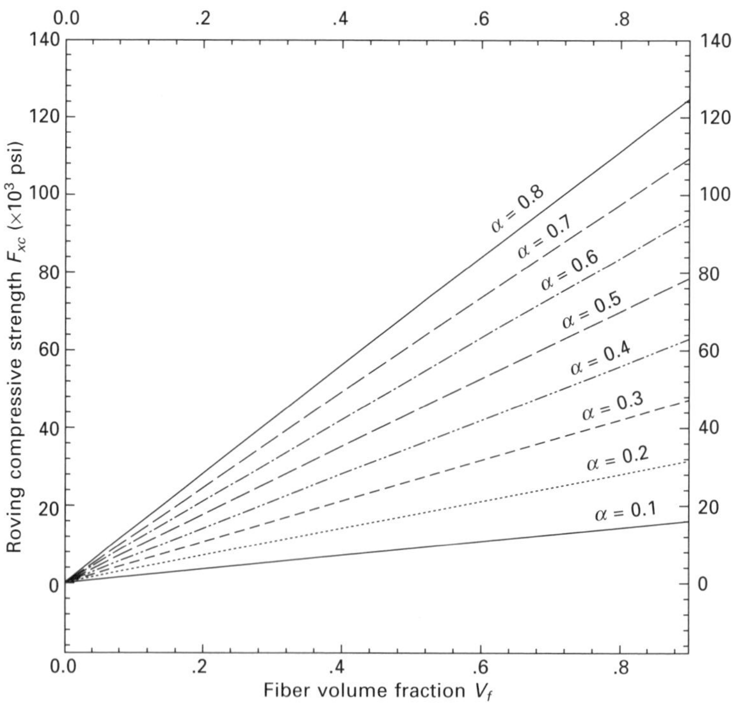

4.18 Roving compressive strength F_{xc} vs. fiber volume fraction V_f.

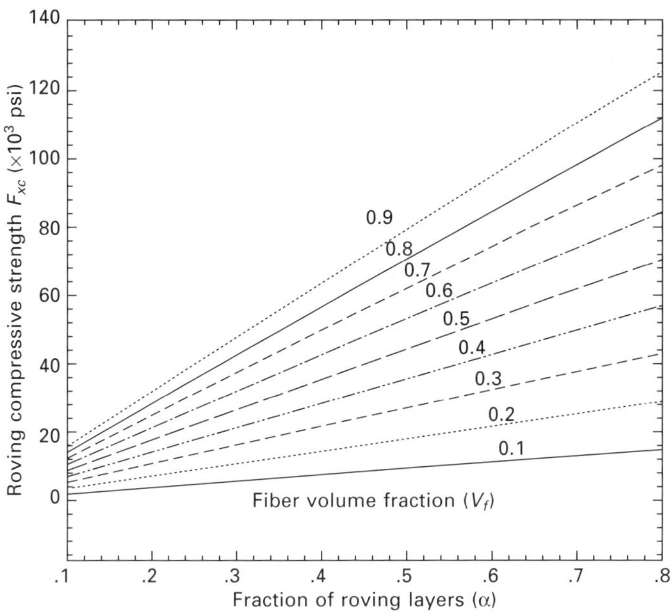

4.19 Roving compressive strength F_{xc} vs. percentage of roving layers α.

panel ultimate compressive strength depends solely on the properties of roving in the panel (e.g., specific roving compressive strength, roving fiber percentage in the panel, and roving fiber volume fraction; see Table 4.4). The specific roving compressive strengths for most common pultruded materials have similar values; therefore, an average ($F_{1c}^* = 173$ ksi, COV = 6.4%) is used in eqn 4.16 to produce the carpet plots of Figs 4.18 and 4.19. The plots can be generally applied to predict the panel compressive strength for most pultruded FRP shapes.

An alternative approach to produce carpet plots for compressive strength F_{xc} is using conventional failure criteria (e.g., Tsai-Wu criterion) by predicting the first ply failure (FPF) load and fiber failure load (FF). Based on Tsai-Wu criterion and the CADEC program (Barbero 1999), by applying $-N_x$ to the laminate, we can obtain the strength ratio R, which is defined as the strength over applied stress as

$$R = \frac{F_{xc}}{(\sigma_c)_{\text{applied}}} \tag{4.24}$$

therefore,

$$F_{xc} = R(\sigma_c)_{\text{applied}} \tag{4.25}$$

where $(\sigma_c)_{\text{applied}} = -N_x/t$ and t is the thickness of the laminate.

Using the CADEC program (Barbero 1999), the compressive strength of an E-glass/polyester composite panel with $V_f = 20\%$ and also $V_f = 50\%$ are computed, and the carpet plots are shown in Figs 4.20 to 4.23 for the compressive strength based on FPF and FF loads. The design for compressive strength is based on an indirect approach using the computer software (CADEC) for simplified design of the pultruded panels with various fiber volume fractions ($V_f = 20\% \sim V_f = 50\%$). The ultimate compressive strength can then be approximated by estimating a value between the compressive strength values at $V_f = 20\%$ and $V_f = 50\%$ (see Fig. 4.24).

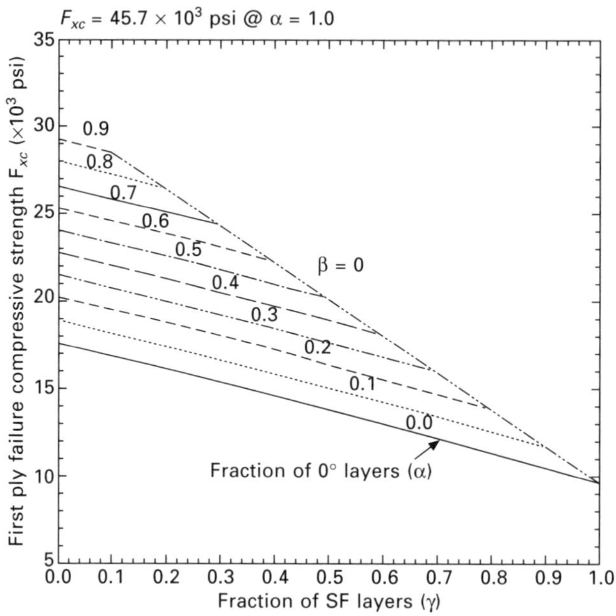

4.20 Carpet plot of FPF compression strength F_{xc} for $V_f = 20\%$.

Shear strength

Similar to the panel compressive strength prediction, the approach to produce carpet plots for shear strength F_{xy} is based on the Tsai-Wu criterion, and the FPF shear and FF shear strength ratios are predicted using CADEC. By applying N_{xy} to the laminate, we can obtain the strength ratio R as

$$R = \frac{F_{xy}}{(\sigma_{xy})_{\text{applied}}} \qquad 4.26$$

therefore,

$$F_{xy} = R(\sigma_{xy})_{\text{applied}}$$

where $(\sigma_{xy})_{\text{applied}} = N_{xy}/t$. 4.27

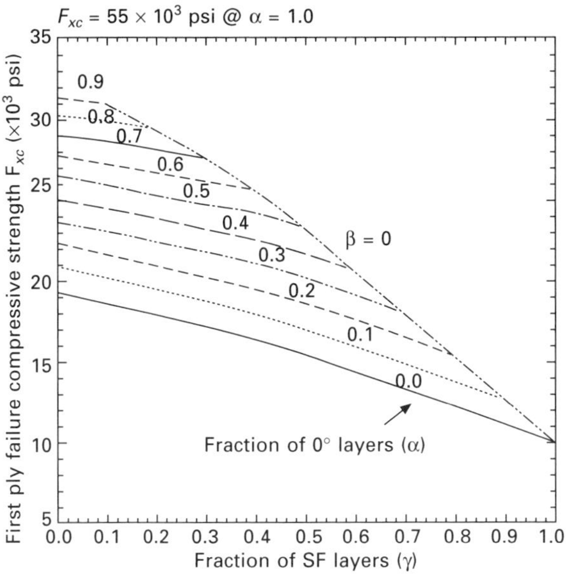

4.21 Carpet plot of FPF compression strength F_{xc} for V_f = 50%.

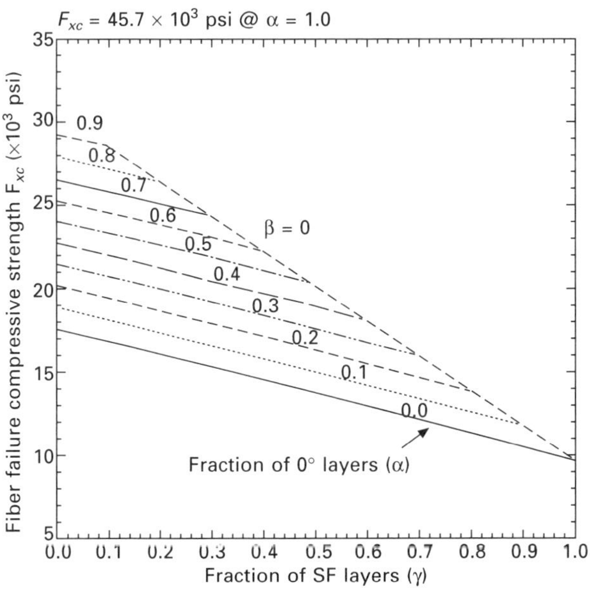

4.22 Carpet plot of FF compression strength F_{xc} for V_f = 20%.

Based on eqn 4.27, carpet plots for FPF shear and FF shear strengths can be generated for an E-glass/polyester composite panel with various fiber volume fractions. Figs 4.25 to 4.26 and Figs 4.27 to 4.28 show the carpet plots for $V_f = 20\%$ and $V_f = 50\%$, respectively. Once again, the ultimate shear

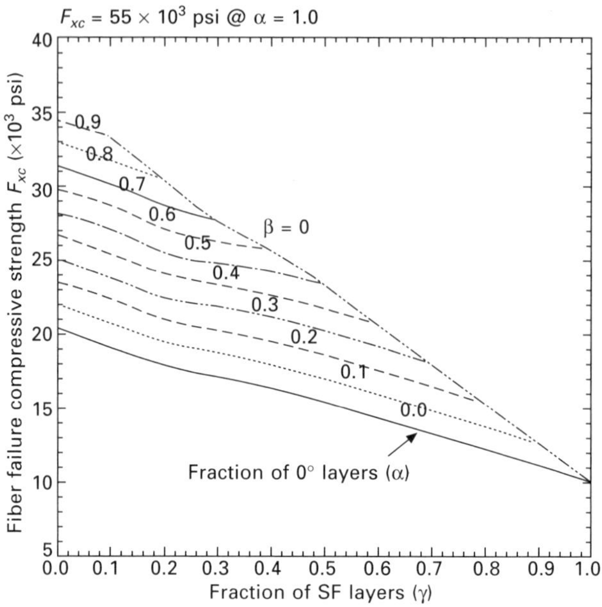

4.23 Carpet plot of FF compression strength F_{xc} for V_f = 50%.

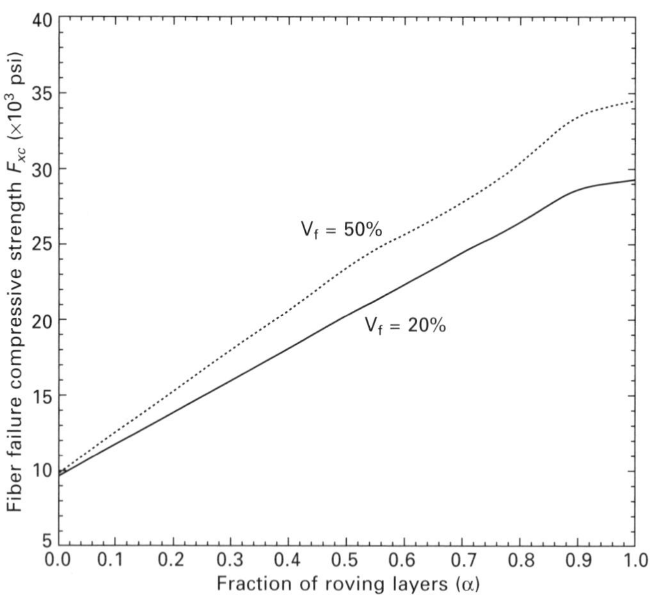

4.24 FF compression strength for V_f = 20% and V_f = 50%.

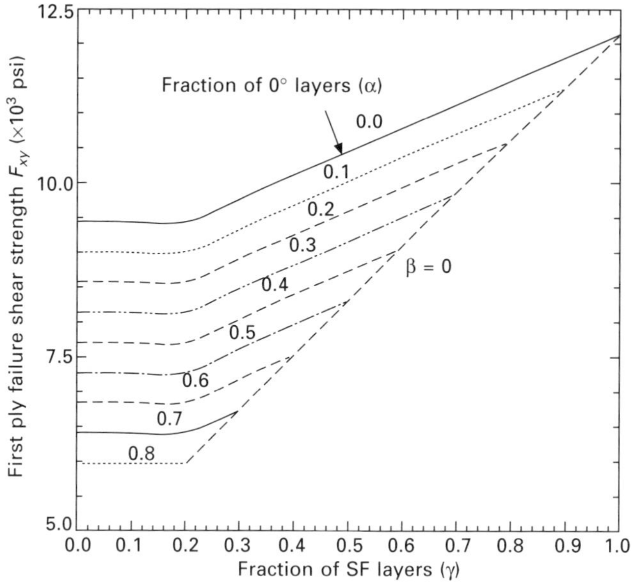

4.25 Carpet plot of FPF shear strength F_{xy} for $V_f = 20\%$.

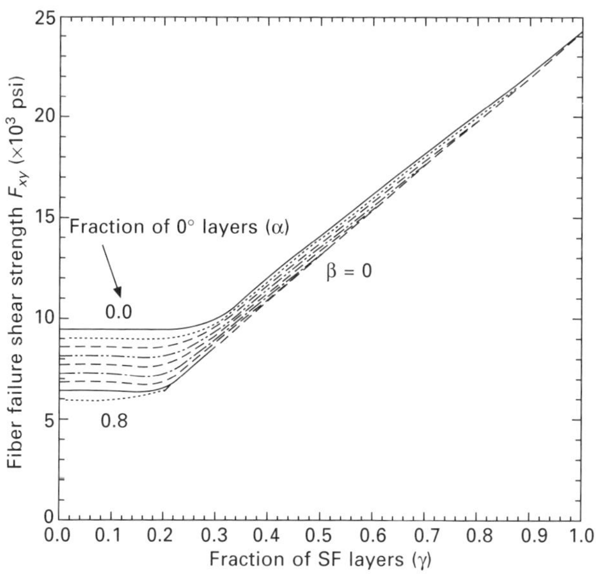

4.26 Carpet plot of FF shear strength F_{xy} for $V_f = 20\%$.

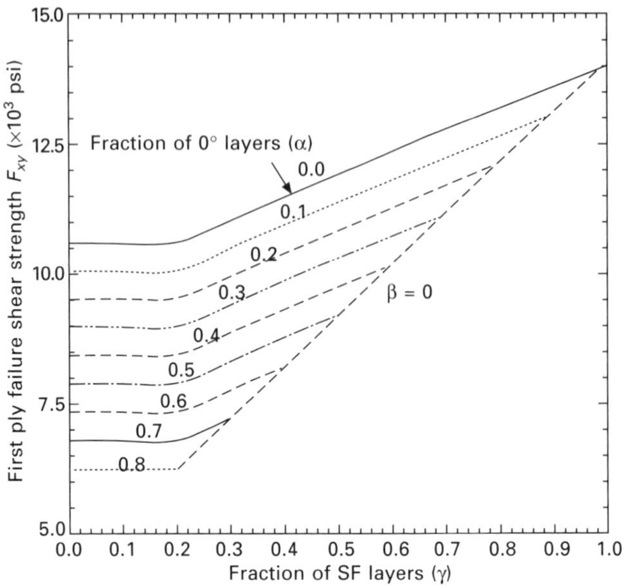

4.27 Carpet plot of FPF shear strength F_{xy} for $V_f = 50\%$.

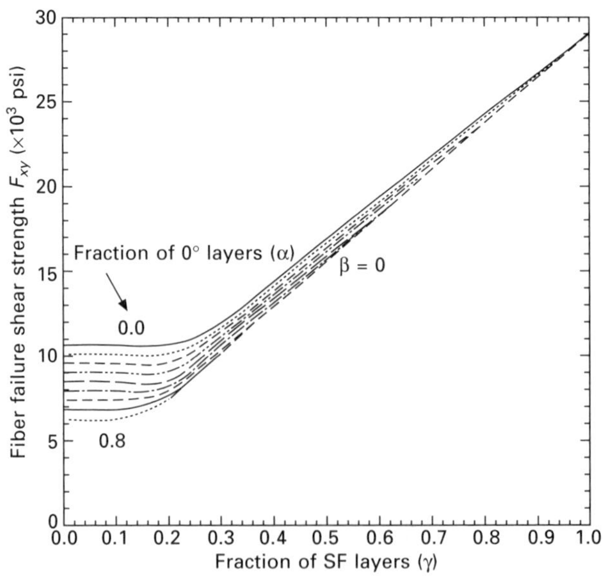

4.28 Carpet plot of FF shear strength F_{xy} for $V_f = 50\%$.

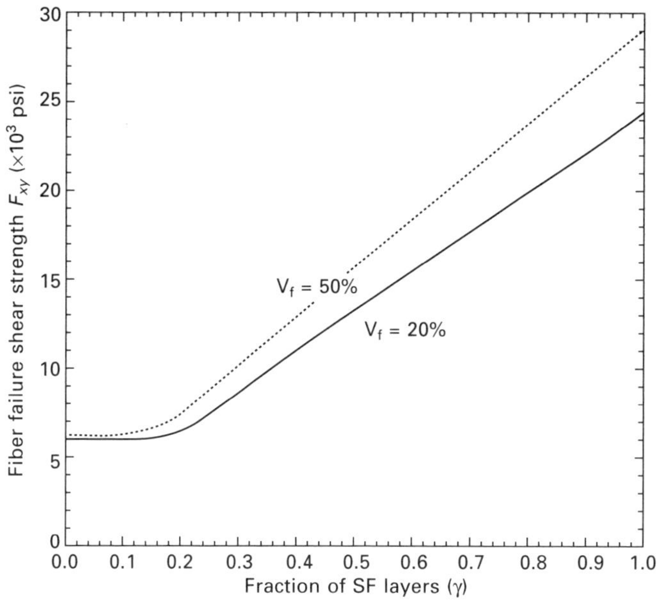

4.29 FF shear strength F_{xy} for $V_f = 20\%$ and $V_f = 50\%$.

strength of the panels with various fiber volume fractions can be approximated by the plot in Fig. 4.29.

4.3.3 Member stiffness properties

FRP composite shapes are used as structural members (e.g., beams, columns and deck panels) due to their high strength and stiffness to weight ratios, corrosion resistance, and structural efficiency. Beams and columns are structural members that carry mainly transverse flexural and axial compressive loads, respectively. In this section, the stiffness properties of FRP members are evaluated using a formal engineering approach to the mechanics of thin-walled laminated beams (MLB) (Barbero *et al.* 1993), based on kinematic assumptions consistent with Timoshenko's beam theory.

The MLB approach is adopted in this study to model pultruded structural shapes based on the first-order shear deformation theory for thin- and moderately thick-walled laminated beams with open or closed cross-sections. In this model, the stiffnesses of a beam are computed by adding the contributions of the stiffnesses of the component panels, which in turn are obtained from the effective beam moduli as given in eqn 4.11. The model accounts for membrane stiffness and flexure stiffness of the walls. Warping effects due to non-uniform bending (shear lag) are not included in this model. Therefore, this theory is more appropriate for moderately thick-walled laminated beams than for thin-walled laminated beams. The position of the neutral axis is defined in such a way that the behavior of a thin-walled beam-column with asymmetric material and/or cross-sectional shape is completely described by axial, bending, and shear stiffness coefficients (A_Z, D_Y, F_Y) only. Furthermore, a shear correction factor is obtained from energy equivalence.

The basic kinematic assumptions in MLB (Barbero *et al.* 1993) are: (i) the contour does not deform in its own plane, and (ii) a plane section originally normal to the beam axis remains plane, but not necessarily normal due to shear deformation. Straight FRP beams with at least one axis of geometric and material symmetry are considered. The pultruded sections are modeled as assemblies of flat walls. The compliance matrices $[\alpha]_{3\times3}$, $[\beta]_{3\times3}$, $[\delta]_{3\times3}$ of the individual panels (see eqn 4.10) are obtained from classical lamination theory (CLT) (Tsai 1988). For each wall panel, the position of the middle surface is defined as (Fig. 4.30):

$$y(s_i) = s_i \sin \phi_i + \bar{y}_i \quad \text{for} \quad -\frac{b_i}{2} \leq s_i \leq \frac{b_i}{2} \qquad 4.28$$

where b_i is the wall width and $\bar{y}_i$ is the position of the wall centroid.

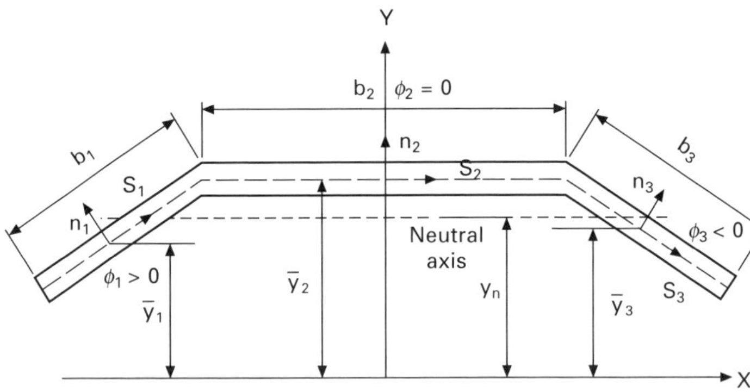

4.30 Global (beam) and local (panel) coordinator systems used in MLB.

By balancing the off-axis plies symmetrically, the shear-extension (α_{16}) and shear-bending (β_{16}) coupling coefficients in eqn 4.10 vanish ($\alpha_{16} = \beta_{16} = 0$). Then using the beam theory assumptions ($\overline{N}_y = \overline{M}_y = 0$), the compliance equations (eqn 4.10) can be reduced to

$$\begin{Bmatrix} \bar{\varepsilon}_x \\ \bar{\kappa}_x \\ \bar{\gamma}_{xy} \\ \bar{\kappa}_{xy} \end{Bmatrix} = \begin{bmatrix} \alpha_{11} & \beta_{11} & 0 & 0 \\ \beta_{11} & \delta_{11} & 0 & 0 \\ 0 & 0 & \alpha_{66} & \beta_{66} \\ 0 & 0 & \beta_{66} & \delta_{66} \end{bmatrix} \begin{Bmatrix} \overline{N}_x \\ \overline{M}_x \\ \overline{N}_{xy} \\ \overline{M}_{xy} \end{Bmatrix} \qquad 4.29$$

Inverting eqn 4.29, the reduced constitutive equation for the ith panel is obtained as

$$\begin{Bmatrix} \overline{N}_x \\ \overline{M}_x \\ \overline{N}_{xy} \\ \overline{M}_{xy} \end{Bmatrix} = \begin{bmatrix} \overline{A}_i & \overline{B}_i & 0 & 0 \\ \overline{B}_i & \overline{D}_i & 0 & 0 \\ 0 & 0 & \overline{F}_i & \overline{C}_i \\ 0 & 0 & \overline{C}_i & \overline{H}_i \end{bmatrix} \begin{Bmatrix} \overline{\varepsilon}_x \\ \overline{\kappa}_x \\ \overline{\gamma}_{xy} \\ \overline{\kappa}_{xy} \end{Bmatrix} \qquad 4.30$$

where $\overline{A}_i$, $\overline{B}_i$, $\overline{D}_i$, $\overline{F}_i$, $\overline{C}_i$ and $\overline{H}_i$ are, respectively, the ith panel extensional, bending-extension, bending, shear, twisting-shear and twisting stiffnesses; they can be expressed in terms of panel engineering properties (see equation 4.11) and given as

$$\overline{A}_i = (\delta_{11}\Delta_1^{-1})_i,\quad \overline{B}_i = (-\beta_{11}\Delta_1^{-1})_i,\quad \overline{D}_i = (\alpha_{11}\Delta_1^{-1})_i$$

$$\overline{F}_i = (\delta_{66}\Delta_2^{-1})_i,\quad \overline{H}_i = (\alpha_{66}\Delta_2^{-1})_i,\quad \overline{C}_i = (-\beta_{66}\Delta_2^{-1})_i \qquad 4.31$$

$$\Delta_1 = \alpha_{11}\delta_{11} - \beta_{11}^2,\quad \Delta_2 = \alpha_{66}\delta_{66} - \beta_{66}^2$$

General expressions for the axial, bending and transverse shear stiffness coefficients of FRP members are derived from the beam variational problem (Barbero *et al.* 1993) while the torsional stiffness is found using energy balance between the work done by the external torque and the strain energy due to shear (Barbero 1999). Hence, the axial (A_Z), bending (D_X or D_Y), shear (F_X or F_Y), and torsional (D_t^{open} for open cross-section and D_t^{close} for close-cross section) stiffnesses that account for the contribution of all the walls can be computed as

$$A_Z = \sum_{i=1}^{n} \overline{A}_i b_i$$

$$B_Y = \sum_{i=1}^{n} [\overline{A}_i(\overline{y}_i - y_n) + \overline{B}_i \cos\phi_i] b_i$$

$$D_Y = \sum_{i=1}^{n} \left[\overline{A}_i \left((\overline{y}_i - y_n)^2 + \frac{b_i^2}{12}\sin^2\phi_i \right) + 2\overline{B}_i(\overline{y}_i - y_n)\cos\phi_i + \overline{D}_i \cos^2\phi_i \right] b_i \qquad 4.32$$

$$F_Y = \sum_{i=1}^{n} F_i b_i \sin^2\phi_i$$

$$D_t^{\text{open}} = \text{GJ} = 4\sum_{i=1}^{n} \overline{H}_i b_i$$

$$D_t^{\text{close}} = \frac{(2\Gamma_s)^2}{\sum_{i=1}^{n} b/\overline{F}_i} + \frac{3}{4}\left(4\sum_{i=1}^{n} \overline{H}_i b_i \right)$$

where Γ_s is the area enclosed by the contour of shear flow. The beam bending-extension coupling coefficient (B_X or B_Y) can be eliminated by defining the location of the neutral axis of bending (x_n or y_n) as

$$y_n = \frac{\sum_{i=1}^{n}(\bar{y}_i\bar{A}_i + \cos\phi_i\bar{B}_i)b_i}{A_z} \tag{4.33}$$

By introducing the coordinate $y' = y - y_n$, we are able to decouple the extensional and bending responses (i.e., $B_Y = 0$). An explicit expression for the static shear correction factor (K_X or K_Y) is derived from energy equivalence. As an approximation in design, the shear correction factor for pultruded sections can be taken as 1.0. General equations of the shear correction factor for various laminates are given by Madabhusi-Raman and Davalos (1996), and for FRP sections are presented by Lopez-Anido (1994).

For each laminated wall (e.g., a flange or a web), the stiffness values are obtained either by the micro/macromechanics approach (see eqn 4.11) or from coupon tests. If we incorporate stress resultant assumptions compatible with beam theory, and we assume that the off-axis plies of pultruded panels are balanced symmetric (no extension-shear and bending-twist couplings are present), the extensional, bending, shear, and twisting stiffnesses of the ith panel are expressed in terms of panel apparent moduli as:

$$\bar{A}_i = (E_x)_i t_i,\quad \bar{D}_i = (E_x)_i t_i^3/12,\quad \bar{F}_i = (G_{xy})_i t_i,\quad \bar{H}_i = \frac{(G_{xy})_i t_i^3}{12} \tag{4.34}$$

where the engineering properties (E_x, E_y, ν_{xy}, and G_{xy}) of the panel are computed by using micro/macromechanics approach and plotted in parametric forms as carpet plots (see section 4.3.2).

For most common thin-walled structures (e.g., I- and Box-sections), the beam centroid is the neutral axis of bending (no beam bending-extension coupling), and the axial (A), bending (D), shear (F) and torsional (D_t) stiffnesses of the beam (that account for the contribution of all the panels) can be simplified as

$$A = \sum_{i=1}^{n}(E_x)_i t_i b_i,$$

$$D = \sum_{i=1}^{n}\left[(E_x)_i t_i\left(\frac{b_i^2}{12}\sin^2\phi_i\right) + \frac{(E_x)_i t_i^3\cos^2\phi_i}{12}\right]b_i$$

$$F = \sum_{i=1}^{n}(G_{xy})_i t_i b_i \sin^2\phi_i \qquad D_t^{\text{open}} = \frac{1}{3}\sum_{i=1}^{n}(G_{xy})_i t_i^3 b_i$$

$$D_t^{\text{close}} = \frac{(2\Gamma_s)^2}{\sum_{i=1}^{n} b_i/((G_{xy})_i t_i)} + \frac{1}{4}\left(\sum_{i=1}^{n}(G_{xy})_i t_i^3 b_i\right) \tag{4.35}$$

where b_i is the panel width, and ϕ_i is the cross-sectional orientation of the ith panel with respect to the x-axis.

The preceding stiffness formulas are readily applied in engineering design to various structural shapes. For example, the beam bending (D) and shear (F) stiffnesses (as shown in eqn 4.35) are simplified for I- and box beams (Fig. 4.31) as follows.

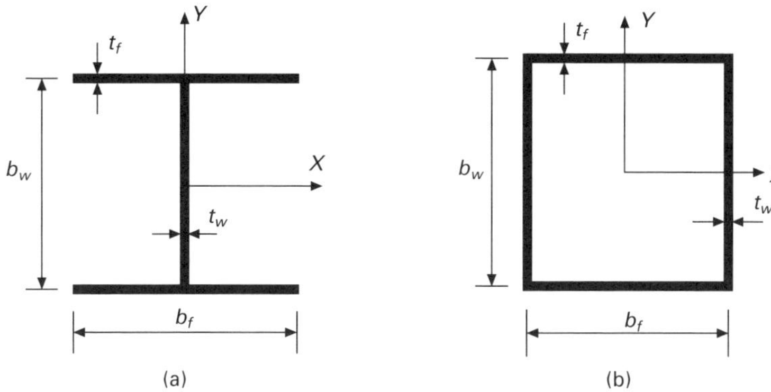

4.31 Geometry of I- and box-sections: (a) I-beam; (b) box beam.

For I section in Fig. 4.31(a):

$$\text{strong axis: } D_Y = \frac{1}{2}(E_x)_f t_f b_w^2 b_f + \frac{1}{12}(E_x)_w t_w b_w^3 + \frac{1}{6}(E_x)_f t_f^3 b_f,\ F_Y = (G_{xy})_w t_w b_w \qquad 4.36a$$

$$\text{weak axis: } D_X = \frac{1}{12}(E_x)_w t_w^3 b_w + \frac{1}{6}(E_x)_f t_f b_f^3$$

$$F_X = 2(G_{xy})_f t_f b_f \qquad 4.36b$$

where E_x and G_{xy} are the longitudinal and shear moduli, respectively. The subscripts f and w represent the quantities relating to the flange and web panels, respectively.

For box section in Fig. 4.31(b):

strong axis:

$$D_Y = \frac{1}{2}(E_x)_f t_f b_w^2 b_f + \frac{1}{6}(E_x)_w t_w b_w^3 + \frac{1}{6}(E_x)_f t_f^3 b_f$$

$$F_Y = 2(G_{xy})_w t_w b_w \qquad 4.37a$$

weak axis:

$$D_X = \frac{1}{2}(E_x)_w t_w b_f^2 b_w + \frac{1}{6}(E_x)_f t_f b_f^3 + \frac{1}{6}(E_x)_w t_w^3 b_w$$

$$F_X = 2(G_{xy})_f t_f b_f \qquad 4.37b$$

Table 4.5 lists the bending and shear stiffnesses of four selected beams based on eqns 4.36 and 4.37.

Table 4.5 Beam bending and shear stiffness properties

FRP Shapes	$D = EI$ ($\times 10^8$ psi-in^4)		$F = GA$ (10^6 psi-in^2)	
	Strong-axis	Weak-axis	Strong-axis	Weak-axis
WF-Beam 6″ × 6″ × 3/8″ **(WF6 × 6)**	1.776	0.570	1.292	3.066
I-Beam 4″ × 8″ × 3/8″ **(I4 × 8)**	2.558	0.199	1.772	2.379
WF-Beam 4″ × 4″ × 1/4″ **(WF4 × 4)**	0.334	0.111	0.585	1.351
Square Tube 4″ × 4″ × 1/4″ **(Box 4 × 4)**	0.364	0.338	1.100	1.176

4.3.4 Mechanical behaviors of FRP shapes

Most FRP shapes are thin-walled structures and made of E-glass fiber and polyester or vinylester resins. Due to the relatively low stiffness of FRP composites and thin-walled sectional geometry, problems with large deformation including shear deformation and structural stability need to be considered in design and analysis (Qiao *et al.* 1999).

Inability to meet service requirements due to large deflections is usually the dominant design limit for FRP shapes subjected to transverse loading. Critical global and local buckling due to the thin-walled structure and/or large slenderness ratios of component panels is another important criterion in design of FRP shapes. Further, potential material failure due to the relatively low compressive and shear strengths of composites should also be considered. Thus, the mechanical behaviors of FRP shapes associated with large deformation, elastic stability, and material failure are addressed in this section.

Elastic deflections

According to Timoshenko beam theory, the deflection of a beam is caused by bending and shear when subjected to transverse loading and can be obtained by solving the equilibrium equations. Deflections at discrete locations can be computed by employing energy methods that incorporate the beam bending and shear stiffnesses. For some conventional materials, the shear deformation can be neglected because of the higher shear modulus. But for FRP materials, the shear modulus of usually $E/10$, compared to $E/2.5$ for metallic beams, is relatively lower; therefore, the shear deformation must be considered. Formulas for maximum bending and shear deflections for typical beam loading and boundary conditions are given in Table 4.6, as

$$\delta_{\text{Total}} = \delta_{\text{Bending}} + \delta_{\text{Shear}} \tag{4.38}$$

where in Table 4.6, P is the concentrated load, q is the uniformly distributed load, M is the transverse bending moment, L is the span length, K is the shear correction factor ($K = 1.0$ can be assumed), and D and F are the beam bending and shear stiffnesses (see eqn 4.35).

Table 4.6 Bending and shear deflections of Timoshenko's beams

Beam loading + B.C.)	δ_{Bending}	δ_{Shear}
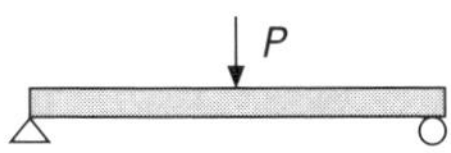	$\frac{1}{48}\frac{PL^3}{D}$	$\frac{1}{4}\frac{PL}{KF}$
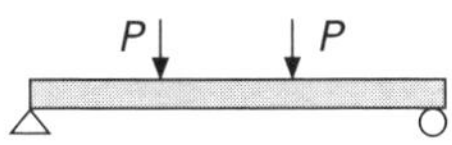	$\frac{23}{648}\frac{PL^3}{D}$	$\frac{1}{3}\frac{PL}{KF}$
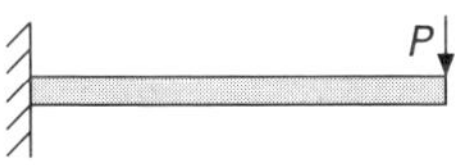	$\frac{1}{3}\frac{PL^3}{D}$	$\frac{PL}{KF}$
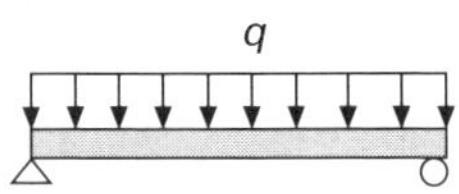	$\frac{5}{384}\frac{qL^4}{D}$	$\frac{1}{8}\frac{qL^2}{KF}$
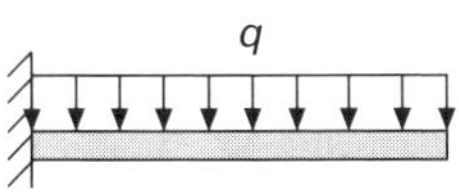	$\frac{1}{8}\frac{qL^4}{D}$	$\frac{1}{2}\frac{qL^2}{KF}$
	$\frac{1}{8}\frac{ML^2}{D}$	0

Elastic strains and stresses

For the ith panel (Fig. 4.30), the mid-surface strains and curvatures in terms of the beam resultant forces and moments are calculated as

$$\bar{\varepsilon}_x = \frac{N_Z}{A_Z} + (\bar{Y}_i - Y_n)\frac{M_Y}{D_Y}, \quad \bar{\kappa}_x = \frac{M_Y}{D_Y}\cos\phi_i, \quad \bar{\gamma}_{xy} = \frac{V_Y}{K_Y F_Y}\sin\phi_i \tag{4.39}$$

where N_Z, M_Y, and V_Y are, respectively, the resultant internal axial force, bending moment, and transverse shear force acting on the beam in the global coordinate system. The subscripts x, y, and z refer to the local coordinate system of individual panels defined similarly as those in MLB approach. Then, applying eqn 4.30, we can obtain the resultant forces and moments ($\overline{N}_x$, $\overline{M}_x$, and $\overline{N}_{xy}$) acting on the ith panel. Combining the constitutive relations of eqn 4.10 with the assumptions of $\overline{N}_y = \overline{M}_y = 0$ and $\overline{M}_{xy} = 0$, the mid-surface strains and curvatures on the ith panel are obtained as:

$$\begin{Bmatrix} \bar{\varepsilon}_x^0 \\ \bar{\varepsilon}_y^0 \\ \bar{\gamma}_{xy}^0 \\ \bar{\kappa}_x \\ \bar{\kappa}_y \\ \bar{\kappa}_{xy} \end{Bmatrix} = \begin{bmatrix} \alpha_{11} & \alpha_{16} & \beta_{11} \\ \alpha_{12} & \alpha_{26} & \beta_{12} \\ \alpha_{16} & \alpha_{66} & \beta_{16} \\ \beta_{11} & \beta_{16} & \delta_{11} \\ \beta_{12} & \beta_{26} & \delta_{12} \\ \beta_{16} & \beta_{66} & \delta_{16} \end{bmatrix} \begin{Bmatrix} \overline{N}_x \\ \overline{N}_{xy} \\ \overline{M}_x \end{Bmatrix} \qquad 4.40$$

where the overbar identifies a panel quantity. Based on classical lamination theory (CLT), the ply strains (ε_x, ε_y, and ε_{xy}) and stresses (σ_x, σ_y, and σ_{xy}) can be correspondingly obtained through the thickness of each panel. Using coordinate transformations, the ply strains (ε_1, ε_2, and γ_{12}) and stresses (σ_1, σ_2, and τ_{12}) can be computed in principal material directions. Table 4.7 shows comparisons between predictions from eqns 4.38 and 4.39 and experimental measurements of four selected beams.

Global buckling

Euler buckling of FRP columns

For an elastic column with pin-pin boundaries at the ends and under axial load P, the Euler buckling load can be easily obtained by applying beam theory, and it is defined as

$$P_E = \frac{n^2 \pi^2 D}{L^2} \qquad 4.41$$

where D is the bending stiffness, n is the buckled wave number, and L is the span length. When the load exceeds the critical value P_E for $n = 1$, the column becomes unstable. For a column made of composite materials, the bending stiffness (D) is computed from beam theory using eqn 4.35. It is important to note that D is computed assuming that the cross-section does not deform during bending.

Table 4.7 Beam deflections and strains (L = 12.0 ft)

FRP Shapes	Axis of load	Deflection δ (in/kip)		Strain ε_{Top} ($\mu\varepsilon$/kip)		Strain ε_{Bottom} ($\mu\varepsilon$/kip)	
		Test	Design	Test	Design	Test	Design
WF-Beam	Strong	0.388	0.378	–616.9	–608.2	668.6	608.2
6″ × 6″ × 3/8″ **(WF6 × 6)**	Weak	1.169	1.106	–1902.7	–1900.0	1823.5	1900.0
I-Beam	Strong	0.271	0.264	–576.6	–566.0	594.1	566.0
4″ × 8″ × 3/8″ **(I4 × 8)**	Weak	3.511	3.152	–3646.8	–3630.0	3557.6	3630.0
WF-Beam	Strong	1.833	1.926	–2081.3	–2160.0	2121.8	2160.0
4″ × 4″ × 1/4″ **(WF4 × 4)**	Weak	5.769	5.627	–5879.6	–6480.0	5913.9	6480.0
Square Tube	Strong	1.886	1.742	–2139.8	–1990.0	2141.7	1990.0
4″ × 4″ × 1/4″ **(Box4 × 4)**	Weak	1.947	1.873	–1944.4	–2140.0	1903.9	2140.0

Influence of the end-conditions

Equation 4.41 can be generalized for various end-conditions by identifying the inflection points of the deflected shape in Fig. 4.32. The effective length is the distance between the inflection points or the end pinned supports, and it is represented by L in eqn 4.41, where its value is conveniently computed as $L_e = \xi L$, in terms of tabulated end-restraint coefficient ξ (Table 4.8). The ideal boundary conditions in Fig. 4.32 cannot be realized in practice because of flexibility of the connections. Therefore, experimentally adjusted values of the end-restraint coefficient are used, as given in Table 4.8. Unfortunately, there is no experimental data for FRP composite columns. The values for wood columns are often used in design since the material properties of wood are relatively similar to those of FRP composites. Then, the Euler buckling load in eqn 4.41 becomes

$$P_E = \frac{\pi^2 D}{(\xi L)^2} \tag{4.42}$$

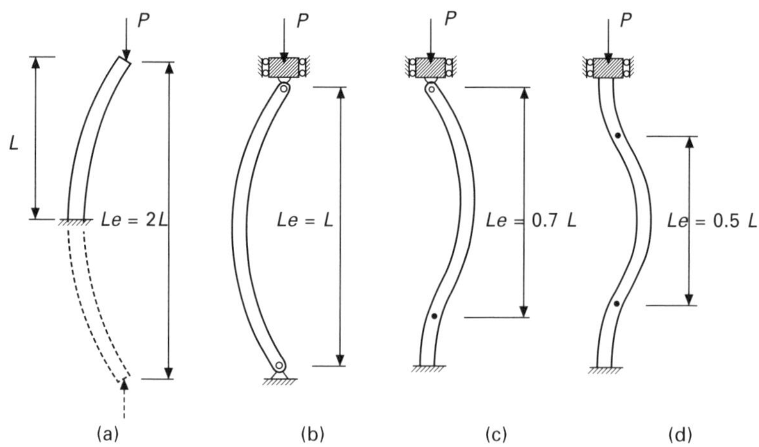

4.32 Effective length of column for various end-conditions: (a) one fixed, and the other free; (b) both ends pinned; (c) one fixed end, and the other pinned; (d) both ends fixed.

Table 4.8 End-restraint coefficients for long column buckling

End-restraint	ξ_{theory}	ξ_{steel}	ξ_{wood}
Pinned-pinned	1.0	1.0	1.0
Clamped-clamped	0.5	0.65	0.65
Pinned-clamped	0.7	0.8	0.8
Clamped-free	2.0	2.1	2.4

If the end-conditions are the same with respect to the weak and strong axes of bending, buckling will occur with respect to the weak-axis of bending, because the stabilizing effect is weaker with respect to this axis. But if the end conditions are different, not only the values of (D) but also the values of the end restraint condition will determine the direction of buckling. In those cases, the bending stiffness (D) and coefficient ξ with respect to both the weak and strong axes are needed.

Influence of shear deformation

The shear deformation increases the deflection of a member when undergoing buckling, and it can be interpreted as a reduction of the stabilizing effect brought by the bending stiffness of the column. Shear deformation beam theory can be used to derive the equation for critical load, and the result is a reduction of the critical load given in eqn 4.42, and is modified as

$$P_{ES} = \frac{P_E}{1 + P_E / F} \qquad 4.43$$

The above correction to eqn 4.42 is negligible for practical cases. The shear stiffness (F) affects the beam deflections only for short beams, and the same holds for columns. But short thin-walled columns fail by local buckling at loads lower than P_E. Therefore, the correction of P_E for shear has very limited application.

Flexural-torsional buckling of FRP beams

In this section, the flexural-torsional buckling of pultruded FRP composite I- and C-section beams is analyzed using the second variational total potential energy principle and Rayleigh-Ritz method (Qiao *et al.* 2003; Shan and Qiao 2005). The total potential energy of FRP shapes based on nonlinear plate theory is derived, which includes shear effect and bending-twisting coupling.

Theoretical background

The energy method is used to analyze the flexural-torsional buckling of FRP beams. Each beam can be separated into top flange, bottom flange and web panels (Fig. 4.33). The total potential energy equation governing instability is derived by treating the web panel as a plate and the flange panels as beams and plates.

The total potential energy (Π) of a system is the sum of the strain energy (U) and the work (W) done by the external loads, and it is expressed as

$$\Pi = U + W \qquad 4.44$$

where $W = -\sum P_i q_i$, and $U = U(\varepsilon_{ij})$. For linear elastic problems, the strain energy is given as

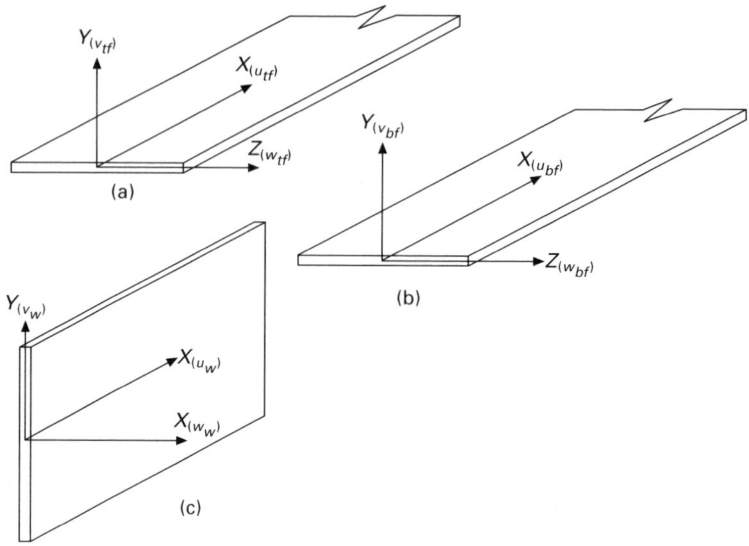

4.33 Coordinate system of I- or C-beam: (a) top flange; (b) bottom flange; (c) web.

$$U = \frac{1}{2}\int_V \sigma_{ij}\varepsilon_{ij}dV \tag{4.45}$$

For a structure in an equilibrium state, the total potential energy attains a stationary value when the first variation of the total potential energy ($\delta\Pi$) is zero. Then, the condition for the state of equilibrium is expressed as

$$\delta\Pi = -\Sigma P_i\delta q_i + \int_V \sigma_{ij}\delta\varepsilon_{ij}dV = 0 \tag{4.46}$$

The structure is in a stable equilibrium state if the value of the potential energy is a relative minimum. It is possible to infer whether a stationary value of a functional Π is a maximum or a minimum by observing the sign of $\delta^2\Pi$. If $\delta^2\Pi$ is positively defined, Π is a minimum. Thus, the condition for the state of stability is characterized by the inequality (Shan and Qiao 2005)

$$\delta^2\Pi = -\Sigma P_i\delta^2 q_i + \int_V (\sigma_{ij}\delta^2\varepsilon_{ij} + \delta\sigma_{ij}\delta\varepsilon_{ij})dV > 0 \tag{4.47}$$

Because q_i is usually being expressed as linear functions of displacement variables, $\delta^2 q_i$ in eqn 4.47 vanishes. Therefore, the critical condition for stability analysis becomes

$$\delta^2\Pi = \delta^2 U = \int_V (\sigma_{ij}\delta^2\varepsilon_{ij} + \delta\sigma_{ij}\delta\varepsilon_{ij})dV = 0 \tag{4.48}$$

The second variation of the total strain energy of the whole beam can be obtained by summing the contributions from web, top and bottom flanges as

$$\delta^2 U = \delta^2 U^{tf} + \delta^2 U^{bf} + \delta^2 U^{w} \tag{4.49}$$

and the critical condition (instability) is defined as

$$\delta^2 \Pi = \delta^2 U = 0 \tag{4.50}$$

which can be solved by employing the Rayleigh-Ritz method. The formulations of total strain energy, displacement field, and stress resultants for I- and C-section FRP beams, respectively, are given in details by Qiao *et al.* (2003) and Shan and Qiao (2005).

Explicit solutions

For the global (flexural-torsional) buckling of I- or C-section beams, the cross-section of the beam is considered as undistorted. As the web panel is not allowed to distort and remains straight in flexural-torsional buckling, the sideways deflection and rotation of the web are coupled. For example, the shape functions of buckling deformation for both the sideways deflection and rotation of the web, which satisfy the cantilever beam boundary conditions, can be selected as exact transcendental functions as (Qiao *et al.* 2003)

$$\begin{Bmatrix} w \\ \theta \end{Bmatrix} = \begin{Bmatrix} \bar{w} \\ \bar{\theta} \end{Bmatrix} \sum_{m=1,2,3,\ldots} \times \left\{ \sin\left(\frac{\lambda_m x}{L}\right) - \sinh\left(\frac{\lambda_m x}{L}\right) - \beta_m \left[\cos\left(\frac{\lambda_m x}{L}\right) - \cosh\left(\frac{\lambda_m x}{L}\right) \right] \right\} \tag{4.51}$$

where $\beta_m = \dfrac{\sinh(\lambda_m) + \sin(\lambda_m)}{\cos(\lambda_m) + \cosh(\lambda_m)}$, and λ_m satisfies the following transcendental equation

$$\cos(\lambda_m)\cosh(\lambda_m) - 1 = 0 \tag{4.52}$$

with $\lambda_1 = 1.875104$, $\lambda_2 = 4.694091$, $\lambda_3 = 7.854757$, ...

By applying the Rayleigh-Ritz method and solving for the eigenvalue of the potential energy equilibrium equation (eqn 4.50), the flexural-torsional buckling load, P_{cr}, for a cantilever I-section with a point load applied at the centroid of the free-end is obtained (Qiao *et al.* 2003)

$$P_{cr} = \Psi_1 \cdot \{ b_w L \Psi_2 + (\sqrt{\Psi_3 + \Psi_4 + \Psi_5 + \Psi_6 + \Psi_7}) / b_w \} \tag{4.53}$$

where $\Psi_1 = (6b_f + b_w)/[2L^3 \cdot (76.5b_f^2 - 6.96 b_f b_w + 0.16 b_w^2)]$

$$\Psi_2 = (123b_f - 5.6b_w)D_{16}$$

$$\Psi_3 = a_{11}b_f^3(279.5b_f^2 - 25.5b_f b_w + 0.6b_w^2)$$

$$\Psi_4 = b_f b_w^5(62.7L^2 d_{66} D_{11} - 305.4b_w^2 D_{11}^2$$

$$- 1377.4L^2 D_{16}^2 - 5511L^2 D_{11} D_{66})$$

$$\Psi_5 = b_w^6(7b_w^2 D_{11}^2 + 31.4L^2 D_{16}^2 + 125.5L^2 D_{11} D_{66})$$

$$\Psi_6 = a_{11}b_f^3 b_w^2(1118b_f^5 d_{11} - 101.8b_f^4 b_w d_{11} + 2.3b_f^3 b_w^2 d_{11}$$

$$+ 5043.5b_f^3 L^2 d_{66} + 4.64b_w^5 D_{11} + 20.9b_w^3 L^2 D_{66})$$

$$\Psi_7 = a_{11}b_f^4 b_w^3[b_w(-203.6b_w^2 D_{11} + 10.5L^2 d_{66} - 918.5L^2 D_{66})$$

$$+ b_f(2235.8b_w^2 D_{11} - 459.5L^2 d_{66} + 10087L^2 D_{66})]$$

and the following material parameters are defined as:

$$a_{11} = 1/\alpha_{11},\ a_{66} = 1/\alpha_{66},\ d_{11} = 1/\delta_{11},\ d_{66} = 1/\delta_{66} \qquad 4.54$$

In a similar fashion, the solution for global buckling load of cantilever C-section beams was recently obtained by Shan and Qiao (2005).

Again, as an illustration of the proposed solution to global buckling of FRP beams, several FRP cantilever I- and C-section beams are analyzed, and their predictions are compared both with the experimental data and finite element analyses (see Table 4.9) (Qiao *et al.* 2003; Shan and Qiao 2005). As indicated in Table 4.9, the favorable agreement of analytical solutions with experimental results and finite element eigenvalue analyses demonstrates the validity of the present variational principle methodology for global buckling analyses.

Table 4.9 Comparisons for flexural-torsional buckling loads of I- and C-beams

Section	Dimensions (in)	Analytical solution P_{cr} (lb)	Finite element P_{cr} (lb)	Experimental data P_{cr} (lb)
I-	4 × 8 × 1/4	717.6	664.5	661.6
(L = 12 ft)	6 × 6 × 1/4	1262.1	1298.1	1231.1
C-	1-1/8 × 4 × 1/4	17.6	17.7	14.1
(L = 12 ft)	1-5/8 × 6 × 1/4	48.5	49.0	42.8

For design purpose, we adopt the simplified engineering equations for flexural-torsional buckling of I-beams developed by Pandey *et al.* (1995) for several commonly used loading conditions (Table 4.10); these formulas are

Table 4.10 Critical flexural-torsional buckling loads of I-beams under different loading conditions

Beam (loading + B.C.)	The critical buckling load
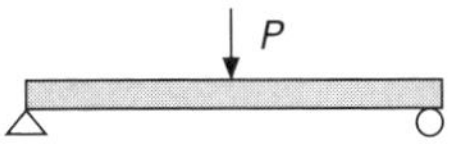	$P_{cr} = \frac{17.17}{L^2}\sqrt{D_y \cdot JG}\sqrt{1 + \frac{\pi^2}{L^2}\frac{I_{ww}}{JG}}$
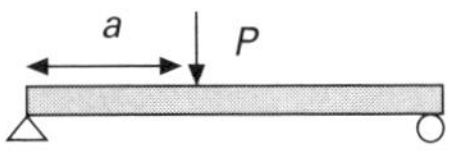	$P_{cr} = \frac{4.342}{L^2}\sqrt{D_y \cdot JG}\, f(\xi)$
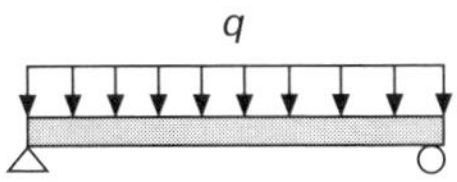	$q_{cr} = \frac{28.46}{L^3}\sqrt{D_y \cdot JG}\left[\sqrt{\frac{\pi^2}{L^2}\left(\frac{I_{ww}}{JG} + 0.212k^2\frac{D_y}{JG}\right) + 1} - \frac{1.44k}{L}\sqrt{\frac{D_y}{JG}}\right]$
	$P_{cr} = 5.08\sqrt{1 + \frac{19.3}{k}\left[\frac{(k+13)(k+3)}{(k+10)^2}\right]}\,\frac{\sqrt{D_y JG}}{L^2}$
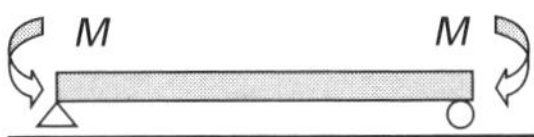	$M_{cr} = \sqrt{A_t\frac{\pi^2}{L^2}D_y JG + A_w\frac{\pi^4}{L^4}I_{ww}D_y}$

1. where D_y is the bending stiffness around the strong axis;

$$JG = \frac{2(G_{xy})_f t_f^3 b_f}{3} + \frac{(G_{xy})_w t_w^3 b_w}{3};$$

$$I_{ww} = \frac{(E_x)_f t_f b_w^2 b_f^3}{24} + \frac{(E_x)_f t_f^3 b_f^3}{36} + \frac{(E_x)_w t_w^3 b_w^3}{144};$$

2. k is $h_w/2$, 0, $h_w/2$ for loads applied at the level of the centroid of the bottom flange, the section centroid and the centroid of the top flange, respectively;

3. $$f(\xi) = \frac{1}{\xi(1-\xi)}\sqrt{\frac{2.33 - 1.75\xi - 2.8\xi^2 + 3.5\xi^3 - \xi^4 + ((7/\xi) + 2.8\xi^2 - 7\xi)(I_{ww}/L^2 JG)}{\xi^6 - 4.4\xi^5 + 4.89\xi^4 + 2.75\xi^3 - 6.286\xi^2 + 2.2}}$$ and $\xi = a/L$;

4. A_t and A_w are the buckling coefficients which depend upon the bending moment distribution and support conditions defined by Sherbourne and Pandey (1989).

based on Vlasov's theory. As an example, the flexural-torsional buckling loads are calculated for different lengths of cantilever FRP I beam I6 × 3 (6″ × 3″ × 3/8″) and channel-beam C6 × 2-B (6″ × 1-11/16″ × 3/8″), each subjected to a tip load at the free-end. Based on the lay-up information provided by the manufacturer and a micro/macromechanics approach, the panel stiffness coefficients are given in Table 4.11. In Fig. 4.34, the classical solution refers to the predictions by Vlasov's theory (see Table 4.10) (Pandey *et al.* 1995). The flexural-torsional critical buckling loads (P_{cr}) versus the span lengths (L) for the geometries of I-beam I6 ×3 (6″ × 3″ × 3/8″) and

Table 4.11 Panel stiffness coefficients for I- and C-section beams

Section	D_{11} (lb-in)	D_{12} (lb-in)	D_{22} (lb-in)	D_{66} (lb-in)	a_{11} (lb/in)	a_{66} (lb/in)	d_{11} (lb-in)	d_{66} (lb-in)
16 × 3	12,993	2,548	6,076	2,918	1,978,576	307,778	18,587	3,618
C6 × 2-B	14,599	3,050	8,210	2,640	1,234,568	213,675	13,496	2,639

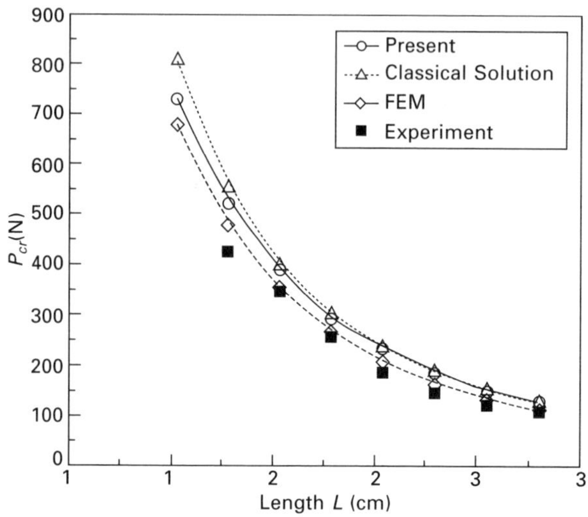

4.34 Flexural-torsional buckling load for an FRP I-beam (6″ × 3″ × 3/8″).

channel-beam C6 × 2-B (6″ × 1-11/16″ × 3/8″) are shown in Figs 4.34 and 4.35, respectively.

Local buckling of FRP shapes

For short-span FRP shapes, local buckling occurs more readily (see Fig. 4.36). In general, the local buckling analyses of FRP shapes are accomplished by modeling the flanges and webs individually and considering the flexibility of the flange-web connections (Qiao *et al.* 2001). In this type of simulation, each part of FRP shapes (Fig. 4.37) is modeled as a composite plate subjected to elastic restraints along the unloaded edges (i.e., the flange-web connections) (Qiao *et al.* 2001).

The first variational total potential energy approach is hereby applied to local buckling analysis of FRP plates, and the Ritz method is used to establish an eigenvalue problem. Explicit solutions for local buckling problems of two types of elastically restrained plates (see Fig. 4.37) were obtained by Qiao and Zou (2002, 2003). By incorporating the discrete plate solutions and

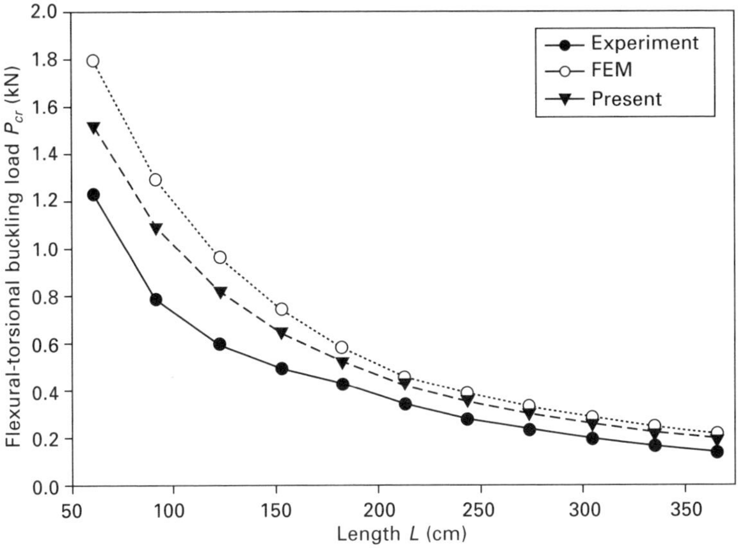

4.35 Flexural-torsional buckling load for an FRP channel beam (6″ × 1-11/16″ × 3/8″).

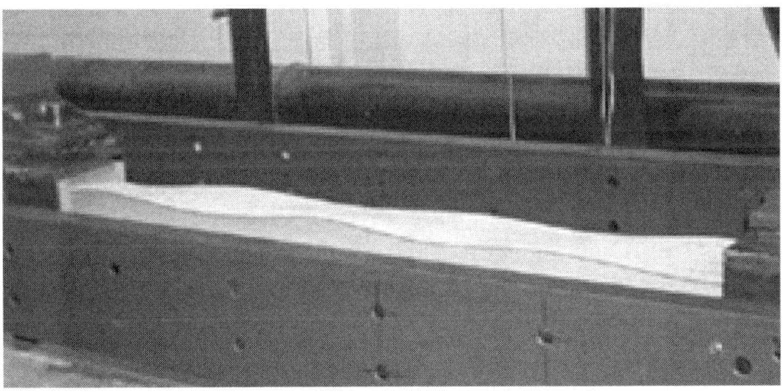

4.36 Local buckling of an FRP column (WF 6″ × 6″ × 1/4″).

considering the rotational restraint stiffness at the flange-web connection, the local buckling of FRP structural shapes is then determined.

Theoretical background

The local buckling of an orthotropic plate subjected to uniform in-plane axial load along the simply supported edges and rotationally restrained either at two unloaded edges (*RR* plate in Fig. 4.37(a)) or at one unloaded edge with the other free (*RF* plate in Fig. 4.37(b)) is briefly presented in this section. A variational formulation of the Ritz method is herein used to analyze

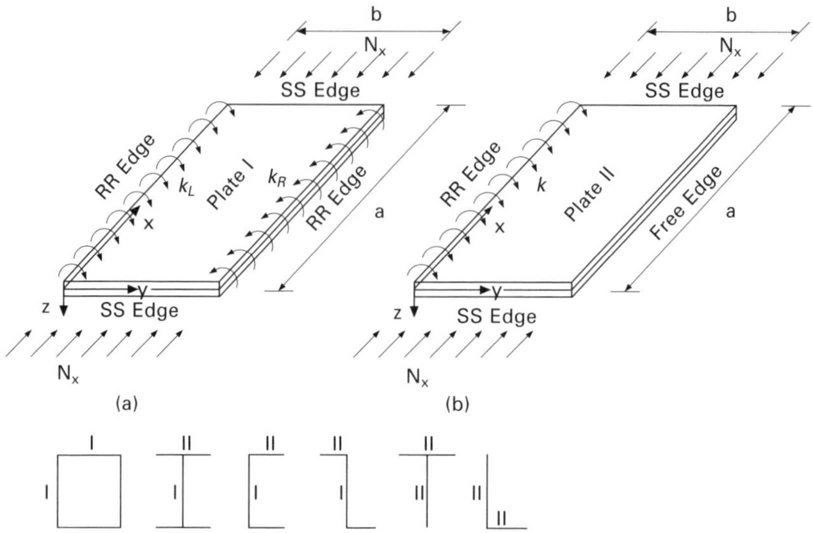

4.37 Geometry of orthotropic plate elements of FRP shapes: (a) RR unloaded edges; (b) RF unloaded edges (RR – rotationally restrained; SS-simply supported).

the elastic buckling of an orthotropic plate with the boundary conditions shown in Fig. 4.37. In the variational form of the Ritz method, the first variations of the elastic strain energy stored in the plate (δU_e), the strain energy stored in the elastic restraints along the rotationally restrained boundaries of the plate (δU_Γ), and the work done by the axial in-plane force (δV) are computed by properly choosing the out-of-plane buckling displacement functions (w) (Qiao and Zou 2002; 2003).

The elastic strain energy of an orthotropic plate (U_e) is given as

$$U_e = \frac{1}{2}\iint_{\Omega} \{D_{11}w_{,xx}^2 + D_{22}w_{,yy}^2 + 2D_{12}w_{,xx}w_{,yy} + 4D_{66}w_{,xy}^2\}dxdy \tag{4.55}$$

where D_{ij} (i, j = 1, 2, 6) are the plate bending stiffness coefficients (Jones 1999) and Ω is the area of the plate.

For the plate with rotational restraints distributed along the unloaded boundary edges, the strain energy (U_Γ) stored in equivalent elastic springs at the flange-web connections for the *RR* plate and *RF* plate are given, respectively, as

$$U_\Gamma = \frac{1}{2}\int_\Gamma k_L\left(\frac{\partial w}{\partial y}\Big|_{y=0}\right)^2 d\Gamma + \frac{1}{2}\int_\Gamma k_R\left(\frac{\partial w}{\partial y}\Big|_{y=b}\right)^2 d\Gamma \tag{4.56a}$$

for *RR* plate

$$U_\Gamma = \frac{1}{2}\int_\Gamma k\left(\frac{\partial w}{\partial y}|_{y=0}\right)^2 d\Gamma \text{ for } RF \text{ plate} \quad 4.56b$$

where k_L and k_R in eqn 4.56a are the rotational restraint stiffness at the restrained edges of $y = 0$ and b, respectively (Fig. 4.37(a)); and k in eqn 4.56b is the rotational restraint stiffness at the restrained edge of $y = 0$ (Fig. 4.37(b)).

The work (V) done by the in-plane uniformly distributed compressive force (N_x) can be written as

$$V = \frac{1}{2}N_x \iint_\Omega w_{,x}^2\, dxdy \quad 4.57$$

where N_x is defined as the uniform compressive force per unit length at the simply supported boundary of $x = 0$ and a.

Using the equilibrium condition of the first variational principle of the total potential energy

$$\delta\Pi = \delta U_e + \delta U_\Gamma - \delta V = 0 \quad 4.58$$

and substituting the proper out-of-plane displacement function (w) into eqn 4.58, the standard buckling eigenvalue problem can be solved by the Ritz method.

Explicit solutions

To solve the eigenvalue problem, it is important to choose the proper out-of-plane buckling displacement function (w). In this study, to explicitly obtain the analytical solutions for local buckling of two types of representative plates as shown in Fig. 4.37, the unique buckling displacement fields are proposed as follows (Qiao and Zou 2002, 2003; Qiao and Shan 2005).

$$w(x, y) = \left\{\frac{y}{b} + \psi_1\left(\frac{y}{b}\right)^2 + \psi_2\left(\frac{y}{b}\right)^3 + \psi_3\left(\frac{y}{b}\right)^4\right\}\sum_{m=1}^{\infty}\alpha_m \sin\frac{m\pi x}{a}$$

for RR Plate 4.59a

$$w(x, y) = \left\{(1 - \omega)\frac{y}{b} + \omega\left[\frac{3}{2}\left(\frac{y}{b}\right)^2 - \frac{1}{2}\left(\frac{y}{b}\right)^3\right]\right\}\sum_{m=1}^{\infty}\alpha_m \sin\frac{m\pi x}{a}$$

for RF Plate 4.59b

where ψ_1, ψ_2, ψ_3 and ω are the unknown constants which can be solved by satisfying the boundary conditions (see details in Qiao and Shan (2005)). Then the assumed displacement function for Plate I and Plate II shown in Fig. 4.37 can be obtained, respectively, as

$$w(x, y) = \left\{\frac{y}{b} + \frac{k_L b}{2D_{22}}\left(\frac{y}{b}\right)^2 - \frac{12D_{22}^2 + D_{22}(5k_L + 3k_R)b + k_L k_R b^2}{6D_{22}^2 + D_{22}k_R b}\left(\frac{y}{b}\right)^3 + \frac{12D_{22}^2 + D_{22}(4k_L + 4k_R)b + k_L k_R b^2}{12D_{22}^2 + 2D_{22}k_R b}\left(\frac{y}{b}\right)^4\right\} \times \sum_{m=1}^{\infty} \alpha_m \sin\frac{m\pi x}{a} \quad \text{for } RR \text{ Plate} \qquad 4.60a$$

$$w(x, y) = \left\{\left(1 - \frac{bk}{3D_{22} + bk}\right)\frac{y}{b} + \frac{bk}{3D_{22} + bk}\left[\frac{3}{2}\left(\frac{y}{b}\right)^2 - \frac{1}{2}\left(\frac{y}{b}\right)^3\right]\right\} \times \sum_{m=1}^{\infty} \alpha_m \sin\frac{m\pi x}{a} \quad \text{for RF Plate} \qquad 4.60b$$

Noting that either k_L, k_R or $k = 0$ corresponds to the simply supported boundary condition; whereas, either k_L, k_R or $k = \infty$ represents the clamped (built-in) boundary condition at rotationally restrained edges.

By substituting eqn (4.60) into the equilibrium condition of the first variational principle of the total potential energy (eqn 4.58, the solution of an eigenvalue problem for the *RR* plate and *RF* plate (Fig. 4.37) is obtained (Qiao and Shan 2005).

For the *RR* plate when $k_L = k_R = k$, the local buckling stress resultant is simplified to

$$N_{cr}^{RR} = \frac{24}{b^2}\left\{1.871\sqrt{\frac{\tau_2}{\tau_1}}\sqrt{D_{11}D_{22}} + \frac{\tau_3}{\tau_1}(D_{12} + 2D_{66})\right\} \qquad 4.61$$

where, the coefficients τ_1, τ_2, and τ_3 are functions of the rotational restraint stiffness k, and defined as

$$\tau_1 = 124 + 22\frac{kb}{D_{22}} + \frac{k^2b^2}{D_{22}^2},$$

$$\tau_2 = 24 + 14\frac{kb}{D_{22}} + \frac{k^2b^2}{D_{22}^2},\ \tau_3 = 102 + 18\frac{kb}{D_{22}} + \frac{k^2b^2}{D_{22}^2}$$

and the resulting critical aspect ratio for the *RR* plate is given as

$$\gamma_{cr}^{RR} = 0.663\left\{\frac{m^4(\eta_1 k^2b^2 + \eta_2 D_{22}kb + 4\eta_3 D_{22}^2)D_{11}}{(\eta_{12}k^2b^2 + \eta_{13}D_{22}kb + 36\eta_{14}D_{22}^2)D_{22}}\right\}^{\frac{1}{4}} \tag{4.62}$$

where

$$\eta_1 = \frac{76D_{22}^2 + 17D_{22}kb + k^2b^2}{D_{22}^2}$$

$$\eta_2 = \frac{1{,}140D_{22}^2 + 272D_{22}kb + 17k^2b^2}{D_{22}^2}$$

$$\eta_3 = \frac{1{,}116D_{22}^2 + 285D_{22}kb + 19k^2b^2}{D_{22}^2}$$

$$\eta_{12} = \frac{36D_{22}^2 + 13D_{22}kb + k^2b^2}{D_{22}^2}$$

$$\eta_{13} = \frac{396D_{22}^2 + 156D_{22}kb + 13k^2b^2}{D_{22}^2}$$

$$\eta_{14} = \frac{24D_{22}^2 + 11D_{22}kb + k^2b^2}{D_{22}^2}$$

while for the *RF* plate, the local buckling stress resultant and the critical aspect ratio are obtained, respectively, as

$$N_{cr}^{RF} = \frac{112(15D_{22}^2 + 10D_{22}kb + 2k^2b^2)D_{66} - 28(5D_{22}kb + k^2b^2)D_{12}}{b^2(140D_{22}^2 + 77D_{22}kb + 11k^2b^2)} + \frac{4\sqrt{35D_{11}D_{22}kb(3D_{22} + kb)}}{b^2\sqrt{140D_{22}^2 + 77D_{22}kb + 11k^2b^2}} \tag{4.63}$$

$$\gamma_{cr}^{RF} = 0.9133m\left\{\frac{(140D_{22}^2 + 77D_{22}kb + 11k^2b^2)D_{11}}{(3D_{22} + kb)kbD_{22}}\right\}^{\frac{1}{4}} \tag{4.64}$$

Note that eqns 4.61 and 4.63 are independent of the number of buckling half-wave length (m).

Application to FRP plates

Based on the explicit formulas in eqns 4.61 and 4.63, design formulas of critical local buckling load (N_{cr}) for several common orthotropic plate cases of applications (see Fig. 4.38) and their related critical aspect ratio (γ_{cr}) are summarized in Table 4.12.

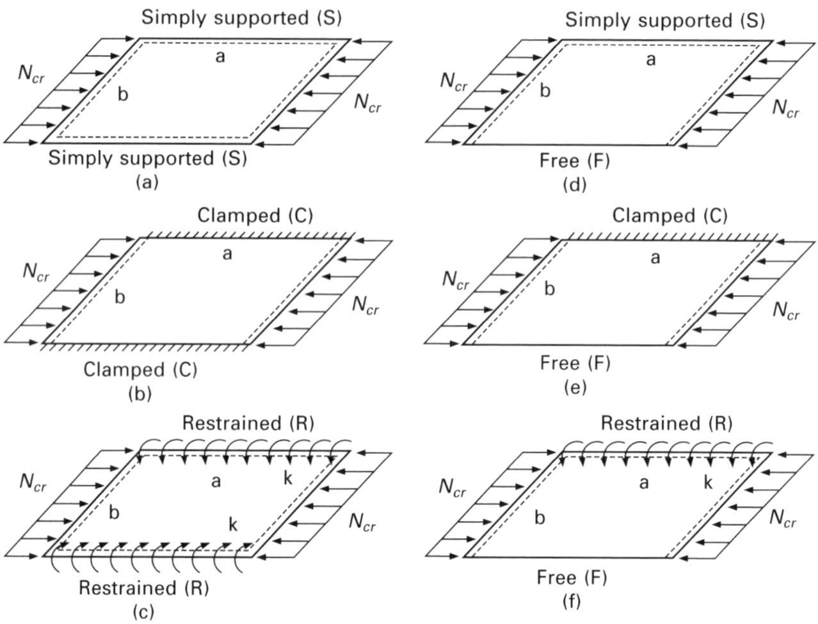

4.38 Common plates with various unloaded edge conditions: (a) Case 1: SS plate; (b) Case 2; CC plate; (c) Case 3; RR plate; (d) Case 4; SF plate; (e) Case 5; CF plate; (f) Case 6: RF plate.

Application to local buckling of FRP thin-walled sections

Once the explicit solutions for elastically restrained plates (Fig. 4.37) are obtained, they can be applied to predict the local buckling of FRP shapes. In the discrete plate analysis of FRP shapes, the rotational restraint stiffness (k) is needed to determine the critical buckling strength. Based on the studies by Bleich (1952) for isotropic materials and Qiao *et al.* (2001) and Qiao and Zou (2002, 2003) for composite materials, the rotational restraint stiffness coefficients (k) for local buckling of different FRP shapes were recently examined by Qiao and Shan (2005).

The explicit formulas for local buckling stress resultants (N_{cr}) and rotational restraint stiffness (k) are summarized in Table 4.13, and they can be used to predict the local buckling of several common FRP profiles. As an illustration of the proposed explicit solutions to local buckling of FRP shapes, several

Table 4.12 Local buckling loads and critical aspect ratios for FRP plates

Plate Type	Local buckling stress resultant (N/cm)	Critical aspect ratio
SS plate (Fig. 4.38(a))	$N_{cr}^{SS} = \frac{2\pi^2}{b^2}\left\{\sqrt{D_{11}D_{22}} + (D_{12} + 2D_{66})\right\}$	$\gamma_{cr}^{SS} = \left(\frac{m^4 D_{11}}{D_{22}}\right)^{1/4}$
CC plate (Fig. 4.38(b))	$N_{cr}^{CC} = \frac{24}{b^2}\left\{\begin{matrix}1.871\sqrt{D_{11}D_{22}} \\ + (D_{12} + 2D_{66})\end{matrix}\right\}$	$\gamma_{cr}^{CC} = 0.663\left(\frac{m^4 D_{11}}{D_{22}}\right)^{1/4}$
RR plate (Fig. 4.38(c))	Equation 4.61	Equation 4.62
SF plate (Fig. 4.33(d))	$N_{cr}^{SF} = \frac{12D_{66}}{b^2} + \frac{\pi^2 D_{11}}{a^2}$	–
CF plate (Fig. 4.38(e))	$N_{cr}^{CF} = \frac{-28D_{12} + 4\sqrt{385D_{11}D_{22}} + 224D_{66}}{11b^2}$	$\gamma_{cr}^{CF} = 1.6633m\left(\frac{D_{11}}{D_{22}}\right)^{1/4}$
RF plate (Fig. 4.38(f))	Equation 4.63	Equation 4.64

FRP box and I-sections are analyzed. The explicit predictions match well with the experimental and finite element data (see Table 4.14).

Ultimate bending and shear failure

Due to the relatively low compressive and shear strength properties of FRP composites, the material failure needs to be evaluated as well. Similar to beam deflection and buckling, the beam bending and shear strengths (ultimate failure loads) for three-point bending can be expressed in terms of panel strength properties as:

$$\text{Bending: } P_{\text{fail}}^{\text{bending}} = \frac{8F_c D}{(E_x)_f (b_w - t_f)L} \qquad 4.65$$

$$\text{Shear: } P_{\text{fail}}^{\text{shear}} = F_{xy} b_w t_w \qquad 4.66$$

where, F_c and F_{xy} are the compressive and shear strengths of FRP panels, and their values are given in Table 4.15. The compressive and shear strength values can also be obtained from the carpet plots given in section 4.3.2. The design values and experimental results are given in Table 4.16.

4.3.5 Equivalent analysis of FRP cellular decks

A multicellular FRP composite bridge deck can be modeled as an orthotropic plate, with equivalent stiffnesses that account for the size, shape, and constituent

Table 4.13 Rotational restraint stiffness (k) and critical local buckling stress resultant (N_{cr}) of different FRP shapes

FRP section	Buckled plate[(a)]	Critical local buckling stress resultant N_{cr}	Rotational restraint stiffness k
Box-	Flange	Equation 4.61 with D^f_{if}	$k = \dfrac{D^{w*}_{22}}{b_f \rho_1\left(\dfrac{b_w}{b_f}\right)}\left(1 - \dfrac{b_w^2}{b_f^2}\dfrac{\sqrt{D^f_{11}D^f_{22}} + D^f_{12} + 2D^f_{66}}{\sqrt{D^w_{11}D^w_{22}} + D^w_{12} + 2D^w_{66}}\right)^{(b)}$
	Web	Equation 4.61 with D^w_{ij}	$k = \dfrac{D^{f*}_{22}}{b_w \rho_1\left(\dfrac{b_f}{b_w}\right)}\left(1 - \dfrac{b_f^2}{b_w^2}\dfrac{\sqrt{D^w_{11}D^w_{22}} + D^w_{12} + 2D^w_{66}}{\sqrt{D^f_{11}D^f_{22}} + D^f_{12} + 2D^f_{66}}\right)^{(b)}$
I-	Flange	Equation 4.63 with D^w_{ij}	$k = \dfrac{D^{w*}_{22}}{b_w}\left(1 - \dfrac{6b_w^2}{\pi^2 b_f^2}\dfrac{D^f_{66}}{\sqrt{D^w_{11}D^w_{22}} + D^w_{12} + 2D^w_{66}}\right)$
	Web	Equation 4.61 with D^w_{ij}	$k = \dfrac{D^{f*}_{22}}{b_w \rho_2\left(\dfrac{b_f}{b_w}\right)}\left(1 - \dfrac{\pi^2 b_f^2}{6b_w^2}\dfrac{\sqrt{D^w_{11}D^w_{22}} + D^w_{12} + 2D^w_{66}}{D^f_{66}}\right)^{(b)}$
Channel and Z-	Flange	Equation 4.63 with D^f_{ij}	$k = \dfrac{2D^{f*}_{22}}{b_w}\left(1 - \dfrac{6b_w^2}{\pi^2 b_f^2}\dfrac{D^f_{66}}{\sqrt{D^w_{11}D^w_{22}} + D^w_{12} + 2D^w_{66}}\right)^{(b)}$
	Web	Equation 4.61 with D^w_{ij}	$k = \dfrac{D^{f*}_{22}}{b_w \rho_2\left(\dfrac{b_f}{b_w}\right)}\left(1 - \dfrac{\pi^2 b_f^2}{6b_w^2}\dfrac{\sqrt{D^w_{11}D^w_{22}} + D^w_{12} + 2D^w_{66}}{D^f_{66}}\right)^{(b)}$

Table 4.13 Continued

FRP section	Buckled plate[(a)]	Critical local buckling stressresultant N_{cr}	Rotational restraint stiffness k
T-	Flange	Equation 4.63 with D_{ij}^{f}	$k = \frac{D_{22}^{w*}}{1.9b_f} e^{-\frac{1}{2}\left(\frac{b_w - \frac{b_f}{2}}{4.5}\right)^2}$
	Web	Equation 4.63 with D_{ij}^{w}	$k = \frac{D_{22}^{f*}}{1.9b_w} e^{-\frac{1}{2}\left(\frac{b_f - \frac{b_w}{2}}{4.5}\right)^2}$
L-	Flange	$N_{cr} = \frac{12D_{66}^{f}}{(b^{f})^2} + \frac{\pi^2 D_{11}^{f}}{a^2}$	$k = 0$[(c)]
	web	$N_{cr} = \frac{12D_{66}^{w}}{(b^{w})^2} + \frac{\pi^2 D_{11}^{w}}{a^2}$	$k = 0$[(c)]

Notes:

(a) Buckled plate refers to the first buckled discrete element (either flange or web) in the FRP shapes.

(b) $\rho_1\left(\frac{b_i}{b_j}\right) = \frac{1}{2\pi} \tanh\frac{\pi b_i}{2b_j}\left\{1 + \frac{\frac{\pi b_i}{b_j}}{\sinh\left(\frac{\pi b_i}{b_j}\right)}\right\}$, $\rho_2\left(\frac{b_i}{b_j}\right) = \frac{1}{4\pi} \frac{3\cosh^2\left(\frac{\pi b_i}{b_j}\right) + \left(\frac{\pi b_i}{b_j}\right)^2 + 1}{\frac{\pi b_i}{b_j} + 3\sinh\left(\frac{\pi b_i}{b_j}\right)\cosh\left(\frac{\pi b_i}{b_j}\right)}$, where b_i (b_j)($i = f$ or w) is the width of flange or web, respectively.

(c) In the L-section, only the case of equal flange and web legs is herein given.

(d) $D_{ij}(i, j = 1, 2, 6)$ are the bending stiffness per unit length and D_{22}^{*} is the transverse bending stiffness of a unit length.

Table 4.14 Comparisons of critical local buckling stress resultants for Box- and I-sections. (Qiao and Zoo 2002, 2003)

Section	Dimensions (in)	Explicit solution (lb/in)	FEM (lb/in)	Experimental results (lb/in)
Box-	4 × 8 × 1/4	2725.5	2746.6	–
	4 × 6 × 1/4	4461.9	4419.7	–
I-	4 × 4 × 1/4	4729.7	4702.3	4600.1
	6 × 6 × 1/4	2309.8	2216.7	2243.0

Table 4.15 Panel strength properties of FRP shapes

FRP Shapes	F_{cx}(×10^3 psi)		F_{xy}(×10^3 psi)
	Compression test	Strength (Barbero *et al.* 1999)	Iosipescu test
WF-Beam 6″ × 6″ × 3/8″ **(WF6 × 6)**	54.498 (COV = 3.76%)	45.55	12.866 (COV = 2.36%)
I-Beam 4″ × 8″ × 3/8″ **(14 × 8)**	61.060 (COV = 2.41%)	56.65	13.022 (COV = 8.13%)
WF-Beam 4″ × 4″ × 1/4″ **(WF4 × 4)**	57.133 (COV = 3.50%)	53.10	13.167 (COV = 29.17%)
Square Tube 4″ × 4″ × 1/4″ **(Box4 × 4)**	60.657 (COV = 5.33%)	47.20	11.138 (COV = 4.84%)

(NOTE: COV = Coefficient of variation).

Table 4.16 Beam ultimate bending and shear loads

FRP Shapes	Bending, $P_{fail}^{bending}$			Shear, P_{fail}^{shear}		
	Span (ft)	Test (kips)	Design (kips)	Span (ft)	Test (kips)	Design (kips)
I-Beam 3″×6″ ×3/8″ **(I3×6)**	9.5	11.52	12.10	2.0	22.00	28.95
I-Beam 4″×8" ×3/8″ **(I4×8)**	9.5	20.28	24.50	–	–	–
WF-Beam 4″×4" ×1/4″ **(WF4×4)**	9.5	5.25	7.52	2.0	12.80	13.11
Square Tube 4″×4″ ×1/4″ **(Box4×4)**	9.5	5.79	7.98	2.0	17.40	22.27

materials of the cellular deck. Thus, the complexity of material anisotropy of the panels and structural orthotropy of the deck system can be reduced to an equivalent orthotropic plate with global elastic properties in two orthogonal directions, parallel and transverse to the longitudinal axis of the deck cell. These equivalent orthotropic plate properties can be directly used in the analysis and design of deck-and-stringer bridge system, as presented in section 4.3.6 and section 4.4, and they can also serve to simplify modeling procedures either in explicit or numerical formulations.

In this section, the development of equivalent stiffness for cellular decks consisting of multiple FRP box beams is presented. Multicell box sections are commonly used in deck construction because of their light weight, efficient geometry, and inherent stiffness in flexure and torsion. Also, this type of deck has the advantage of being relatively easy to build. It can either be assembled from individual box-beams or manufactured as a complete section by pultrusion or vacuum assisted resin injection molding (VARIM) process. The elastic equivalence concept (Troitsky 1987) presented here accounts for out-of-plane shear effects, and the results for a multicell box section are verified experimentally and by finite element analyses (Qiao *et al.* 2000). As an illustrative example, we derive the bending, shear and torsional equivalent stiffnesses for a deck composed of multiple box sections (Fig. 4.39).

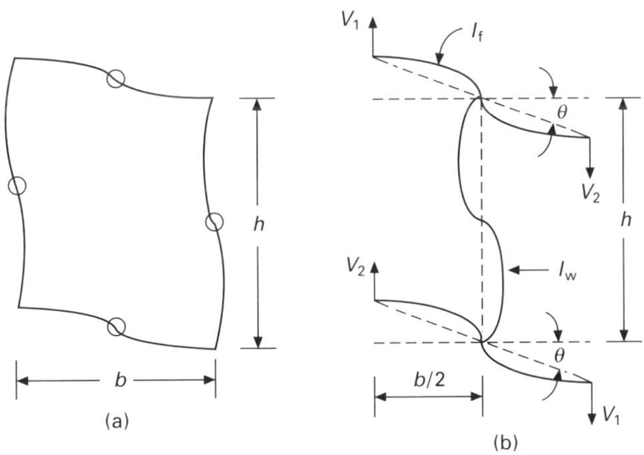

4.39 Vierendeel distortion in multi-cell box-beam (Cusen and Pama 1975): (a) cell distortion; (b) distortion parameters.

Equivalent longitudinal stiffnesses

The bending stiffness of the deck in the longitudinal direction, or x-axis in Fig. 4.40, is expressed as the sum of the bending stiffness of individual box beams (D_b can be obtained by MLB, see section 4.3.3):

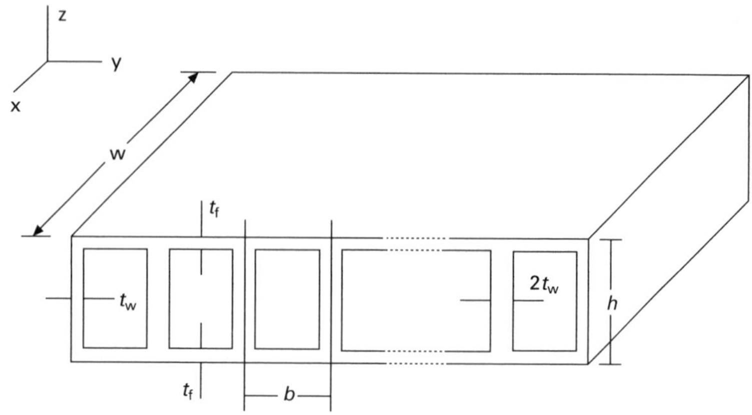

4.40 Geometric notation of multi-cell box deck.

$$D_x = n_c D_b \qquad 4.67$$

where n_c is the number of cells. For the section shown in Fig. 4.40, b is the width of a cell, h is the height of a cell, t_f and t_w are the thicknesses of the flange and web, respectively. If all panels have identical material lay-up and $t_f = t_w = t$, eqn 4.67 becomes

$$D_x = n_c E_x (h^2 + t^2 + 3(h)b)\frac{(h)(t)}{6} \qquad 4.68$$

where E_x is the modulus of elasticity of a panel in the x-direction computed by micro/macromechanics, obtained experimentally, or evaluated from the carpet plots (see Figs 4.10 and 4.14).

The out-of-plane shear stiffness of the deck in the longitudinal direction, F_x, is expressed as a function of the stiffness for the individual beams (F_b):

$$F_x = n_c F_b \qquad 4.69$$

where F_b is the shear stiffness obtained by MLB (see section 4.3.3), and n_c = number of cells. This expression can be further approximated in terms of the in-plane shear modulus of the component panel, G_{xy}, and cross-sectional area of the beam webs:

$$F_x = n_c G_{xy} (2t)h \qquad 4.70$$

where G_{xy} is the in-plane shear modulus of the panel walls, and it can be evaluated from the carpet plots (see Figs 4.12 and 4.16).

Equivalent transverse stiffnesses

An approximate value for the deck bending stiffness in the transverse direction, D_y, may be obtained by neglecting the effect of the transverse diaphragms and the second moment of area of the flanges about their own centroids. For a deck as shown in Fig. 4.45 with $t_f = t$:

$$D_y = \frac{1}{2} E_y (w)(t)h^2 \quad 4.71$$

where w is the length of the deck in the longitudinal direction, and E_y is the modulus of elasticity of the panel in the y-direction, which can be obtained from the carpet plots (Figs 4.11 and 4.15).

For multiple box sections, the simplest way to obtain the deck's out-of-plane transverse shear stiffness is to treat the structure as a Vierendeel frame in the transverse direction (Cusen and Pama 1975). For the Vierendeel frame (Fig. 4.39), the inflection points are assumed at the midway of the top and bottom flanges between the webs. The shear stiffness in the transverse direction, F_y, for the cross-section shown in Fig. 4.40 may be written as

$$F_y = \frac{V}{\theta} = \frac{12E_y}{b\left(\dfrac{h}{I_w} + \dfrac{b}{2I_f}\right)} \quad 4.72$$

where the moments of inertia I are defined as:

$$I_f = \frac{wt_f^3}{12}; I_w = \frac{w(2t_w)^3}{12} \quad 4.73$$

For $t_f = t_w = t$, eqn 4.72 can be simplified as

$$F_y = \frac{2E_y wt^3}{b\left(b + \dfrac{h}{4}\right)} \quad 4.74$$

where E_y is the modulus of elasticity of a panel in the y-direction.

Torsional stiffness of cellular FRP decks

The torsional rigidity of a multi-cell section, GJ, is evaluated by considering the shear flow around the cross-section of a multi-cell deck. For a structure where the webs and flanges are small compared with the overall dimensions of the section, Cusen and Pama (1975) have shown that the torsional rigidity may be written as

$$GJ = \frac{4A^2G_{xy}}{\Sigma \dfrac{ds}{t}} + \Sigma G_{xy}(ds)\frac{t^3}{3} \quad 4.75$$

where A is the area of the deck section including the void area and is defined as $A = n_c bh$, and $\Sigma\ ds/t$ represents the summation of the length-to-thickness ratio taken around the median line of the outside contour of the deck cross-section. For a constant panel thickness t, the torsional rigidity can be simplified as

$$GJ = \frac{2(n_c bh)^2 G_{xy} t}{(n_c b + h)} + \frac{2}{3}(n_c b + h) G_{xy} t^3 \qquad 4.76$$

The above approximate equation is justified by the fact that for a multi-cell deck, the net shear flow through interior webs is negligible, and only the shear flow around the outer webs and top and bottom flanges is significant. The second term in eqn 4.76 is relatively small compared to the first term and can be ignored.

If the deck is treated as an equivalent orthotropic plate, its torsional rigidities depend upon the twist in two orthogonal directions. Thus the torsional stiffness D_{xy} may be taken as one-half of the total torsional rigidity given by eqn 4.76 divided by the total width of the deck:

$$D_{xy} = \frac{GJ}{2n_c b} \qquad 4.77$$

Substituting eqn 4.76 into eqn 4.77 and neglecting the second term in eqn 4.76, we get:

$$D_{xy} = \frac{n_c G_{xy} bh^2 t}{(n_c b + h)} \qquad 4.78$$

where, D_{xy} is the torsional stiffness per unit width (lb-in^4/in).

Equivalent orthotropic material properties

Once the stiffness properties of an actual deck are obtained, it is relatively simple to calculate the equivalent orthotropic plate material properties, which can further simplify the design analysis of deck and deck-and-stringer bridge systems.

To calculate the moduli of elasticity $(E_x)_p$ and $(E_y)_p$ for the equivalent orthotropic plate, the relationship $D = EI$ is used, leading to

$$(E_x)_p = 12 \frac{D_x}{t_p^3 b_p}(1 - \nu_{xy}\nu_{yx}) \qquad 4.79$$

$$(E_y)_p = 12 \frac{D_y}{t_p^3 l_p}(1 - \nu_{xy}\nu_{yx}) \qquad 4.80$$

where the subscript 'p' indicates property related to the equivalent orthotropic plate; t_p is the thickness of the plate (= h for the actual deck, Fig. 4.40), b_p

is the width of the plate ($= n_c b$ for the actual deck), and l_p is the length of the plate ($= w$ for the actual deck). The Poisson's ratios ν_{ij} are defined as

$$\nu_{ij} = -\frac{\varepsilon_j}{\varepsilon_i} \qquad 4.81$$

where ε is the strain in the i or j direction. For orthotropic materials, the Poisson's ratio must conform to the following relationship:

$$\frac{\nu_{ij}}{E_i} = \frac{\nu_{ji}}{E_j} \quad \text{or} \quad \frac{\nu_{xy}}{D_x} = \frac{\nu_{yx}}{D_y} \qquad 4.82$$

In this study, we use the approximation $(\nu_{xy})_p = 0.3$ for the equivalent deck panel, which is typically used for pultruded composites.

To calculate the out-of-plane shear moduli $(G_{xz})_p$ and $(G_{yz})_p$, the relationship $F = GA$ is used, leading to

$$(G_{xz})_p = \frac{F_x}{t_p b_p} \qquad 4.83a$$

$$(G_{yz})_p = \frac{F_y}{t_p l_p} \qquad 4.83b$$

Finally, to calculate the in-plane shear modulus $(G_{xy})_p$, we use

$$(G_{xy})_p = 6\frac{D_{xy}}{t_p^3} \qquad 4.83c$$

As shown in Qiao *et al.* (2000), the equivalent orthotropic deck properties in eqns 4.79 to 4.83c are successfully implemented as material input in finite element (FE) analysis, and the predictions obtained correlated well with results for both FE analysis of actual deck geometry and properties and experimental data. Using these equivalent material properties, it is now possible to implement explicit plate solutions (see section 4.3.6) for analysis and design of cellular decks.

4.3.6 Macro-flexibility analysis of deck-and-stringer bridge systems

The equivalent properties for cellular decks in section 4.3.5 and stiffnesses for FRP beams in section 4.3.3 can be efficiently used to analyze and design deck-and-stringer bridge systems. An overview of a series solution for stiffened orthotropic plates based on first-order shear deformation theory and transverse interaction of forces between the deck and the stringers is presented in this section. The solutions for symmetric and antisymmetric load cases are used to obtain the solution for asymmetric loading. Based on deck-stringer transverse interaction force functions, wheel load distribution factors are derived, which are used later to provide design guidelines for deck-and-stringer bridge systems.

First-order shear deformation theory for FRP composite deck

A first-order shear deformation theory (Reddy 1984) is applied to analyze the behavior of a geometrically orthotropic FRP composite deck. Instead of direct modeling of the actual deck geometry, an equivalent orthotropic plate, as discussed in section 4.3.5, is used to simplify the analysis. The formulas for equivalent orthotropic material properties accounting for deck geometry and panel laminated material properties are given in eqns 4.79–4.83. The equilibrium equations accounting for transverse shear deformation of an orthotropic plate are:

$$A_{55}\frac{\partial}{\partial x}\left(\psi_x + \frac{\partial w_o}{\partial x}\right) + A_{44}\frac{\partial}{\partial y}\left(\psi_y + \frac{\partial w_o}{\partial y}\right) + q(x, y) = 0$$

$$\frac{\partial}{\partial x}\left(D_{11}\frac{\partial \psi_x}{\partial x} + D_{12}\frac{\partial \psi_y}{\partial y}\right)$$

$$+ D_{66}\frac{\partial}{\partial y}\left(\frac{\partial \psi_x}{\partial y} + \frac{\partial \psi_y}{\partial x}\right) - A_{55}\left(\psi_x + \frac{\partial \omega_o}{\partial x}\right) = 0 \qquad 4.84$$

$$D_{66}\frac{\partial}{\partial x}\left(\frac{\partial \psi_x}{\partial y} + \frac{\partial \psi_y}{\partial x}\right)$$

$$+ \frac{\partial}{\partial y}\left(D_{12}\frac{\partial \psi_x}{\partial x} + D_{22}\frac{\partial \psi_y}{\partial y}\right) - A_{44}\left(\psi_y + \frac{\partial w_o}{\partial y}\right) = 0$$

where A_{ij} $(i,j = 4, 5)$ are the intralaminar shear stiffnesses, and D_{ij} $(i, j = 1, 2, 6)$ are the bending stiffnesses for an orthotropic material.

The deck-and-stringer bridge system (Fig. 4.41(a)) can first be analyzed as an orthotropic plate stiffened by edge stringers (or beams), as shown in Fig. 4.41(b) (Salim *et al.* 1995, 1997; Salim 1997). Then, the contributions of interior stringers are accounted for in the formulation by considering the interaction forces (Fig. 4.41(c)) and the comparability conditions along rib lines between the deck and stringers. The analysis is general with respect to: (i) size and stiffness of the deck, and (ii) type of loading (uniform and/or concentrated). The formulation is concerned first with symmetric and antisymmetric loading conditions.

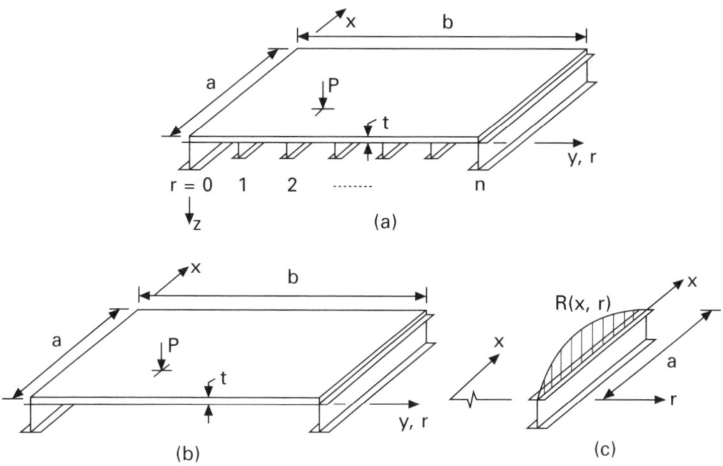

4.41 Deck-and-stringer bridge system: (a) deck-and-stringer system; (b) plate with exterior stringers only; (c) interior stringer.

Bridge system under symmetric loading

A Fourier polynomial series is employed to obtain the solutions for the equilibrium equations (eqn 4.84). The solution for a symmetric loading (Salim *et al.* 1997; Salim 1997; Brown 1998) is given as

$$w_o(x, y) = \sum_{i,j=1}^{\infty} W_{ij} \sin \alpha x(\beta y + W_o)$$

$$\psi_x(x, y) = \sum_{i,j=1}^{\infty} X_{ij} \cos \alpha x(\sin \beta y + X_o) \qquad 4.85$$

$$\psi_y(x, y) = \sum_{i,j=1}^{\infty} X_{ij} \sin \alpha x \cos \beta_y$$

where $\alpha = i\pi/a$ and $\beta = j\pi/b$, and W_{ij}, X_{ij}, and Y_{ij} are the coefficients to be determined to complete the solution. Note that these series approximations satisfy the essential boundary conditions. The generalized loading can be written as the following infinite double series

$$q(x, y) = \sum_{i,j=1}^{\infty} Q_{ij} \sin \alpha x \sin \beta y \qquad 4.86$$

Q_{ij} are the Fourier coefficients in the representation of the load $q(x, y)$.

By substituting the general solution eqns 4.85 and 4.86 into eqn 4.84 and reducing by orthogonality conditions (Salim *et al.* 1995; 1997), we obtain the following system of equations for any number of terms (i, j):

$$\begin{bmatrix} K_{11} & K_{12} & K_{13} \\ K_{21} & K_{22} & K_{23} \\ K_{13} & K_{23} & K_{33} \end{bmatrix} \begin{Bmatrix} W_{ij} \\ X_{ij} \\ Y_{ij} \end{Bmatrix} = \begin{Bmatrix} Q_{ij} \\ 0 \\ 0 \end{Bmatrix} \qquad 4.87$$

where K_{ij} are the deck stiffness coefficients under a symmetric loading (Brown 1998) and given as:

$$K_{11} = \alpha^2 A_{55} + \beta^2 A_{44} + \frac{4\alpha^2}{\pi} A_{55} W_o$$

$$K_{12} = \alpha A_{55} + \frac{4\alpha}{\pi} A_{55} X_o$$

$$K_{13} = \beta A_{44} \qquad 4.88$$

$$K_{21} = \alpha A_{55} + \frac{4\alpha}{\pi} A_{55} W_o$$

$$K_{22} = \alpha^2 D_{11} + \beta^2 D_{66} + A_{55} + \frac{4\alpha^2}{\pi} D_{11} X_o + \frac{4}{\pi} A_{55} X_o$$

$$K_{23} = \alpha\beta\,(D_{12} + D_{66})$$

$$K_{33} = \alpha^2 D_{66} + \beta^2 D_{22} + A_{44}$$

For a one-term approximation, the constants W_o and X_o are obtained by satisfying the boundary conditions of the edge-stiffened orthotropic plate (Fig. 4.41(b)):

$$W_o = A_{44} c \left(\frac{Y_{11}}{W_{11}} + \beta \right) \qquad 4.89a$$

$$X_o = -\frac{A_{44}}{\alpha^3 D} \left(\frac{Y_{11} + \beta W_{11}}{X_{11}} \right) \qquad 4.89b$$

where, $c = \frac{1}{\alpha^2}\left(\frac{1}{\kappa F} + \frac{1}{\alpha^2 D} \right)$; κ = the stringer shear correction factor; F and D are, respectively, the shear and bending stiffnesses of the stringer which are obtained based on MLB (see section 4.3.4). For a bridge deck without stringers, the edge deformation coefficients W_o and X_o are defined by

$$W_o = \frac{\pi}{4} \left[-1 + \frac{A_{44}\beta^2}{A_{55}\alpha^2 D_{12}} \left(\frac{\alpha^2 D_{12}^2 (3D_{66} + D_{12}) - D_{22}(D_{11}\alpha^2 + A_{55})(D_{12} + 2D_{66}) + D_{12} D_{66} \beta^2}{A_{44}(D_{12} + 2D_{66}) + D_{66}(D_{22}\beta^2 - D_{12}\alpha^2)} \right) \right]$$

4.90a

$$X_o = -\frac{\pi D_{66}}{2D_{12}}\left(\frac{D_{12}\alpha^2 - D_{22}\beta^2}{2D_{66}\alpha^2 + D_{22}\beta^2}\right) \qquad 4.90b$$

where the bending (D_{ij}) and shear (A_{ij}) stiffnesses can be evaluated as

$$D_{11} = \frac{D_x}{b_p},\ D_{22} = \frac{D_y}{l_p},\ D_{12} = \frac{\upsilon_{xy} D_x}{l_p},\ D_{66} = \frac{D_{xy}}{2} \qquad 4.91a$$

$$A_{44} = \frac{5}{6}\frac{F_x}{l_p},\ A_{55} = \frac{5}{6}\frac{F_x}{b_p} \qquad 4.91b$$

where l_p and b_p are the length and width of the bridge deck panel, respectively.

For any interior stringer at any location r ($r = 0, 1...n$) (see Fig. 4.41(c)), the generalized deflection function for any symmetric loading is (Brown 1998)

$$w^R(x, r) = R_{11}\frac{1}{\alpha^2}\left(\frac{1}{\kappa F} + \frac{1}{\alpha^2 D}\right)\sin\frac{\pi x}{a}\left(\sin\frac{\pi r}{n} + W_o\right) \qquad 4.92$$

where, $$R_{11} = \frac{Q_{11}}{\frac{1}{\alpha^2}\left(\frac{1}{\kappa F} + \frac{1}{\alpha^2 D}\right)\frac{Q_{11}}{W_{11}} + \frac{n}{b}\left(1 + \frac{4W_o}{\pi}\right)}$$

Bridge system under antisymmetric loading

Analogous to the symmetric case, eqns 4.85 and 4.86 are modified for a first-term approximation of an antisymmetric loading as

$$w_o(x, y) = W_{12}\sin\alpha x\left(\sin 2\beta y + W_1\left(1 - \frac{2y}{b}\right)\right)$$

$$\psi_x(x, y) = X_{12}\cos\alpha x\left(\sin 2\beta y + X_1\left(1 - \frac{2y}{b}\right)\right) \qquad 4.93$$

$$\psi_y(x, y) = Y_{12}\sin ax\cos 2\beta y$$

$$q(x, y) = Q_{12}\sin\alpha x\sin 2\beta y$$

By substituting eqn (4.93) into eqn 4.84, we obtain the stiffness matrix for an orthotropic deck under antisymmetric loading (Brown 1998; Qiao *et al.* 2000). The constants W_1 and X_1 are determined as

$$W_1 = A_{44}c\left(\frac{Y_{12}}{W_{12}} + 2\beta\right)\left(\frac{1}{1 + \frac{2}{b}A_{44}c}\right) \qquad 4.94a$$

$$X_1 = -\frac{A_{44}}{\alpha^3 D}\left(\frac{Y_{12} + 2\beta W_{12}}{X_{12}}\right) \tag{4.94b}$$

where, c is the same as given in eqn 4.89. The generalized deflection function for antisymmetric loading is

$$w^R(x, r) = R_{12}\frac{1}{\alpha^2}\left(\frac{1}{\kappa F} + \frac{1}{\alpha^2 D}\right)\sin\frac{\pi x}{a}\left(\sin\frac{2\pi r}{n} + W_1 \cos\frac{\pi r}{n}\right) \tag{4.95}$$

where, $R_{12} = \dfrac{Q_{12}}{\dfrac{1}{\alpha^2}\left(\dfrac{1}{\kappa F} + \dfrac{1}{\alpha^2 D}\right)\dfrac{Q_{12}}{W_{12}} + \dfrac{4n}{b}\left(1 + \dfrac{2W_1}{3\pi}\right)}$

Bridge system under asymmetric loading

The asymmetric case is obtained by superposition of the symmetric and antisymmetric load conditions. By simply adding the symmetric and antisymmetric responses, the generalized deflection function for an interior stringer under an asymmetric load is written as

$$w^R(x, r) = \left[R_{11}\left(\sin\frac{\pi r}{n} + W_o\right) + R_{12}\left(\sin\frac{2\pi r}{n} + W_1 \cos\frac{\pi r}{n}\right)\right] \times \frac{1}{\alpha^2}\left(\frac{1}{\kappa F} + \frac{1}{\alpha^2 D}\right)\sin\frac{\pi x}{a} \tag{4.96}$$

Wheel load distribution factors

The above solution is used to define wheel-load distribution factors for any of the stringers. The load distribution factor for any interior stringer ith is defined as the ratio of the interaction forces $R(x, r)$ for the ith stringer to the sum of interaction forces for all stringers. The general expressions of load distribution factors in terms of the number of stringers m (where, $m = n + 1$) for symmetric and asymmetric loads (Brown 1998; Qiao *et al.* 2000) are respectively

$$W_f^{\text{Sym}}(r) = \frac{\sin\dfrac{r-1}{m-1}\pi + W_o}{\dfrac{2}{\pi}(m-1) + mW_o} \tag{4.97}$$

$$W_f^{\text{Asym}}(r) = \frac{\left[\begin{array}{l} R_{11}\left(\sin\pi\dfrac{r-1}{m-1} + W_o\right) \\ + R_{12}\left(\sin 2\pi\dfrac{r-1}{m-1} + W_1\left(1 - 2\dfrac{r-1}{m-1}\right)\right)\end{array}\right]}{\sum\limits_{r=1}^{m}\left[\begin{array}{l} R_{11}\left(\sin\pi\dfrac{r-1}{m-1} + W_o\right) \\ + R_{12}\left(\sin 2\pi\dfrac{r-1}{m-1} + W_1\left(1 - 2\dfrac{r-1}{m-1}\right)\right)\end{array}\right]} \quad 4.98$$

The maximum wheel-load distribution factor under symmetric loading occurs when $(r-1)/(m-1) = 1/2$; i.e., $\sin(\pi/2) = 1$. Therefore, the maximum load distribution factor for symmetric loading is given as

$$(W_f^{\text{Sym}})_{\max} = \frac{1 + W_o}{mW_o + \dfrac{2}{\pi}(m-1)} \quad 4.99$$

Similarly, the maximum value of W_f^{Antisym} for antisymmetric loading can be obtained as

$$(W_f^{\text{Antisym}})_{\max} = \frac{1 + 0.5W_1}{(m-1)\left(\dfrac{1}{\pi} + \dfrac{W_1}{4}\right)} \quad 4.100$$

The maximum wheel load distribution factors given in eqns 4.99 and 4.100 can be used in analysis and design of deck-and-stringer bridge systems as illustrated later in design examples (section 4.4.2).

4.4 Design guidelines and examples

4.4.1 Design guidelines for FRP shapes

Based on the aforementioned analytical studies in section 4.3, design guidelines for FRP shapes including members and deck-and-stringer bridge systems are provided in this section, and the proposed step-by-step design analysis procedures follow closely the flow chart illustrated in Fig. 4.4. The following guidelines for analysis and design of FRP shapes are suggested:

1. Based on the manufacturer's information and material lay-ups, the panel stiffness and strength properties are obtained by using either carpet plots (see section 4.3.2) or conducting experimental coupon tests.
2. Based on the panel material properties obtained in step 1, the elastic stiffness properties of FRP beam members and bridge deck are calculated by MLB and equivalent analysis, respectively.

3. Using Timoshenko's beam theory, the deflections of FRP beams with different boundary and loading conditions are calculated.
4. Based on MLB, the elastic strains and stresses of FRP beams are obtained.
5. The factors of safety for material failure are obtained by comparing to the panel strength properties obtained in step 1.
6. Following the explicit formulas of local and global buckling (see section 4.3.4), the stability analysis of FRP beams is implemented.
7. For design and analysis of deck-and-stringer bridge system, the stiffness properties and deformation of the stringers are obtained using MLB and Timoshenko's beam theory; while the material properties of orthotropic deck panels are obtained by performing the elastic equivalency analysis.
8. Based on the limitation on the serviceability of bridge system (e.g., the deflection limit), the number of FRP stringers can be determined by the macro-flexibility analysis.
9. Based on MLB, the maximum normal and shear stresses of the stringers and related factors of safety are calculated to meet the requirements for a safe design.
10. A summary of design calculations and design sketches are used to complete the design.

To illustrate the above design guidelines, three detailed design examples are provided in the next section.

4.4.2 Design examples

In the following, two design examples for FRP beams and one for a deck-and-stringer bridge system are provided.

Example 1: Design of an FRP I-beam (SI units)

Problem statement: analyze a pultruded thin-walled composite material I-beam (Fig. 4.42) based on the following data and the carpet plots shown in section 4.3.2.

Beam dimensions:	0.2 m × 0.4 m × 0.02 m (Fig. 4.42)
Panel material and lay-up:	E-glass fiber with polyester (V_f = 50.0%), unidirectional fiber roving layer (0°): $\alpha = 60\%$, continuous strand mat (CSM) layer: β = 10%, ±45° stitched fiber (SF) layer: $\gamma = 30\%$, uniform panel thickness and material lay-up throughout the beam.
Beam span:	$L = 9.0$ m
Loading type:	Beam under bending with uniformly distributed load q = 4,000 N/m (Fig. 4.42)

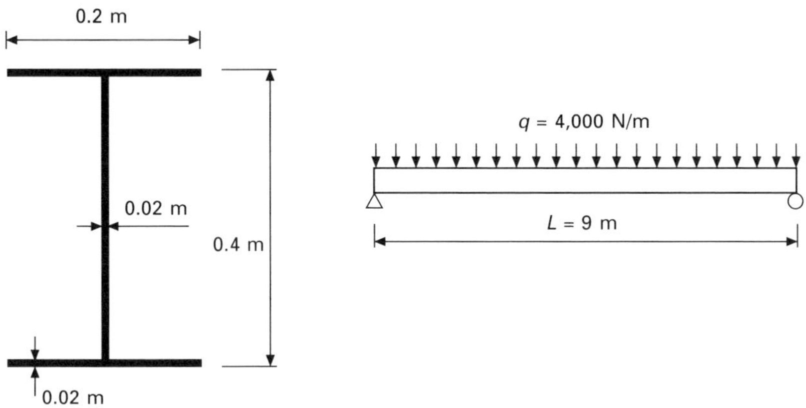

4.42 FRP I-beam dimensions and loading.

Requirements: evaluate this beam as follows:

1. Based on the carpet plots, obtain the panel equivalent moduli and strength values.
2. Compute the beam bending ($D = EI$) and shear ($F = GA$) stiffness values.
3. Compute the beam maximum deflection, and also bending and shear deflection components, respectively.
4. Compute the beam maximum bending and shear stresses and corresponding factors of safety (F.S.).
5. Compute the local buckling strength (σ_x^{cr}) of the beam top flange and the beam local buckling load. Also check the corresponding F.S.
6. Compute the critical global buckling load (P_{cr}^{global}) and obtain the corresponding F.S.

Solution: based on the carpet plots (Figs 4.10–4.13, 4.23, and 4.28) and the design guidelines given in section 4.4.1, the following solutions are obtained.

1. Panel engineering properties:
Carpet plots:

Unidirectional fiber roving (0°) layer:	$\alpha = 60\%$;
CSM layer:	$\beta = 10\%$;
±45° SF layer:	$\gamma = 30\%$.

Panel stiffness properties
$E_x = 4.25 \times 10^6$ psi = 29.3 GPa (Fig. 4.10), $E_y = 2.0 \times 10^6$ psi = 13.8 GPa (Fig. 4.11), $G_{xy} = 0.96 \times 10^6$ psi = 6.6 GPa (Fig. 4.12), $v_{xy} = 0.4$ (Fig. 4.13).

Panel strength properties:
$F_{xc} = 2.6 \times 10^4$ psi = 180 MPa (Fig. 4.23), $F_{xy} = 0.75 \times 10^4$ psi = 51.7 MPa (Fig. 4.28).

2. Beam stiffness coefficients:
The beam bending and shear stiffnesses:

$$D = \frac{1}{2}(E_x)_f t_f b_w^2 b_f + \frac{1}{12}(E_x)_w t_w b_w^3 + \frac{1}{6}(E_x)_f t_f^3 b_f$$

$$F = (G_{xy})_w t_w b_w \qquad \text{... eqn 4.36}$$

where: $(E_x)_f = (E_x)_w = E_x = 29.3$ GPA; $(G_{xy})_f = (G_{xy})_f = G_{xy} = 6.6$ GPa;

$t_f = t_w = 0.02$ m; $b_f = 0.2$ m; $b_w = 0.4$ m

Beam bending stiffness:

$$D = EI = 12.509 \text{ MPa-m}^4$$

Beam shear stiffness:

$$F = GA_w = 5.280 \text{ MPa-m}^2$$

where $A_w = t_w b_w = 0.008 \text{ m}^2$

3. Beam deflections:
Beam maximum deflection (under uniformly distributed load):

$$\delta_{\max} = \delta_b + \delta_s = \frac{qL^4}{384D} + \frac{qL^2}{8KF} \qquad \text{... eqn 4.38 and Table 4.6}$$

$$\delta_{\max} = 0.00546 \text{ m} + 0.00767 \text{ m} = 13.13 \text{ mm}$$

$\delta_{\max} = 13.13 \text{ mm} < L/250 = 36.0 \text{ mm} \Rightarrow$ Safe design for deflection ($\delta < L/250$ is used in this example as a deflection design limit).

Beam deflection due to bending:

$$\delta_b = \frac{qL^4}{384D}; \ \delta_b = 0.00546 \text{ m} = 5.46 \text{ mm}$$

Beam deflection due to shear:

$\delta_s = \dfrac{qL^2}{8KF}$; $\delta_s = 0.00767$ m $= 7.67$ mm; where K is the shear correction factor assumed to be 1.0.

4. Beam stresses
Maximum resultant moment and shear force:
for a beam under uniform load bending:

$$M_{\max} = \frac{qL^2}{8} \text{ and } V_{\max} = \frac{qL}{2};$$

where: $q = 4{,}000$ N/m and $L = 9.0$ m.

$M_{max} = 40{,}500$ N-m and $V_{max} = 18{,}000$ N

Maximum bending and shear strains:

$$\varepsilon_x = \frac{M}{D} y_i \quad \text{and} \quad \gamma_{xy} = \frac{V}{F} \sin \phi_i \qquad \text{... eqn 4.39}$$

where: $y_i = b_w/2 = 0.2$ m (maximum compressive bending strain occurs on the top flange); $\phi_i = 90°$ (maximum shear strain in the web panel);

$$\varepsilon_x = 647.6\mu\varepsilon \text{ and } \gamma_{xy} = 3{,}409\ \mu\varepsilon$$

Approximate maximum bending and shear stresses:

$$\sigma_x \approx E_x \varepsilon_x \text{ and } \tau_{xy} \approx G_{xy}\gamma_{xy}$$

$$\sigma_x \approx 18.975 \text{ MPa}; \ \tau_{xy} \approx 22.499 \text{ MPa}$$

Factors of safety (F.S.) for design:

$$(F.S.)_{\text{bending}} = \frac{F_{xc}}{\sigma_x} \quad \text{and} \quad (F.S.)_{\text{shear}} = \frac{F_{xy}}{\sigma_{xy}}$$

$$(F.S.)_{\text{bending}} = 9.486 \text{ and } (F.S.)_{\text{shear}} = 2.298 \Rightarrow \text{safe design}$$

5. Beam local buckling load:
Material properties:

$$D_{11} = \frac{t^3 E_x}{12(1 - \upsilon_{xy}\upsilon_{yx})} = 21{,}232 \text{ N-m},$$

$$D_{22} = \frac{t^3 E_y}{12(1 - \upsilon_{xy}\upsilon_{yx})} = 10{,}000 \text{ N-m},$$

$$D_{12} = \frac{t^3 E_y \upsilon_{xy}}{12(1 - \upsilon_{xy}\upsilon_{yx})} = 4000 \text{ N-m}, \ D_{66} = \frac{t^3 G_{xy}}{12} = 4{,}400 \text{ N-m}.$$

Top flange local buckling stress resultant

$$N_{cr}^{\text{local}} = 7.8 \times 10^8 \text{ N/m} \qquad \text{... eqn 4.63 and Table 4.13}$$

Critical beam buckling load:

$$\frac{N_{cr}^{\text{local}}}{t} = \frac{My}{I} \Rightarrow q_{cr}^{\text{local}} = \frac{16 D N_{cr}^{\text{local}}}{b_w t E_x L^2}$$

$$= 8222.32 \text{ N/m} > q = 4{,}000 \text{ N/m}$$

where: $M = \dfrac{q_{cr}^{\text{local}} L^2}{8}$ and $y = \dfrac{b_w}{2}$

Factor of safety for local buckling:

$$(F.S.)_{\text{local}} = \frac{q_{cr}^{\text{local}}}{q} = 2.06 \Rightarrow \text{Safe design for local buckling.}$$

6. Beam global buckling load:
Critical beam global buckling load (Table 4.10):

$$q_{cr} = \frac{28.46}{L^3}\sqrt{D_y \cdot JG} \times \left[\sqrt{\frac{\pi^2}{L^2}\left(\frac{I_{ww}}{JG} + 0.212k^2\frac{D_y}{JG}\right) + 1} - \frac{1.44k}{L}\sqrt{\frac{D_y}{JG}}\right]$$

where: $$JG = \frac{2(G_{xy})_f t_f^3 b_f}{3} + \frac{(G_{xy})_w t_w^3 b_w}{3} = 14080 \text{ Pa-m}^4$$

$$I_{ww} = \frac{(E_x)_f t_f b_w^3 b_f^3}{24} + \frac{(E_x)_f t_f^3 b_f^3}{36} + \frac{(E_x)_w t_w^3 b_w^3}{144} = 31410 \text{ Pa-m}^6$$

$$q_{cr}^{\text{global}} = 7882.3 \text{ N/m}$$

Factor of safety for global buckling

$$(F.S.)_{\text{global}} = \frac{q_{cr}^{\text{global}}}{q} = 1.97 \Rightarrow \text{Safe design for global buckling.}$$

7. Summary
Based on the step-by-step procedures shown above for the I-beam under the uniformly distributed load, it is concluded that the beam satisfies the design requirements.

Example 2: design of an FRP box beam (English units)

Problem statement: Analyze an existing pultruded FRP structural box-beam (Box 4 × 4) (Fig. 4.43) based on the following data and the panel engineering properties given in Tables 4.2 (see page 131) and 4.15 (see page 174).

Beam dimensions: 4 in × 4 in × 1/4 in (Fig. 4.43).

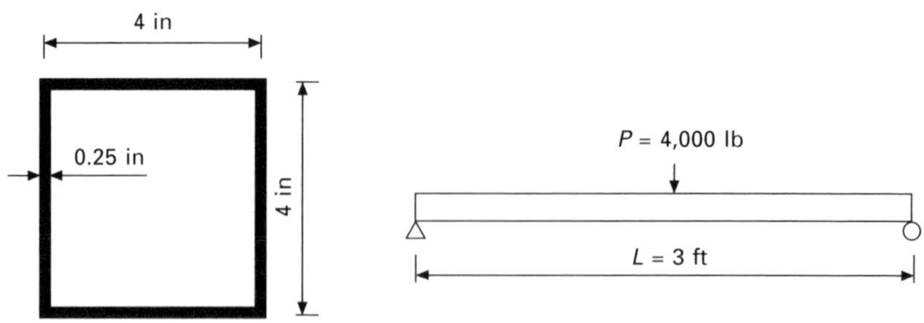

4.43 FRP box-beam dimensions and loading.

Panel material and lay-up: E-glass fiber with polyester resins. Uniform panel thickness and material lay-up throughout the beam, and the experimental coupon test values shown in Tables 4.2 (page 131) and 4.15 (page 174) will be used in the design analysis.

Beam span: $L = 3.0$ ft.

Loading type: Beam under three-point bending with central concentrated load $P = 4{,}000$ lb (Fig. 4.43)

Requirements: based on the design guidelines given in section 4.4.1, perform the following design analysis:

1. Based on the experimental coupon test results of panels, summarize the panel engineering properties.
2. Compute the beam bending ($D = EI$) and shear ($F = GA$) stiffnesses values.
3. Compute the beam maximum deflection.
4. Predict the beam maximum bending and shear strains.
5. Predict the local buckling strength (σ_x^{cr}) of the beam top flange and the beam local buckling load. Also check the corresponding *F.S.*

Design solution: based on the panel experimental test data (Tables 4.2 and 4.15) and design guidelines given in section 4.4.1, the following design analysis for the box-beam is performed:

1. Panel engineering properties:
From panel coupon test data:
Box panel stiffness properties (Table 4.2):

$$E_x = 4.295 \times 10^6 \text{ psi and } G_{xy} = 0.548 \times 10^6 \text{ psi}$$

Box panel strength properties (Table 4.15):

$$F_{xc} = 60.657 \times 10^3 \text{ psi and } F_{xy} = 11.138 \times 10^3 \text{ psi}$$

2. Beam stiffness coefficients:
The beam bending and shear stiffnesses:

$$D = \frac{1}{2}(E_x)_f t_f b_w^2 b_f + \frac{1}{6}(E_x)_w t_w b_w^3 + \frac{1}{6}(E_x)_f t_f^3 b_f$$

$$F = 2(G_{xy})_w t_w b_w \text{ (eqn 4.36)}$$

where: $(E_x)_f = (E_x)_w = E_x = 4.295 \times 10^6$ psi; $(G_{xy})_f = (G_{xy})_f = G_{xy} = 0.548 \times 10^6$ psi; $t_f = t_w = 0.25$ in; $b_f = b_w = 4.0$ in

Beam bending stiffness:

$$D = EI = 4.586 \times 10^7 \text{ psi-in}^4$$

Beam shear stiffness:

$$F = GA_w = 1.096 \times 10^6 \text{ psi-in}^2$$

3. Beam deflections:
Beam maximum deflection (3-point bending):

$$\delta_{\max} = \delta_b + \delta_s = \frac{PL^3}{48D} + \frac{PL}{4KF} \quad \text{eqn 4.38 and Table 4.6}$$

$\delta_{\max} = 0.085 + 0.033 = 0.118$ in $< L/250 = 0.144$ in $\Rightarrow$ safe design for deflection.

4. Beam maximum strains and stresses:
Maximum resultant moment and shear force for a beam under three-point bending:

$$M_{\max} = \frac{PL}{4} \quad \text{and} \quad V_{\max} = \frac{P}{2}$$

where: $P = 4000$ lb and $L = 3.0$ ft
$M_{\max} = 36{,}000$ lb-in and $V_{\max} = 2000$ lb

Maximum bending and shear strains

$$\varepsilon_x = \frac{M}{D} y_i \quad \text{and} \quad \gamma_{xy} = \frac{V}{F} \sin \phi_i \qquad \text{... eqn 4.39}$$

where: $y_i = b_w/2 = 2.0$ in (maximum compressive bending strain occurs on the top flange); $\phi_i = 90°$ (maximum shear strain in the web panel)

$$\varepsilon_x = 1570\ \mu\varepsilon \text{ and } \gamma_{xy} = 1825\ \mu\varepsilon$$

Approximate maximum bending and shear stresses:

$$\sigma_x \approx E_x \varepsilon_x \text{ and } \tau_{xy} \approx G_{xy} \gamma_{xy}$$

$$\sigma_x \approx 6743 \text{ psi; } \tau_{xy} \approx 1000 \text{ psi}$$

Factors of safety (F.S.) for design:

$$(F.S.)_{\text{bending}} = \frac{F_{xc}}{\sigma_x} \quad \text{and} \quad (F.S.)_{\text{shear}} = \frac{F_{xy}}{\sigma_{xy}}$$

$(F.S.)_{\text{bending}} = 9.0$ and $(F.S.)_{\text{shear}} = 11.0 \Rightarrow$ Safe design for material strength.

5. Beam local buckling load:
Material properties

$$D_{11} = \frac{t^3 E_x}{12(1 - \upsilon_{xy}\, \upsilon_{yx})} = 5869 \text{ lb-in}$$

$$D_{22} = \frac{t^3 E_x}{12(1 - \upsilon_{xy}\,\upsilon_{yx})} = 1911 \text{ lb-in}$$

$$D_{12} = \frac{t^3 E_y \upsilon_{xy}}{12(1 - \upsilon_{xy}\upsilon_{yx})} = 688 \text{ lb-in}, \; D_{66} = \frac{t^3 G_{xy}}{12} = 713 \text{ lb-in}$$

Top flange local buckling stress resultant:

$$N_{cr}^{\text{local}} = 6743.4 \text{ lb/in} \qquad \text{... eqn 4.61 and Table 4.13.}$$

Critical beam buckling load:

$$\frac{N_{cr}^{\text{local}}}{t} = \frac{My}{I} \Rightarrow P_{cr}^{\text{local}} = \frac{8DN_{cr}^{\text{local}}}{Lb_w tE_x} = 16{,}000 \text{ lb} > P = 4000 \text{ lb}$$

where: $M = \dfrac{P_{cr}^{\text{local}} L}{4}$ and $y = \dfrac{b_w}{2}$

Factor of safety for local buckling:

$$(F.S.)_{\text{local}} = \frac{P_{cr}^{\text{local}}}{P} = 4.0 \Rightarrow \text{Safe design for local buckling.}$$

6. Summary

Based on the step-by-step procedures shown above for the box-beam under the three-point bending, it is concluded that the beam satisfies the design requirements.

Example 3: design of an FRP deck-and-stringer bridge

Problem statement: design an FRP deck-and-stringer bridge system (Fig. 4.44) using the available box sections and subjected to symmetric loading.

Deck-and-stringer bridge: a single-lane short-span bridge consisting of cellular box sections and wide flange I-beam stringers (15 ft width × 20 ft span) (4.572 m × 6.1 m)

I-beam stringer: 12″ × 12″ × 3/4″ (30.48 cm × 30.48 cm × 1.905 cm). Panel material and lay-up:
E-glass fiber with polyester (V_f = 50.0%).
Unidirectional fiber layer (0°): α = 60%.
CSM layer: β =10%.
±45° SF layer: γ = 30%.
Uniform panel thickness and material lay-up.

Box-beam section: 6″ × 10″ × 3/8″ (15.24 cm × 25.4 cm × 0.953 cm). Panel material and lay-up (see Figs 4.3 and 4.5 as reference).

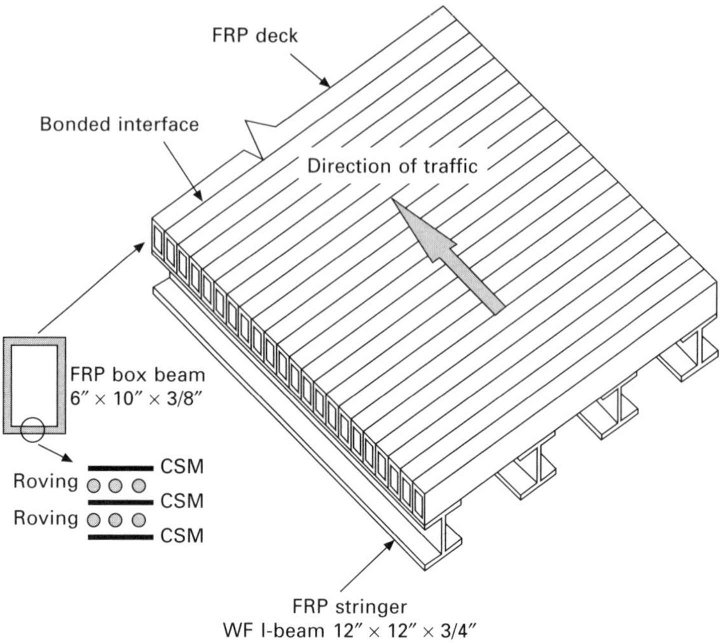

4.44 Schematic of FRP deck on FRP stringers.

	three CSM layers and two roving layers (total: five layers).
	CSM: 3 oz/ft^2 (0.09 g/cm^2); $t_{CSM} = 0.045$ in (0.11cm).
	Roving: 113 Yield, 20 tow/in; t_{roving} = 0.12 in (0.3 cm).
Deflection limit:	$L/500$.
Loading type:	Single-lane with M_{max} =207.4 kip-ft (AASHTO HS-20) (281.2 kN-m).
Requirements:	determine number of the stringers (m) and evaluate their performance.

1. Based on the carpet plots, obtain the panel equivalent moduli and strength values for I-beam stringer and box-beam section.
2. Based on the box-section panel properties, compute the FRP cellular deck stiffness properties (D_x, F_x, D_y, F_y, and D_{xy}).
3. Based on the values in 2, compute the elastic equivalent properties for FRP cellular deck (E_x^p, E_y^p, G_{xy}^p, G_{yz}^p, G_{xz}^p, assuming $\nu_{xy} = 0.3$).
4. Based on the panel properties of I-beam given in 1, compute the stringer bending and shear stiffnesses.
5. Based on the deck stiffness properties obtained from 2, compute deck (an orthotropic plate) bending (D_{11}, D_{22}, D_{12}, D_{66}) and shear (A_{44} and A_{55}) stiffnesses.

6. Compute the deck edge deflection coefficient W_0.
7. Based on the deflection limit, obtain the required number of stringers in the bridge system.
8. Compute the maximum bending and shear stresses in the stringer and check safety factors.
9. Finalize the design.

Design solution:

1. Panel equivalent moduli and strength:

I-beam:

$$\alpha = 60\%,\ \beta = 10\%,\ \gamma = 30\% \text{ and } V_f = 50.0\%$$

From the carpet plots (Figs 4.10–4.13, 4.22, 4.28):

$E_x = 4.25 \times 10^6$ psi (29.3 GPa), $E_y = 2.0 \times 10^6$ psi (13.8 GPa), $G_{xy} = 0.96 \times 10^6$ psi (6.6 GPa), $v_{xy} = 0.4$, $F_{xc} = 2.6 \times 10^4$ psi (180 MPa), $F_{xy} = 0.75 \times 10^4$ psi (51.7 MPa).

Box-beam:

$V_f = ((V_f)_{\text{roving}} \times 2 \times t_{\text{roving}} + (V_f)_{\text{CSM}} \times 3 \times t_{\text{CSM}})/t = 41\%$,

where $(V_f)_{\text{roving}} = n_r A_r / t_r = 45.36\%$, $A_r = 1/(Y\rho_r)$ (eqns 4.3 and 4.4), $(V_f)_{\text{CSM}} = w/\rho_r t_{\text{CSM}} = 32\%$ (eqn 4.2) based on $t_{\text{CSM}} = 0.045$ in, $t_{\text{roving}} = 0.12$ in, $Y = 113$ (yard/lb), $\rho_r = 2.5$ g/cm^3, $n_r = 20$ (in^{-1}), and $w = 3$ oz/ft^2; $\alpha = 2 \times t_{\text{roving}} \times (V_f)_{\text{roving}}/(V_f t) = 70\%$, $\beta = 3 \times t_{\text{CSM}} \times (V_f)_{\text{CSM}}/(V_f t) = 30\%$.

Based on the ratio plots (Figs 4.14–4.17):

$$(E_x)_{41}/(E_x)_{50} = 0.82,\ (E_y)_{41}/(E_y)_{50} = 0.82$$

$$(G_{xy})_{41}/(G_{xy})_{50} = 0.82,\ (v_{xy})_{41}/(v_{xy})_{50} = 1.0$$

From the carpet plots of $V_f = 50\%$ (Figs 4.10–4.13), the panel stiffness properties for $V_f = 41\%$ are obtained as follows:

$$(E_x)_{41} = 0.82 \times (E_x)_{50} = 3.84 \times 10^6 \text{ psi} = 26.4 \text{ GPa}$$

$$(E_y)_{41} = 0.82 \times (E_y)_{50} = 1.83 \times 10^6 \text{ psi} = 12.6 \text{ GPa}$$

$$(G_{xy})_{41} = 0.82 \times (E_x)_{50} = 0.64 \times 10^6 \text{ psi} = 4.4 \text{ GPa}$$

$$(v_{xy})_{41} = 1.0 \times (v_{xy})_{50} = 0.357$$

2. FRP cellular deck stiffness properties

Number of cells:

$$n_c = L/w_{\text{box}} = 40.$$

Bending stiffness:

$$D_x = n_c E_x (h^2 + t^2 + 3(h)b)\frac{(h)(t)}{6} = 77 \text{ MPa-m}^4 \qquad \text{... eqn 4.68}$$

Transverse stiffness:

$$D_y = \frac{1}{2} E_y (w)(t)h^2 = 17.7 \text{ MPa-m}^4 \qquad \text{... eqn 4.71}$$

Shear stiffness in longitudinal direction:

$$F_x = n_c G_{xy} (2t)h = 852.1 \text{ MPa-m}^2 \qquad \text{... eqn 4.70}$$

Shear stiffness in transverse direction

$$F_y = \frac{2E_y wt^3}{b\left(b + \frac{h}{4}\right)} = 3.02 \text{ MPa-m}^2 \qquad \text{... eqn 4.74}$$

Torsional stiffness:

$$D_{xy} = \frac{n_c G_{xy} bh^2 t}{(n_c b + h)} = 2.59 \text{ MPa-m}^2 \qquad \text{... eqn 4.78}$$

3. Elastic equivalent properties for FRP cellular deck

$v_{xy} = 0.3$ then $v_{yx} = E_y \times v_{xy} / E_x = 0.143$

Bending moduli:

$$(E_x)_p = 12\frac{D_x}{t_p^3 b_p}(1 - v_{xy} v_{yx}) = 8.85 \text{ GPa} \qquad \text{... eqn 4.79}$$

$$(E_y)_p = 12\frac{D_y}{t_p^3 l_p}(1 - v_{xy} v_{yx}) = 2.71 \text{ GPa} \qquad \text{... eqn 4.80}$$

Out-of-plane shear moduli:

$$(G_{xz})_p = \frac{F_x}{t_p b_p} = 0.55 \text{ GPa} \qquad \text{... eqn 4.83a}$$

$$(G_{yz})_p = \frac{F_x}{t_p b_p} = 0.0026 \text{ GPa} \qquad \text{... eqn 4.83b}$$

In-plane shear modulus:

$$(G_{xy})_p = 6\frac{D_{xy}}{t_p^3} = 0.97 \text{ GPa} \qquad \text{... eqn 4.83c}$$

4. Stringer bending and shear stiffnesses

$$D = \frac{1}{2}(E_x)_f t_f b_w^2 b_f + \frac{1}{12}(E_x)_w t_w b_w^3$$

$$+\frac{1}{6}(E_x)_f t_f^3 b_f = 9.23 \text{ MPa-m}^4$$

$$F = (G_{xy})_w t_w b_w = 38.33 \text{ MPa-m}^2$$ ... eqn 4.36a

5. Bending and shear stiffnesses of the bridge deck

$$D_{11} = \frac{D_x}{b_p} = 12.63 \text{ MPa-m}^3,\ D_{22} = \frac{D_y}{l_p} = 3.88 \text{ MPa-m}^3$$

$$D_{12} = \frac{\upsilon_{xy} D_x}{l_p} = 1.164 \text{ MPa-m}^3,\ D_{66} = \frac{D_{xy}}{2} = 1.3 \text{ MPa-m}^3$$

$$A_{44} = \frac{5}{6}\frac{F_x}{l_p} = 0.55 \text{ MPa-m},\ A_{55} = \frac{5}{6}\frac{F_x}{b_p} = 116.53 \text{ MPa-m}$$

... eqn. 4.91

6. Deck edge deflection coefficient

$\alpha = \pi/l_p = 0.687$, $\beta = \pi/b_p = 0.515$ and

$$W_o = \frac{\pi}{4}\left[\begin{matrix} -1 + \dfrac{A_{44}\beta^2}{A_{55}\alpha^2 D_{12}} \\ \left(\dfrac{\alpha^2 D_{12}^2(3D_{66} + D_{12}) - D_{22}[(D_{11}\alpha^2 + A_{55}) \times (D_{12} + 2D_{66}) + D_{12}D_{66}\beta^2}{A_{44}(D_{12} + 2D_{66}) + D_{66}(D_{22}\beta^2 - D_{12}\alpha^2)}\right)\end{matrix}\right] = -1.97$$

... eqn 4.90a

7. Number of stringers in the bridge system

$M_{max} = 207.4$ kip-ft $= 281.197$ kN-m

Deflection limit

The dimensions of the deck are used to evaluate the maximum allowable moment per lane (M_{max}) according to AASHTO (1989). Then the equivalent concentrated load (P_e) at mid-span is calculated as

$$P_e = \frac{4M_{max}}{L} = 184.5 \text{ kN} \qquad 4.101$$

where L is the length of a stringer (span of the bridge) and is 20 feet (6.096 m) in this example. The maximum load distribution factor for symmetric loading is given in eqn 4.99, as

$$(W_f^{Sym})_{max} = \frac{1 + W_o}{mW_o + \frac{2}{\pi}(m - 1)}$$

The design load (P_d) is found from

$$P_d = P_e N_L (W_f^{\text{sym}})_{\max} \tag{4.102}$$

where N_L is the number of lanes and $(W_f^{\text{sym}})_{\max}$ is found from the above equation.

The live load is defined as

$$P_{LL} = P_d(1 + DLA) \tag{4.103}$$

where the dynamic load allowance (DLA) factor for short span bridge is taken as 0.20.

The midspan deflection δ_{LL} of a stringer is evaluated as

$$\delta_{LL} = P_{LL}\left(\frac{L^3}{48D} + \frac{L}{4\kappa F}\right) \tag{4.104}$$

Based on eqns 4.101 to 4.104, the deflection limit can be written as a function of the number of stringers m:

$$\delta_{LL} = \frac{L}{500}$$

$$= M_{\max}\left(\frac{4}{L}\right)N_L\left(\frac{1 + W_o}{mW_o + (m - 1)\dfrac{2}{\pi}}\right)\left(\frac{L^3}{48D} + \frac{L}{4\kappa F}\right)(1 + DLA) \tag{4.105}$$

In this case, $N_L = 1.0$, and κ is taken as 1.0. By solving eqn 4.105 for the variable m, the result is 7.28, and therefore, $m = 8$ is taken as the number of the stringers.

8. Evaluation and checking

Maximum load distribution factor for symmetric loading:

$(w_f^{\text{sym}})_{\max} = 0.086$ ... eqn 4.99

Design load:

$P_d = 15.8$ kN ... eqn 4.100

Approximate maximum compression (or tension) stress on stringer due to live load (P_{LL}):

$$\varepsilon_x = \frac{M_{LL}}{D}y_i \text{ where } P_{LL} = 18.96 \text{ kN and } M_{LL} = \frac{P_{LL}L}{4} = 28.9 \text{ kN-m}$$

$$\sigma_x = E_x\varepsilon_x = 14.0 \text{ MPa}$$

Approximate maximum shear stress on stringer due to live load (P_{LL}):

$$\gamma_{xy} = \frac{V_{LL}}{F} \text{ where } V_{LL} = \frac{P_{LL}}{2} = 9.48 \text{ kN}$$

$$\tau = G_{xy}\gamma_{xy} = 1.63 \text{ MPa}$$

Safety factors:

$$(F.S.)_{\sigma} = F_{xc}/\sigma_c = 180/14 = 12.85 > 1 \Rightarrow \text{Safe design}$$

$$(F.S.)_{\tau} = F_{xy}/\,\tau = 51.7/1.63 = 31.7 > 1 \Rightarrow \text{Safe design}$$

9. Design summary

Deck: 40 Box-beams (6″ × 10″ × 3/8″) are used as the deck panel (Fig. 4.44)

Stringer: Eight I-beams (12″ × 12″ × 3/4″) are used as bridge stringers with center-to-center spacing of 15 × 12/(8-1) = 25.7″ (65.3 cm).

4.5 Conclusions

A systematic approach for material characterization, analysis and design of all fiber-reinforced polymer (FRP) composite structures is presented in this chapter. First, the advanced research and applications of FRP composites in civil infrastructure are briefly reviewed with the focus on FRP bridge systems, followed by the introduction of one of the most prevalent manufacturing processes, pultrusion, which is commonly used in the manufacturing of civil engineering products. Following the design procedure, several approaches (e.g., micro/macromechanics and carpet plots) are given to obtain the material properties including constituent materials and ply properties, laminated panel engineering properties, and member stiffness properties. The mechanical behaviors of FRP shapes (e.g., bending and shear deformation, local/global buckling, and material failure) are discussed with reasonable details.

An elastic equivalence analysis is used to obtain the apparent properties of cellular deck panels; while a first-order shear deformation macro-flexibility analysis is presented for composite deck-and-stringer systems, which account for load distribution factors under various loading cases. The summarized step-by-step design guidelines for FRP shapes are useful for practicing engineers concerned with design of FRP composite structures. The simplified design analysis procedures are illustrated by three typical design examples which include two FRP beams (I- and box-sections) and one bridge deck-and-stringer system. The present approach described in this chapter can be used efficiently to analyze and design FRP shapes and bridge systems and also develop new design concepts for all composite structures.

4.6 Acknowledgments

The authors would like to acknowledge the valuable contributions provided by Prof. Ever J. Barbero of West Virginia University, Dr Hani A. Salim of University of Missouri, Columbia, and Dr Roberto Lopez-Anido of University of Maine. The contributions by former graduate students Brian Brown of West Virginia University and Xuezhi Li of University of Akron, are greatly appreciated.

4.7 References

AASHTO, *Standard Specifications for Highway Bridges*, The American Association of State Highway and Transportation Officials (AASHTO), Washington, DC, 1989.

Ahmad, S.H. and Plecnik, J.M., *Transfer of Composite Technology to Design and Construction of Bridges*. US DOT Report, September, 1989.

Bakeri, B. and Sunder, S.S., Concepts for Hybrid FRP Bridge Deck System. *Proceedings of the 1st Materials Engineering Congress, ASCE*, Denver, CO. 1990, Vol. 2, pp.1006–1014.

Bakis, C.E., Bank, L.C., Brown, V.L., Cosenza, E., Davalos, J.F., Lesko, J.J., Machida, A., Rizkalla, S.H. and Triantafillou, T.C., Fiber reinforced Polymer Composites for Construction–State-of-the-Art Review, *J. of Composite for Construction*, 2002, 2, 73–87.

Barbero, E.J., *Introduction to Composite Materials Design*. Taylor & Francis, Philadelphia, PA, 1999.

Barbero, E.J., Lopez-Anido, R. and Davalos, J.F., On the Mechanics of Thin-walled Laminated Composite Beams. *J. Composite Materials*, 1993, 27(8): 806–829.

Barbero, E.J Makkapati, S. and Tomblin, J.S., Experimental Determination of The Compressive Strength of Pultruded Structural Shapes, *Composites Science and Technology*, 59 (13): 2047–2054, 1999.

Bleich, F., *Buckling Strength of Metal Structures*, McGraw-Hill Book Company, Inc., New York, NY, 1952.

Brown, B., *Experimental and Analytical Study of FRP Deck-and-Stringer Short-Span Bridges*. Master of Science Thesis, West Virginia University, Morgantown, WV, 1998.

Burnside, P., Barbero, E.J., Davalos, J.F. and GangaRao, H.V.S., Design Optimization of an All-FRP Bridge, *Proceedings of 38th Int. SAMPE Symposium*, 1993.

Chamis, C.C., *Simplified Composites Micromechanics Equations for Strength, Fracture Toughness, and Environmental Effects*. NASA TM-83696,1984.

CP Design Manual: The Pultex Pultrusion Global Design Manual of Standard and Custom Fiber Reinforced Polymer Structural Profiles, Creative Pultrusions, Inc., Alum Bank, PA, 2000.

Cusen, A.R. and Pama, R.P., *Bridge Deck Analysis*, John Wiley & Sons, 1975.

Davalos, J.F. and Qiao, P.Z., *A Computational Approach for Analysis and Optimal Design of FRP Beams, Computers and Structures*, 1999, 70(2): 169–183.

Davalos, J.F., Qiao, P. and Barbero, E.J., Multiobjective Material Architecture Optimization of Pultruded FRP I-beams, *Composite Structures*, 1996a, 35: 271–281.

Davalos, J.F., Salim, H.A., Qiao, P., Lopez-Anido, R. and Barbero, E.J., Analysis and Design of Pultruded FRP Shapes under Bending. Composites, Part B: *Engineering J.*, 1996b, 27(3-4): 295–305.

GangaRao, H.V.S. and Sotiropoulos, S.N., *Development of FRP Bridge Superstructural Systems*. US DOT Report, 1991.

Harris, J.S. and Barbero, E.J., Prediction of Creep Properties of Laminated Composites from Matrix Creep Data, *J. of Reinforced Plastics and Composites*, 1998, 17 (4): 361–379.

Head, P.R., Advanced Composites in Civil Engineering – A Critical Overview at This High Interest, Low Use Stage of Development, *Proceedings of ACMBS*, M. El-Badry, ed., Montreal, Quebec, Canada, 1996, pp. 3–15.

Henry, J.A., *Deck Girders System for Highway Bridges Using Fiber Reinforced Plastics.* M.S. Thesis, North Carolina State University, 1985.

Hull, D., *An Introduction to Composite Materials*, Cambridge University Press, Cambridge, UK, 1981.

Jones, R.M., *Mechanics of Composite Materials.* Taylor & Francis, Inc., Philadelphia, PA, 1999.

Kim, Y., Davalos, J.F. and Barbero, E.J., Progressive Failure Analysis of Laminated Composite Beams, *J. Composite Materials*, 1996, 30, 536–560.

Li, X.Z., *Simplified Analysis and Design of Fiber Reinforced Plastic Structural Shapes*, Engineering Report for MS degree, Department of Civil Engineering, the University of Akron, Akron, OH, 2000.

Lopez-Anido, R., *Analysis and Design of Orthotropic Plates Stiffened by Laminated Beams for Bridge Superstructures.* Ph.D. Dissertation, Department of Civil and Environmental Engineering, West Virginia University, 1994.

Lopez-Anido, R., Troutman, D.L. and Busel, J.P., Fabrication and Installation of Modular FRP Composite Bridge Deck. *Proceedings of Int. Composites Expo '98*, Composites Institute, 1998a, pp. 4-A (1–6).

Lopez-Anido, R., Howdyshell, P.A., Stephenson, L.D. and GangaRao, H.V.S., Fatigue and Failure Evaluation of Modular FRP Composite Bridge Deck. *Proceedings of Int. Composites Expo '98*, Composites Institute, 1998b, pp. 4-B (1–6).

Luciano, R. and Barbero, E.J., Formulas for the Stiffness of Composites with Periodic Microstructure, *Int. J. of Solids and Structures*, 1994, 31(21): 2933–2944.

Madabhusi-Raman, P. and Davalos, J.F., Static Shear Correction Factor for Laminated Rectangular Beams, *Composites Part B: Engineering*, 1996, 27B, 285–293 (Special edition on Composites for Civil Infrastructure).

Mongi, A.N.K., *Theoretical and Experimental Behavior of FRP Floor System*, M.S. Thesis, West Virginia University, Morgantown, WV. 1991.

Nemat-Nasser, S. and Taya, M., On Effective Moduli of an Elastic Body Containing Periodically Distributed Voids, *Quarterly Applied Mathematics*, 1981, 39: 43–59.

Pandey, M.D., Kabir, M.Z. and Sherbourne, A.N., Flexural-torsional Stability of Thin-walled Composite I-section Beams, *Composites Engineering*, 1995, 5(3), 321–342.

Plecnik, J.M. and Azar, W.A., Structural Components, Highway Bridge Deck Applications, *International Encyclopedia of Composites*, edited by I. Lee and M. Stuart, 1991, 6, pp. 430–445.

Plunkett, J.D., *Fiber Reinforcement Polymer Honeycomb Short Span Bridge for Rapid Installation.* IDEA Project Report, November 1997.

Qiao, P.Z., *Analysis and Design Optimization of Fiber-reinforced Plastic (FRP) Structural Beams.* Ph.D. Dissertation, Department of Civil and Environmental Engineering, West Virginia University, 1997.

Qiao, P., Davalos, J.F. and Barbero, E.J., Design Optimization of Fiber-reinforced Plastic Composite Shapes, *J. of Composite Materials*, 1998, 32(2), 177–196.

Qiao, P.Z., Davalos, J.F., Barbero, E.J. and Troutman, D., Equations Facilitate Composite Designs, *Modern Plastics Magazine*, 1999, 76(11): 77–80.

Qiao, P.Z., Davalos, J.F. and Brown, B., A Systematic Analysis and Design Approach for Single-span FRP Deck/Stringer Bridges, Composites, Part B: *Engineering J.*, 2000, 31: 593–609.

Qiao, P.Z., Davalos, J.F. and Wang, J.L., Local Buckling of Composite FRP Shapes by Discrete Plate Analysis, *Journal of Structural Engineering*, 2001, 127(3), 245–255.

Qiao, P.Z. and Zou, G.P., Local Buckling of Elastically Restrained Fiber-reinforced Plastic Plates and its Application to Box Sections, *Journal of Engineering Mechanics*, 2002, 128(12), 1324–1330.

Qiao, P.Z. and Zou, G.P., Local Buckling of Composite Fiber-reinforced Plastic Wide-flange Sections, *Journal of Structural Engineering*, 2003, 129(1), 125–129.

Qiao, P.Z., Zou G.P. and Davalos J.F., Flexural-torsional Buckling of Fiber-reinforced Plastic Composite Cantilever I-beams, *Composite Structures*, 2003, 60, 205–217.

Qiao, P.Z. and Shan, L.Y., Explicit Local Buckling Analysis and Design of Fiber-reinforced Plastic Composite Shapes, *Composite Structures*, 2005, 70(4), 468–483.

Reddy, J.N., *Energy and Variational Methods in Applied Mechanics*, John Wiley, New York, 1984.

Salim, H.A., Davalos, J.F., GangaRao, H.V.S. and Raju, P., An Approximate Series Solution for Design of Deck-and-Stringer Bridges, *Int. J. of Engineering Analysis*, 1995, 2: 15–31.

Salim, H.A., *Modeling and Application of Thin-walled Composite Beams in Bending and Torsion*, Ph.D. Dissertation, West Virginia University, Morgantown, WV, 1997.

Salim, H.A., Davalos, J.F., Qiao, P. and Kiger, S.A., Analysis and Design of Fiber Reinforced Plastic Composite Deck-and-Stringer Bridges, *Composite Structures*, 1997, 38: 295–307.

Shan, L.Y. and Qiao, P.Z., Flexural-torsional Buckling of Fiber-reinforced Plastic Composite Open Channel Beams, *Composite Structures*, 2005, 68(2): 211–224.

Sherbourne, A.N. and Pandey, M.D., Elastic, Lateral-Torsional Stability of Beams; Moment Modification Factor, *J. of Construction Steel Research*, 1989, 13, 337–356.

Tomblin, J.S., *Compressive Strength Models for Pultruded Glass Fiber Reinforced Composites*. Ph.D. Dissertation, Department of Mechanical and Aerospace Engineering, West Virginia University, Morgantown, WV, 1994.

Troitsky, M.S., *Orthotropic Bridges, Theory and Design*, The James F. Lincoln ARC Welding Foundation, Cleveland, Ohio, 1987.

Tsai, S.W., *Composites Design*. Dayton, OH: Think Composites, 1988.

Zureick, A., *Fiber-reinforced polymeric bridge decks*, Seminar Note, National Seminar on Advanced Composite Material Bridges, FHWY, Washington, DC, 1997.

5
Rehabilitation of civil structures using advanced polymer composites

V M KARBHARI, University of California, San Diego, USA

5.1 Introduction

Worldwide, civil infrastructure facilities represent an investment of trillions of dollars and serve as the basis for the comfort, security and advancement of mankind. This infrastructure, which was significantly expanded over the last 50 years, is now reaching a critical age with increasing signs of deterioration and reduced functionality. Deficiencies in the current bridge inventory, for example, range from those related to wear, environmental deterioration, and aging of structural components, to increased traffic demands and changing patterns, and from insufficient detailing at the time of original design and construction to the original use of substandard materials and methods of construction. In addition aspects related to infrequent inspection and inadequate maintenance, as well as changes in design codes, have created the need for substantial rehabilitation and renewal. These deficiencies are not restricted to just bridges and other transportation related structures but are endemic to the built environment. Major metropolitan areas are facing the unintended results of previously unrestricted urban development which led to rapid construction often with substandard materials or without proper detailing, such as in the rapid urban development in South-East Asia. In addition to deterioration and code related deficiencies there is need for rehabilitation measures to handle increased and changed needs related to occupancy and critical services. These aspects have a tremendous impact on society in terms of socio-economic losses resulting from lack of useable space to delays and accidents/collapses.

Conventional materials such as steel, concrete, and timber have a number of advantages, not the least of which is the relatively low cost of raw materials. However, it is clear that conventional materials and technologies, although suitable in some cases and with a history of good applicability, lack in longevity in some cases, and in others are susceptible to rapid deterioration, emphasizing the need for better grades of these materials or the use of newer technologies to supplement the palette of materials and technologies available to the civil designer and engineer. It should also be noted that in a number of cases design alternatives are constrained by the limitations of current materials used such as the length of the span of a bridge due to weight considerations, or the unsupported surface of the dome of a sports facility due to stiffness and weight considerations, or the size of a column due to restrictions of design space. In other cases, constraints such as dead load restrict the expansion of previously built structures to meet current needs (such as the widening of a bridge, or increasing the height of a building, or the pressure of water in a pipeline) or the carriage of higher levels of traffic (both for bridges and in buildings). In all such cases there is a critical need for the use of new and emerging materials and technologies with the end goal of facilitating greater functionality and efficiency.

The history of mankind can be written in terms of the use of materials of construction starting from caves built in rock and the use of animal skins as rough shelters from the environment, through the use of adobe and straw-reinforced mud bricks, to the more modern use of cast- and wrought-iron which initiated the industrial age, and the use of specially designed alloys of steel and titanium, and the increasing use of polymers and fiber reinforced polymer composites. The concept of combining materials to create new systems having unique characteristics is not new to the civil engineering arena and examples such as plywood (a laminated composite consisting of adhesive bonded veneers), specially designed alloys, reinforced concrete, and fiber reinforced concrete, abound in everyday construction.

The use of fiber reinforced polymer (FRP) composites takes this one step further enabling the synergistic combination of reinforcing fibers, with appropriate fillers/additives, in a polymeric resin matrix. The fibrous reinforcement carries load in predetermined directions (indicated by fiber orientation) and the polymer acts as a medium to transfer stresses between adjoining fibers through adhesion, and also provides toughness and protection for the material. The combination of the matrix phase with the fiber reinforcement thus creates a new material which is conceptually analogous to steel rebar reinforced concrete although the reinforcing fractions vary considerably (i.e. reinforced concrete in general rarely contains more than 5% reinforcement, whereas in FRP composites the reinforcing volume fraction can range from 25% to 70% depending on process used). The matrix has both tensile and compression capabilities (unlike cement which is strong only in compression) and unlike concrete, polymeric resins used in the civil infrastructure area impart a high degree of toughness and damage tolerance as well.

The attractiveness of FRP composites as construction materials derives from a set of advantages that includes high specific stiffness and strength characteristics, low weight, corrosion resistance and potentially high durability, tailorability of mechanical properties in the form of an anisotropic material, the ability to have low conductivity and to act as an insulator (in the case of glass fiber reinforced composites), and to be electromagnetically transparent. In addition the wide variety of manufacturing processes available enable these materials to be fabricated both in the field under highly uncontrolled environments (such as in the case of wet layup used in external strengthening applications) and to very precise dimensional tolerances under highly controlled factory conditions (such as in the case of autoclave cured composites).

While a variety of fiber and matrix materials can be used as constituents, for the purposes of the current chapter the discussion will be restricted to aramid, carbon, and glass fibers, and thermoset resins. The interested reader is referred to the excellent works by Mallick (1994), Jones (1999), and the ASM Handbook (2001) amongst others for more details on the materials, manufacturing methods and mechanics pertaining to these materials. Aspects related to durability can be found in works by Springer (1981), Pritchard (2000) and Karbhari *et al.* (2001) amongst others. Again, although FRP composites can, and have been used in a variety of civil engineering applications ranging from reinforcing bars and tendons, bridge decks, structural profiles, for rehabilitation, as aesthetic fascia, etc., the present chapter will only focus on the use of the above mentioned classes of FRP composites to rehabilitation of structural components.

5.2 Rehabilitation and FRP composites

Throughout this chapter it is essential that the terms repair, strengthening and retrofit, which are often used interchangeably, be differentiated since they refer to three different structural conditions, all covered by the general term – rehabilitation (Fig. 5.1). In 'repairing' a structure, the FRP composite material is used to fix a structural deficiency, such as a crack, in order to restore the component to its original level of performance. In contrast, the 'strengthening' of structures or components is specific to those cases whwerein the addition of the FRP composite enhances the existing as-designed performance level thereby providing additional performance. This applies to the strengthening of bridge decks wherein the application of external bonded FRP composites (Fig. 5.2) can result in the addition of load-carrying capacity. The term 'retrofit' is specifically used in this context to represent the seismic

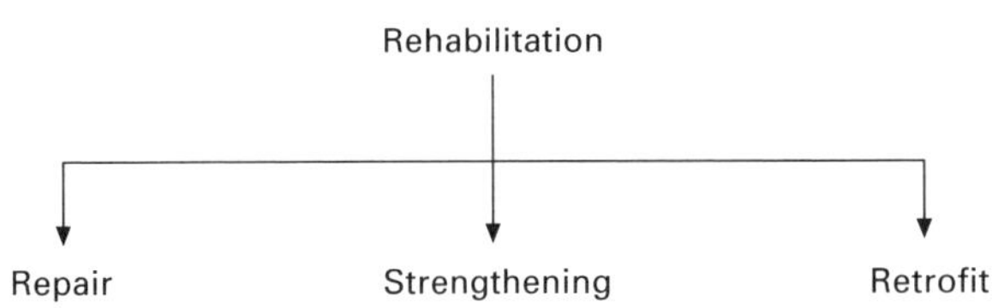

5.1 Categories of rehabilitation.

5.2 Application of composite strips to the soffit of a bridge deck.

upgrade of facilities, such as in the case of the confinement of columns through application of FRP composite jackets. The differentiation is important not only from the perspective of functionality but also since the selection of materials and processes needs to be based on the specific requirements, including those of durability and expected service-life, of the application.

For the purposes of this chapter, discussion related to rehabilitation through the use of FRP composites will be restricted to those cases wherein the FRP composite is bonded onto, or otherwise brought into close contact with, the appropriate portion of the concrete substrate. The methods of seismic retrofitting essentially derive from the earlier work on steel and concrete based jackets (Priestley *et al.*, 1996) where additional material is added to ensure that the concrete core is confined. In the case of strengthening and repair, the methodology is derived from the bolting or bonding of steel plates that was developed in South Africa and France in the 1960s and 1970s (L'Hermite, 1967; Bresson, 1971, Dussek, 1974) as a means of rapid strengthening of concrete and masonry structures. In these instances steel plates were adhesively bonded and/or bolted onto the substrate to provide additional strength and stiffness. Although the methodology was extremely successful and is in fact still used it has a few critical disadvantages, namely,

- Steel plates are heavy and the added weight of the plates to an already weak structure can cause additional damage and in some cases inadvertent failure.
- The weight of the plates necessitates special measures to ensure that they stay attached to the substrate, often necessitating time consuming anchoring.

- The weight of the plates makes handling in the field difficult and requires use of lifting equipment, equipment to hold the plate in intimate contact with the substrate during cure of the adhesive, and makes positioning in areas with existing utilities fairly difficult.
- Due to weight and stiffness of the material it is difficult to use long continuous lengths.
- Field welding, especially butt welding to join shorter lengths can result in damage to the existing structure, serve as a locus for future corrosion, and lead to issues with nondestructive evaluation.
- Long-term durability, especially in moist environments is suspect.
- There are concerns related to bonding during traffic movement due to sagging under self weight.
- The weight and difficulty in handling results in an involved and time-consuming process.

In comparison, FRP composites are extremely light, can be applied without significant equipment, have potentially very good durability, and provide a rapid and flexible means for application in the field with minimal disruption of traffic. In the mid-1980s, driven by the disadvantages listed above, experiments were conducted on the use of FRP as a means of externally bonded rehabilitation both in Europe and Japan, and it was shown that if done properly there was a significant weight saving, little to no disruption of traffic, and the added advantage of being able to maintain the external aesthetics of a structure since configuration could be followed fairly easily (Meier *et al.*, 1992; Taljsten, 1994; McKenna and Erki, 1994; Karbhari, 1996). In addition there is the added advantage of use of the anisotropy of the material which enables increase in stiffness in desired directions without increases in undesired directions, such as in seismic retrofit of columns where one desires confinement in the hoop direction without increase in stiffness in the longitudinal direction.

5.3 Materials and manufacturing processes

While the performance characteristics of conventional materials of construction such as concrete, steel, and timber are fairly well know, those related to the emerging class of FRP composites are still not widely known in the civil engineering community. Moreover, there are as yet very few standardized handbooks that provide these properties, both because the use of FRP composites for the purposes of rehabilitation is still an emerging field, and because the characteristics of FRP composites are variable. It must be emphasized that the properties of FRP composites depend on the properties of the material constituents (i.e. reinforcing fibers, resin matrices, and fillers), the corresponding volume fractions, and the orientation of the fibers. Thus,

FRP composites provide a class of designer materials, for which properties can be tailored, within reason, based on the needs of the application. This provides both a designer's dream (due to the ability to not be constrained by isotropy and a few standardized grades) and a nightmare (since at times the lack of standardization and the availability of almost an infinite number of choices can make routine design difficult) at the same time. It is thus important that the reader develop a basic understanding of the intrinsic characteristics of this new class of materials prior to examining how these materials can be applied to concrete substrates for the purposes of rehabilitation.

5.3.1 Materials overview

For the purpose of this chapter FRP composites will be restricted to the class of materials formed by the combination of strong and stiff fiber assemblies with polymeric resins through processes which provide material characteristics that are different from those of the constituents, and in which the constituents can be physically identified and separated. The focus of the current discussion will remain on material systems, and their constituents, that are commonly used in rehabilitation, rather than on systems which could be used in the future. Thus the description of fibers is restricted to glass, carbon and aramid fibers, and that of the matrices is restricted to epoxy and vinylester type resins.

A comparison of typical fiber properties with those of grade 60 structural steel is shown in Table 5.1. Glass fibers are perhaps the most common

Table 5.1 Properties of typical reinforcing fibers

Fiber type	Designation	Density (g/cc)	Tensile modulus (GPa)	Tensile strength (MPa)	Ultimate strain (%)
Glass	E-glass	2.55	72.4	3450	4.8
	S-glass	2.49	86.9	4800	5.2
Carbon (PAN based)	T-300	1.76	230	3530	1.5
	AS4	1.80	221	3930	1.7
	IM6	1.74	275	5240	1.7
	UHM	1.85	441	3445	0.8
Carbon (pitch based)	P-55	2.00	380	1900	0.5
	P-100	2.15	758	2410	0.32
	UHM	2.20	965	2410	0.27
Aramid	Kevlar 49	1.45	131	3620	2.8
	Kevlar 149	1.47	179	3500	2.0
	Technora	1.39	74	3500	4.6
Steel	Grade 60	7.80	200	Yield = 415 Ultimate = 620	4.5–12

reinforcement type used in composites, accounting for, perhaps, about 90% of all fibers used, and were initially widely used in rehabilitation. Glass is made by fusing silicates with potash, lime, or various metallic oxides. The manufacture of glass filaments, which are between typically 3–20 μm in diameter, begins with the blending of silica and select minerals in a furnace where the melt is homogenized. Once refined, the melt is drawn into fibers by extrusion through platinum-rhodium bushings. The filaments are rapidly cooled to prevent crystallization. These filaments are highly abrasive and are susceptible to moisture-induced degradation hence treated with a sizing/binder soon after forming and prior to gathering into strands. Glass fibers are differentiated based on chemical composition and have letter designations implying specific characteristics and uses (Gupta, 1988). E- and S-Glass fibers are commonly used in structural applications, with rehabilitation applications being almost completely restricted to E-glass fibers. They have good strength and modulus and are easy to form, but are, however, susceptible to degradation in the presence of moisture and alkalis in the bare state (i.e. when not protected by the use of an appropriately selected resin matrix). Further they undergo creep and stress-rupture and therefore should not be used under high levels of sustained load.

The first use of carbon fibers is often ascribed to Thomas Alva Edison who carbonized cotton and bamboo for use as filaments in incandescent lamps (Edison, 1880). However, practical forms for use in structural applications did not appear till the work of Bacon (1960) and Bacon and Tang (1964) based on the use of rayon and cellulose as precursors. The use of isotropic pitch as a precursor was initially pursued by Union Carbide (Bacon and Smith, 1965). Later Shindo in Japan (1961) and Watt in the UK (Watt *et al.*, 1986) both developed carbon fibers from polyacrylonitrile (PAN). Otani later developed high modulus fibers from pitch (Otani *et al.*, 1972). Precursors are materials from which carbon fibers are derived. The three general types used currently are rayon, polyacrylonitrile (PAN) and pitch. Properties of the resulting fiber, as well as the economics of the process, change based on choice of precursor. Rayon based precursors are derived from cellulose and have a low conversion efficiency. PAN precursors are the basis for a majority of commercially available carbon fibers and generally yield the highest tensile strengths. Although most fibers are circular in cross-section the use of PAN as a precursor enables formation of rectangular, dog-boned and 'x' type cross-sections, which can yield closer fiber packing. Pitch precursors are of low cost and are a complex mixture of aromatic hydrocarbons generally derived from petroleum, coal tar, or Polyvinylchloride.

Mesophase pitch, which has long highly oriented molecules, yields very high modulus fibers. The use of pitch, depending on process, can yield either very low modulus or high modulus fibers and until recently had associated problems related to batch-to-batch variation. Carbon fibers are available with a large range of moduli and strength values and therefore need to be carefully specified and referred to in order to avoid the use of erroneous data in design. Low modulus (LM) fibers are attained from pitch and generally have strengths between 350–1000 MPa and a modulus of less than 100 GPa. Intermediate modulus (IM) fibers have moduli up to 300 GPa and a strength to modulus ratio greater than 0.01. High modulus (HM) fibers have moduli higher than 300 GPa and a strength to modulus ratio of less than 0.01. Ultra-high modulus (UHM) fibers have moduli in excess of 500 GP and generally have strengths in the range of 1700–2600 MPa. These fibers can be both from pitch and PAN precursors, however, fibers at the higher range of modulus (>700 GPa) are formed exclusively from mesophase pitch. High tensile (HT) strength fibers have a strength higher than 3000 MPa and a strength to modulus ratio of between 0.015 and 0.02. Irrespective of precursor used the process for formation of the fibers is similar. The PAN precursor is first spun into filament form and then stretched during heating at 200–300 °C to cause orientation and cross-linking of molecules such that decomposition does not take place in subsequent steps. The stretching is essential for attainment of an oriented molecular structure for high strength and stiffness. In the case of the pitch precursor the fiber is spun and then stabilized in similar fashion. Once stabilization is completed, carbonization at 1000–1500 °C causes precursor pyrolysis to about 95% carbon content. Restraint on shrinkage during this stage and graphitization (1500–2800 °C) results in higher orientation and attainment of high tensile modulus and enhanced tensile strength.

It is essential at this stage to emphasize that although the terms carbon and graphite are often used interchangeably, there is a major point of distinction as related to chemical composition. Graphite fibers are subjected to a much higher degree of pyrolysis than carbon fibers resulting in carbon content being about 99% compared to a 95% content for conventional carbon fibers. These fibers are increasingly the reinforcement of choice for purposes of rehabilitation, especially as related to flexural strengthening of beams, girders, and slabs.

Aramid fibers are organic fibers consisting of aromatic polyamides generally manufactured by the extrusion of a polymer solution through a spinneret. These liquid crystalline polymers have an extended chain structure containing aromatic rings, which provide high levels of thermal stability, and amide (-NH-) and carbonyl (-CO-) bonds which are highly resistant to rotation and provide high strength and modulus. The microstructure of aramid fibers, due to the alignment of long, parallel polymer chains is fibrillar and gives an anisotropic nature with higher strength and modulus in the axial direction as compared to the radial (transverse) direction. Because of the fibrillar structure, the fibers are susceptible to the formation of kink-bands or microbuckling in compression. Fiber response is linear in tension but plastic in compression with yield at levels of compressive strain as low as 0.3–0.5%. A number of varieties of aramid (Kevlar, Technora, Twaron) are available and although Kevlar is often mistakenly used as a generic name for aramids it must be noted that it refers specifically to a type of poly para-phenylenetece phthalimide (PPD-T) developed by the Du Pont Company (Blades, 1973; Kwolek *et al.*, 1977).

Aramids have high tensile strength and stiffness, very good impact and abrasion resistance and damage tolerance. However, they do swell in the presence of moisture with equilibrium moisture content at 60% RH being between 1.5–5% depending on fiber type with diameter increasing by 0.5% with a change in 1% moisture content (Chang, 2001). The fiber is susceptible to creep but has a higher creep rupture threshold than E-glass, and shows degradation in the presence of sunlight due to ultraviolet radiation. While not commonly used in rehabilitation, they have been used in specific cases where the characteristics of abrasion and impact resistance are important, or when there is a need for the non-conductive nature of the fibers in addition to the higher modulus in tension.

Although the basic reinforcing element of a FRP composite is a fiber, it is almost never used in that form during fabrication. Rather, a collection of filaments (in the form of tows, yarns, bundles, etc.) is used in a fibrous assembly or in fabric form. The unidirectional architecture is the principal building block of a set of oriented, and non-woven, fabrics. In this form roving, yarn or tow, with fibers oriented parallel to each other are held in place with the use of transversely oriented threads which either stitch the assembly together or are bonded to the assembly using a heating process. Although the 'stitch-bonded' assemblies are easier to handle the points of bond serve as local points of weakness and crack initiation in the composite since, in general, a good bond with the matrix is not achieved at these sites. In most cases of rehabilitation, unidirectional textile fabrics are the preferred form of reinforcement.

Woven fabrics, which are used only in specific cases, are formed by interlocking two or more sets of bundles at prescribed oriented. Biaxial weaves (where the sets of bundles are perpendicular to each other) and triaxial weaves are the most common forms of this type. Although plain weaves, wherein bundles alternate in the weave pattern, are the most common form of assembly, other architectures of yarns perpendicular to each other, yet interwoven at specified intervals (every 3rd, 4th or 5th bundle for example) are also used as needed for specific property ratios and for ease of conformance. These assemblies are often known as 'satins.' Although woven structures enable a high degree of conformance and are easy to handle the interwoven structure involves undulation of fibers, which results in a loss of effective modulus. An alternative to the use of this is the use of stitched, non-woven assemblies wherein layers of unidirectionals are placed at different orientations in a stack and then stitched together.

Irrespective of fiber type and fibrous assembly used, it should, however, be noted that the final properties of FRP composites are dependent on the manufacturing process and generally have properties lower than those of structural steel (with the exception of the higher modulus carbon fibers). A generic comparison of these properties using E-glass, standard modulus carbon, and aramid fibers in epoxy and vinylester type resins is given in Table 5.2.

Table 5.2 Comparison of typical range of FRP composite properties with those of structural steel

Property	Range	Comparison with steel
Modulus	20–138 GPa	1/10th to 2/3rd that of steel
Strength	340–1700 MPa	1–5 times the yield strength of steel
Ultimate strain	1–3%	1/10th to 1/3rd that of steel
density	1.4–2.0 g/cc	4 to 6 times lighter

5.3.2 Manufacturing processes

In the context of this discussion, the term 'manufacturing processes' will be taken to imply the subset of processes used *vis-à-vis* rehabilitation of concrete using externally bonded FRP composites. Based on specifics of application it is useful to divide this into two categories: (i) processes used for the seismic retrofit/strengthening of columns, and (ii) processes used for the flexural and shear strengthening/repair of beams, girders, slabs and walls. Since processing in FRP composites cannot be considered as being distinct from aspects related to materials and configuration, the discussion in the following sections, while focused on manufacturing, will touch on these aspects as well, emphasizing the integrated nature of the three considerations.

Seismic retrofitting

Concrete columns that need to be retrofitted are commonly deficient in flexural ductility, shear strength, bar buckling restraint, and/or lap splice clamping. The use of appropriately designed jackets/wraps around the columns induces lateral confining stresses in the concrete as it expands laterally in compression as a function of the high axial compressive strains, or in the tension zone as a function of dilation of the lap splices, or through the development of diagonal shear cracks. For reinforced concrete columns, conventional retrofit measures include the external confinement of the core by heavily reinforced external concrete sections, the use of steel cables wound helically around the existing column at close spacing, which are then covered by concrete, and the use of steel shells or casings that are welded together in the field, confining the existing columns.

Although some of these methods are very effective, they are time-consuming, needing days for installation; can cause significant traffic disruption due to access and space requirements for heavy equipment; rely on field welding, the quality and uniformity of which are often suspect; and are susceptible to degradation due to corrosion. Also, due to the isotropic nature of the material in steel casings or jackets, the jacket not only provides the needed confinement, but also causes an increase in stiffness and strength capacity of the retrofitted column. Both of these attributes are not desirable, because higher seismic force levels are typically transmitted to adjacent structural elements. The use of fiber-reinforced composites not only provides a means for confinement without the attendant increase in stiffness (through the use of hoop reinforcement only, i.e., no axial reinforcement), but also enables the rapid fabrication of cost-effective and durable jackets, with little to no traffic disruption in a large number of cases. Generically, fiber-reinforced composite wraps (or jackets) can be classified into six basic categories, as a function of process and/or materials form, as shown in Fig. 5.3.

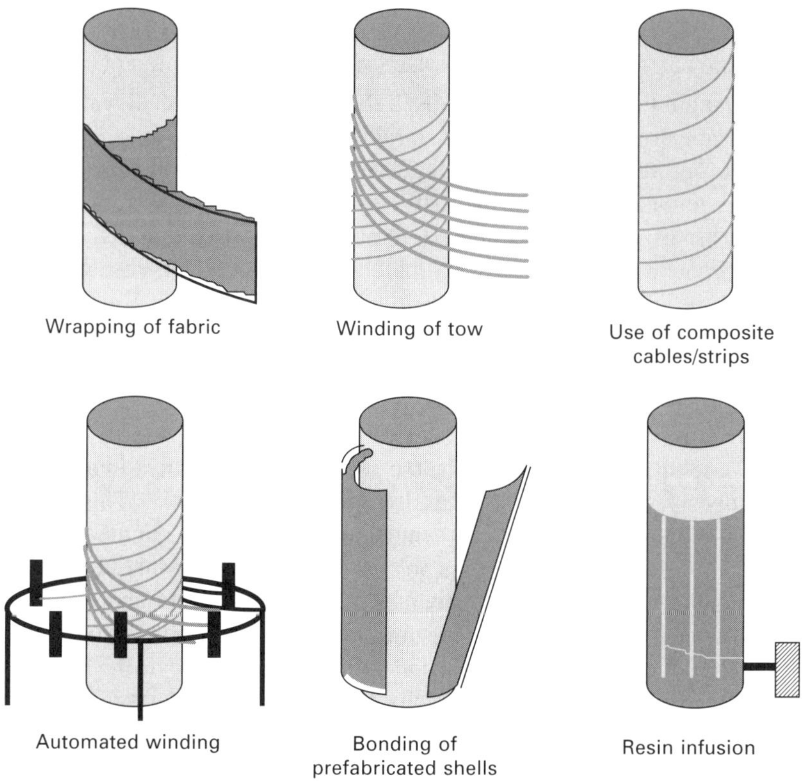

5.3 Methods of application of FRP composites for seismic retrofit and strengthening of columns.

In the wet lay-up process, fabric is impregnated on site and wrapped around a column, with cure generally taking place under ambient conditions. In wet winding, the process of fabrication is automated, but essentially follows the same schema, with the difference being that the ensuing jacket has a nominal prestress due to the use of winding tension. The wet layup process is generally associated with manual application and the use of ambient cure, although it is possible to heat the system after application to achieve higher cure temperatures and hence higher T_g (glass transition temperature). The process affords considerable flexibility of use especially in restricted spaces, but there are concerns related to the quality control of the resin mix, attainment of good wet-out of fibers with uniform resin impregnation without entrapment of excessive voids, good compaction of fibers without excessive wrinkling of the predominantly hoop directed fibers, control of cure kinetics and achievement of full cure, and aspects related to environmental durability during and after cure. In the case of wet winding of tow or tape, the process may be automated, although resin impregnation is still through the use of a wet bath and/or spray, and many of the concerns are the same as those described for the wet layup process.

The use of prepreg material generically entails an elevated cure, with the winding process for tow and tape being automated, and the fabric process being manual in terms of lay-down. Both these variants use an elevated cure and hence concerns related to T_g and durability are assuaged, to an extent, as long as appropriate fabrication techniques are followed and anhydride-based systems are not used (due to the moisture sensitivity of these systems during cure). The use of a prepreg based system potentially provides a higher level of quality control over the incoming material, albeit at added cost, but also necessitates the adherence to a good cure schedule without which all advantages gained through material form are likely to be lost. Further, the necessity of curing under elevated temperatures (usually in the range of 175–300 °F) can cause problems if the substrate concrete is very moist resulting in water vapor driven blistering during the curing of the FRP composite jacket.

In the case of prefabricated shells, the sections are fabricated in a factory and are adhesively bonded in the field so as to form the jacket. This can be done using either two sections or a single section, which is stretched apart, so as to encapsulate the column. This process affords a high level of incoming product quality control due to controlled factory-based fabrication of the shells. However, the efficiency and durability of the system rests on the ability of the adhesive to transfer load, and hence is dependent on the integrity of the bond which is constructed in the field and on the durability of the adhesive in a harsh and widely varying environment which could include excessive moisture (or immersion in water, in the case of a flood plain) and large temperature gradients. Typically the adhesives used to date in such applications are cured under ambient conditions and are either polyurethanes or epoxies. It should be noted that polyurethanes generically contain moisture sensitive components and may undergo reversion in the presence of heat and/or moisture. Although they have good properties at low temperatures they typically show significant reductions in shear performance at elevated temperatures. Epoxies in comparison have a higher range and show less creep and shrinkage, although they have a lower overall peel strength. Unfortunately, to date, very little attention has been paid to the temperature and moisture related deleterious effects on bond performance. It should be noted that degradation of the adhesive either due to excessive moisture absorption or due to combined effects of thermal cycling and moisture absorption can cause debonding of the layers resulting in failure of the overall system, even though there may in fact be very little degradation in the actual factory fabricated FRP composite shells themselves.

In the case of resin infusion, the dry fabric is applied manually and resin is then infused using vacuum with cure being under ambient conditions. The last method, involving the use of cables or prefabricated strips, has not been investigated to a large extent as yet.

Strengthening/repair

Degradation due to corrosion of steel reinforcement, spalling of concrete cover, extensive cracking of concrete due to excessive carbonation and/or freeze-thaw action, effects of alkali-silica reaction, and rapidly changing traffic needs (both in terms of number of vehicles and load levels) have created a critical need for methods of repair and strengthening of beams and girders, as well as deck soffits and bridge deck slabs. The need for increasing load capacity/rating of bridges due to requirements for uniform postings over a region, such as in Europe, or the need to enhance capacity to ensure safe carriage of increased truck axle weights, as in Australia and Japan, provide significant impetus for the adoption of technology that is capable of achieving this both in a cost-effective manner and without significant distress to traffic. In buildings, materials degradation, changing need of building occupancy (e.g., residential to office), and upgrades to facilities (e.g., cutting of floor slabs in older buildings to facilitate the construction of shafts for elevator installation) necessitate the strengthening of existing slabs and beams.

Since the composite element is bonded onto the concrete substrate, the efficacy of the method depends on the combined action of the entire system with emphasis on the integrity of the bond and the interface layers. Thus the composite-adhesive/resin-concrete system has to be considered as a complete system and materials aspects of each of the constituents, and interactions thereof between themselves and with the external environment could have a significant effect on overall efficiency, renewal capacity, and durability.

The 'plate' (of external reinforcement) can itself be fabricated in three generic ways (Karbhari and Zhao, 1998) using adhesive bonding of prefabricated elements, wet layup of fabric, or resin infusion. Of these, the premanufactured alternative (with the strip or plate fabricated through FRP composite processes such as pultrusion, pullforming, or even wet layup using a vacuum bag and controlled factory conditions) shows the highest degree of uniformity and quality control, since it is fabricated under controlled conditions.

In the adhesive bonding approach, the composite strip/panel/plate is prefabricated and then bonded onto the concrete substrate using an adhesive under pressure. Application is rapid, but the efficiency in use is predicated by the use of an appropriate adhesive and through the achievement of a good bond between the concrete substrate and the composite adherend. Care must be taken to ensure that the adhesive is chosen to match as closely as possible both the concrete and composite in relation to their elastic moduli and coefficients of thermal expansion, while providing an interlayer to reduce mismatch induced stresses. Aspects related to environmental effects follow those discussed earlier. Although strips and plates can be premanufactured using a variety of fabric forms, commercially available strips are currently fabricated using unidirectional carbon fiber reinforcement, which is pultruded to preset thicknesses and widths.

The use of a unidirectional form does provide the maximum theoretical efficiency of the fiber, but necessitates the use of thicknesses in excess of those required on the basis of mechanical properties in order to avoid longitudinal splitting during handling, and can suffer from premature interlaminar cracking between layers of filaments. Figure 5.4 shows an example of this wherein the failure surface has moved from one within the concrete

5.4 Close-up showing splitting within the prefabricated strip.

to one within the composite strip itself. In this case the change in failure path is due to the use of a pultruded unidirectional carbon/epoxy strip as the external reinforcement, with the composite having a fiber volume fraction of about 70%. This translates to a relatively low resin content, resulting in microscopic dry spots and/or voids between individual filaments creating weak zones along which fracture can easily propagate. Further, the resin used was itself fairly brittle and hence the energetics of fracture favor a path of propagation through this region once a crack has reached the adhesive-composite interface (as from a flexural crack in the concrete that carries through the adhesive).

The wet layup process entails the application of resin to the concrete substrate followed by the impregnation of layers of fabric, which are bonded onto the substrate using the resin itself. Both the composite and the bond are formed at the same time in the field. The process affords the maximum flexibility in the field but has the disadvantage of field mixing and fabrication. There is concern related to the mixing of the resin system and the absorption of moisture and/or inclusion of impurities. Also, the process inherently bears with it the potential for nonuniform wet-out of the fabric, and wrinkling/shearing of the fabric, both of which affect final properties. The wet layup process is perhaps the most used currently and gives the maximum flexibility for field application, and is probably also the cheapest alternative. However, it presents the most variability, and necessitates the use of excessive resin, and could result in the wrinkling or shear deformation of the fabric used, decreasing its designed strengthening efficiency (Kaiser and Karbhari, 2003). Also the process carries with it the intrinsic entrapment of air voids, and the resulting potential for deterioration with time. The composite formed in the case is generally cured under ambient conditions. At present both plain-weave and unidirectional fabrics are available commercially, with specific types of unidirectionals with paper backing having been developed in Japan specifically to facilitate ease of use in rehabilitation in the field, mimicking the application of wallpaper.

The *in-situ* resin infusion method is a fairly new variant and is capable of achieving uniformity and good fabric compaction, while making it easier for the reinforcement to be placed without excessive unintended deformation. However, this scheme is difficult to apply over large areas and necessitates application of vacuum, which may be difficult to hold on a severely cracked surface. In the resin infusion process, the reinforcing fabric is first formed into a preform, which is attached to the substrate using a vacuum bag. Resin is infused into the fibrous assembly under vacuum to form the composites. As in the wet layup process, the composite and bond are formed at the same time. In a variant of the process, the outer layer of the preform assembly is actually impregnated and partially cured prior to placement in order to provide a good outer surface finish. The process also has the potential to leave dry spots or otherwise not completely fill the preform due to local irregularities, surface conditions, or the inability to hold full vacuum.

As noted earlier, the efficacy of the external composite is largely dependent on the bond between the composite and the concrete. Although significant research has been conducted in other areas of bonding (primarily associated with the bonding of metals) on the critical aspect of surface characterization and treatment, unfortunately very little attention, beyond the level of sand blasting and abrasion, has been paid as related to concrete strengthening. A number of possible variations and interfaces exist for the three cases of wet layup, adhesive bonding, and resin infusion. In most cases local irregularities and cracks in the concrete itself are first filled with a putty, paste or filler. This is likely to be a 'softer' and more compliant layer in relation to crack propagation and fracture. In the case of resin infusion, the infusing resin can itself act in a fashion similar to crack injection, filling all cracks, and hence significantly alter the substrate material. It should be noted that if the surface irregularities are deep or of nonuniform shape/configuration, some will not be filled completely when an adhesive or primer layer is the first one brushed or rolled onto the concrete surface, thereby potentially trapping air voids, hollows, or gaps, which can substantially affect both durability and fracture behavior.

In the wet layup and resin infusion methods discussed above, the function of the adhesive is taken by the resin itself with the bond to the concrete substrate being formed simultaneously with the fabrication of the composite. This is both an advantage and a disadvantage, since the elimination of the third phase in the system, the adhesive, results in the formation of fewer interfaces at which failure could occur, and also eliminates the use of a more compliant layer. The formation of cracks at these interfaces and/or the changes in properties of the adherents as a function of times and environment can result in premature failure of these rehabilitation strategies.

Lack of appropriate surface preparation can lead to defects at the bond line leading to non-optimal strengthening and even premature failure of the system (Kaiser *et al.*, 2003, Kaiser and Karbhari, 2003, Mirmiran *et al.*, 2004). This can be both time consuming and difficult in specific circumstances. In an attempt to ease this problem the use of near surface mounted (NSM) FRP provides a means of obviating this difficulty in specific instances through the placement of the FRP composite reinforcement in a groove cut into the concrete itself. The method is actually a modification of the conventional method of rebar slotting (Johnson, 1965). In the case of rebar, a groove is cut to the appropriate depth and width, into which the rebar is inserted and grouted in using an adhesive. The move to FRP reinforcement provides the advantages of corrosion resistance, light weight, tailorable thermal properties, and optimized use of the reinforcement. Since the FRP is non-corrosive the groove depth is minimized since cover requirements are alleviated. In addition the FRP composite shape can be optimized to provide maximum surface area for shear transfer (as in the case of flat reinforcement) thereby not only decreasing the level of interfacial stress but also increasing the effectiveness of stress transfer between the substrate and the FRP.

In the case of NSM FRP a slot is cut into the concrete, such that the width is about 1.5–2 times the dimension of the reinforcement, the slot is cleaned to remove debris, and then the NSM FRP reinforcement is placed in the groove with adhesive. The adhesive is required to completely encapsulate the FRP reinforcement and it provides protection from the environment and from accidental damage. If required a cosmetic surface can then be added on top. Since the contact between the FRP reinforcement and the concrete substrate is on three sides, development lengths for NSM strengthening are usually shorter than those for wet layup and external adhesive bonding, and in addition the greater operative surface area results in more efficient use of the FRP with peeling forces being less critical as compared to externally bonded FRP plates/strips. The method has been used on bridges and silos in Europe (Carolin, 2001) as well as in the United States (Alkhrdaji, 1999) and has been shown to be an effective means of rehabilitation and is often faster than the competing methods of wet layup and external bonding. The method has particular applicability for strengthening in the negative moment region where the external bonding of FRP could result in local disbondment due to compressive stresses, and on top surfaces of bridge decks where the temperature gradients seen during placement of the asphaltic wear surfaces would cause degradation of the externally bonded FRP. In the case of NSM FRP the top layer of adhesive and concrete actually shields the FRP composite and the FRP composite-concrete bonds from the higher temperatures.

Lack of uniform bond between the FRP reinforcement and concrete can cause premature failure due to peeling and disbondment and hence anchors are often used to ensure continuity of contact between the FRP and the substrate. The form of anchors can range from the use of strips of FRP placed transversely over the bonded FRP to increase the shear area in contact and the use of stirrups bonded over flexural reinforcement, also termed as 'U' wrap anchors (Spadea *et al.*, 1998), to the use of embedded bolt type systems with bearing plates (Sharif *et al.*, 1994; Garden and Hollaway, 1998), and the use of fiber anchors (Teng *et al.*, 2000; Eng, 2004; Schuman, 2004). In all these cases the 'anchors' provide further assurance of integrity of the connection and in some cases can even increase load capacity and ductility of the rehabilitation measure (Eng, 2004; Schuman, 2004).

Due to the concerns related to the time taken to effect a good bond to the substrate, as well as concerns related to overall durability, a recent method of strengthening has focused on the development of a method for the mechanical anchorage of FRP strips to the substrate using multiple small distributed powder-actuated fastners (Lamanna *et al.*, 2004). The method has seen limited application to date but has significant potential for further development especially in cases where immediate use of a rehabilitated structure may be of importance. In this sense the method has tremendous applicability for emergency repairs.

5.4 Characteristics and properties

The characteristics of FRP composites are directly derived from the constituent materials, configuration (which includes fabric architecture, layup and orientation), and processing method (Wilkins and Karbhari, 1991), and in general the characteristics of FRP composites used for strengthening follow those of the generic processes used. Tables 5.3, 5.4 and 5.5 provide typical properties for fabrics and resins used in the wet layup process, and the resulting composites, respectively. Typical properties of prefabricated strips are listed in Table 5.6. As expected, the tensile strength and modulus of the prefabricated strips is significantly higher than those from the wet layup processed composites due to the higher fiber volume fractions possible in these materials. It should also be noted that the wet layup based materials show significantly higher levels of variability, which ultimately has an effect on uniformity of the rehabilitation in the field, on the selection of design allowables for the material (Aradero *et al.*, 2005), and the resulting factors of safety which could be different from those assumed during design (Helbling *et al.*, 2005).

Since FRP composites are still relatively unknown to the practicing civil engineer and infrastructure systems planner, there are heightened concerns related to the overall durability of these materials, especially as related to

Table 5.3 Characteristics of typical fabrics used in rehabilitation

Fabric type	Style (warp/weft)	Aerial weight (g/m^2)	Weight %	Heat set	Theoretical thickness (mm)
		Glass fabric			
100	Weave (E-glass/polymide)	920	98%/2%	Yes	0.36
107	Weave (E-glass/aramid)	920	98%/2%	Yes	0.36
		Carbon fabric			
103	Weave (24 k carbon/polyamide)	610	99%/1%	Heat set	0.34
117	Weave (12 k carbon/polyamide)	300	99%/1%	Heat set	0.17
230	Weave (12 k carbon/polyamide)	230	99%/1%	Heat set	0.12

Table 5.4 Characteristics of typical resins used in rehabilitation

Characteristic/property	HEX-3R Epoxy 300	HEX-3R Epoxy 306
Service temperature range (°C)	−40 °C to +60°	−40 °C to +60°
Application temperature range (°C)	10 °C to 40°	10 °C to 40°
Shelf life (years)	2 years (unopened)	2 years (unopened)
Pot life (hours)	4 hours at 21 °C	3 hours at 21 °C
Cure time (days)	5 days at 21 °C	5 days at 21 °C
Gel time (hours)	14 hours at 21 °C	5 hours at 21 °C
Tack free time (hours)	20 hours at 21 °C	16 hours at 21 °C
Viscosity (cps)	Part A: 13000 cps at 25 °C Part B: 17 cps at 25 °C	Part A: 44000 cps at 25 °C Part B: 18 cps at 25 °C
Glass transition temperature (°C)	51 °C after 7 days at 23 °C 79 °C after 5 days at 23 °C and 48 hours at 60 °C	48 °C after 7 days at 25 °C 80 °C after 5 days at 25 °C and 48 hours at 60 °C
Tensile strength (MPa)	68	65
Tensile modulus (GPa)	3.4	3.56
Elongation at failure (%)	3.1	3.5
Flexural strength (MPa)	116	103
Flexural modulus (GPa)	3.19	2.92
Cured specific gravity	1.16	1.14

their capacity for sustained performance under harsh and changing environmental conditions under load. Although FRP composites have been successfully used in the industrial, automotive, marine and aerospace sectors, there are critical differences in loading, environment and even the types of materials and processes used in these applications as compared to the materials-process-load combinations likely to be used in civil infrastructure. While anecdotal evidence provides substantial reason to believe that if appropriately designed and fabricated, these systems can provide longer lifetimes and lower maintenance costs than equivalent structures fabricated from conventional materials. However, actual data on durability is sparse, not well documented, and in cases where available – not easily accessible to the civil engineer (Karbhari *et al.*, 2003a).

Substantial work is currently ongoing through both laboratory and field testing to assess the durability and longevity of these materials (Green *et al.*, 2000; Murthy *et al.*, 2002; Zhang *et al.*, 2003; Karbhari *et al.*, 2003b; El-Hacha *et al.*, 2004; Abanilla *et al.*, 2006). Figures 5.5(a) and (b), for example, show changes in characteristics of two column wrap systems exposed to field environments in Seattle Washington for a nominal period of exposure

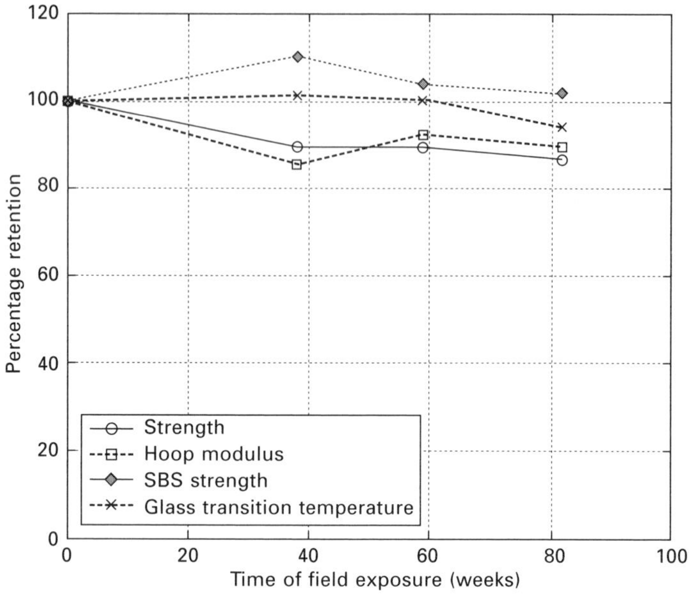

(a)

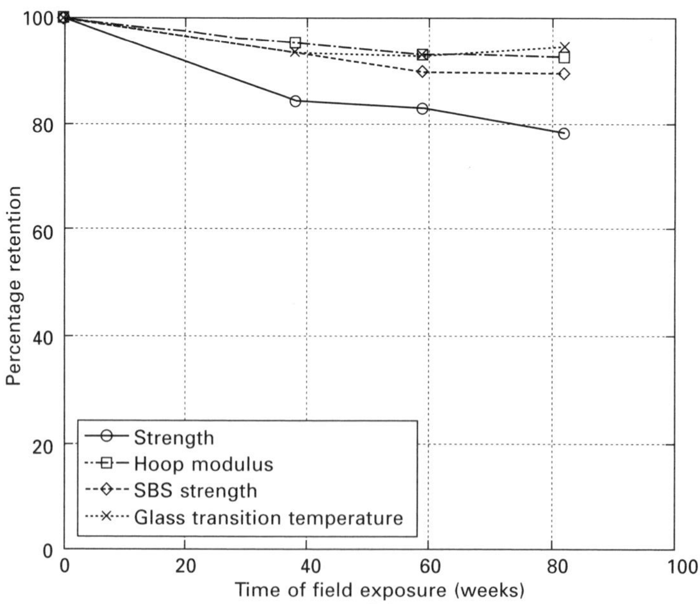

(b)

5.5 (a) Percentage retention of properties as a function of exposure time for an adhesively bonded E-glass/polyester prefabricated shell system; (b) percentage retention of properties as a function of exposure time for a wet layup based E-glass/epoxy column wrap system.

Table 5.5 Laminate properties for typical FRP composites fabricated for rehabilitation using the wet layup method (measured after cure at 70–75 °F for 5 days and 48 hours at 140°F)

Property	Type 100 E-glass fabric[1]		Type 103 carbon fabric[1]	
	Average	Design[2]	Average	Design[2]
Tensile strength (MPa)*	612	558	849	717
Tensile modulus (GPa)*	26.12	24.44	70.55	65.09
% Elongation*	2.45	2.23	1.12	0.98
Tensile strength @ 140°F (MPa)	551	531	847	699
Tensile modulus @ 140°F (GPa)	25.69	23.36	69.84	63.09
% Elongation @ 140ºF	2.28	2.14	1.13	0.97
Compressive strength (MPa)	597	542	779	715
Compressive modulus (GPa)	29.72	25.49	67.01	61.53
90º Tensile strength (MPa)	30	30	24	16
90º Tensile modulus (GPa)	6.65	6.32	4.86	3.97
90° Elongation (ultimate percent strain)	0.46	0.34	0.45	0.33
±45° In-plane shear strength (MPa)	40	34	52	46
±45° In-plane shear modulus (GPa)	2.31	2.11	2.50	2.39
Ply thickness (mm)	1.016		1.016	

*24 coupons per test series; all other values based on six coupon test series
[1]Using the resin listed in Table 5.4
[2]Design value = average value –2 (standard deviation)

Table 5.6 Typical properties of the prefabricated carbon fiber/epoxy strips used in adhesively bonded rehabilitation

Characteristic Description	Type S Standard modulus	Type M Intermediate modulus	Type H High modulus
Longitudinal tensile modulus (GPa)	165	210	300
Longitudinal tensile strength (GPa)	2.8	2.4	1.3
Ultimate strain (%)	>1.7%	>1.2%	>0.45%
Density (g/cm³)	1.5	1.6	1.6

of 18 months (Zhang *et al.*, 2003). A comparison of Fig. 5.5(a) and (b) shows that system B undergoes a higher drop in tensile strength at each of the exposure periods, although the final change in glass transition temperature is about the same. It should be noted that moisture absorption causes both a depression in glass transition temperature and changes in mechanical properties. In both systems standard deviation of results is seen to substantially increase over the second period of exposure, which can be correlated to non-uniform through-thickness moisture absorption levels and gradients. The prolonged exposure through the third period results in the attainment of a more uniform level of moisture through the thickness resulting in a decrease in scatter bounds.

Although strength depends largely on the intrinsic properties of the composite itself, in the configuration used in seismic retrofit, the hoop modulus is largely affected by inter-layer effects. Since the prefabricated adhesively bonded shell system depends on the adhesive for stress transfer between layers of the factory-fabricated shell segments the change in adhesive properties as a result of exposure causes greater flexibility and inter-layer slip resulting in a greater drop in hoop modulus over the period of exposure. This flexibilization also results in a higher apparent toughness resulting in an almost negligible change in inter-laminar shear strength as measured through short-beam-shear testing. It is noted, however, that this is an indication of adhesive deterioration, and a higher level of overall performance degradation, including disbond initiation could reasonably be expected with a high probability after further exposure, especially if a higher degree of moisture were absorbed by the adhesive.

While the use of safety factors, or partial safety factors, in approaches prescribed by ACI-440 (2002) and TR-55 (2000) provide a means of defining 'safe' levels of performance for FRP composites used in rehabilitation, they are based on factors rather than on estimates of reliability. Thus when these are combined with predictive approaches to determine lengths of 'safe' service life they can provide a false sense of security since the designer could conceivably assume that there is a zero probability (or equally low probability) of failure till the threshold value is reached. A more rigorous approach would be to estimate probability of failure using reliability methods and describing material characteristics through appropriate statistical distributions such as the Weibull distribution.

As an example, Fig. 5.6 shows changes in the probability of failure of a two-layer carbon/epoxy wet layup FRP composite immersed in deionized water at 23 °C and tested in flexure. As can be seen, the probability of failure at a specific value of flexural strength increases with time of exposure. It is of interest to assess the ACI-440 (2002) threshold value of 379.64 MPa for the design ultimate strength in this context. Taken simply as a factor of safety, this threshold is reached in 22.62 years and hence one could reasonably expect that till that point in time the composite could be safely used at levels equal to or below the threshold. It is noted, however, that at this threshold the probability of failure is 0.21 as early as five years and 0.48 at the 15 year

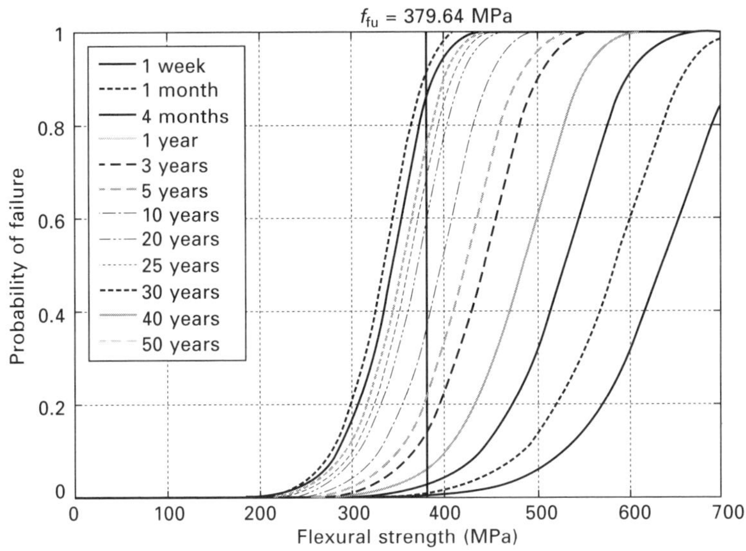

5.6 Estimation of probability of failure for flexural strength response of two layered carbon/epoxy specimens immersed in deionized water as a function of time.

level. This, then, provides a better assessment of safety through consideration of statistical variation in material characteristics and thereby allows the designer to better estimate risk and select design envelopes based on requirements of reliability. It is critical that methods like this are further developed to enable appropriate prediction of service-life of not just the FRP composite used in the rehabilitation, but also that of the rehabilitated structure itself.

5.5 Applications

From its initial development through research projects and field demonstrations, today a significant number of rehabilitation projects have been successfully completed using FRP composites bonded onto concrete substrates. In the area of seismic retrofit, these range from the retrofit of single column bents using the wet layup process to a few thousand columns on the Yolo Causeway in Sacramento, CA using adhesively bonded shells. The application of externally bonded FRP composites for strengthening has found the most applicability in the flexural strengthening of beams and girders through the bonding of FRP reinforcement to the lower surface of the beam/girder or around the web area to provide shear strengthening. A carbon-fiber strengthening system consisting of stabilized unidirectional fabric impregnated in the field using the wet layup process, for example, was used for the repair and strengthening of shear capacity of over 5000 precast prestressed double-tee beam stems at the Pittsburgh airport parking complex, consuming over 34,000 square meters of material. The project was completed in under four months with minimal closures and disruption in use of the garage, and demonstrated the significant advantages of FRP rehabilitation systems over more conventional methods. In another interesting application, over 3000 m of pultruded carbon/epoxy strips were used at LaGuardia airport in 1998 to strengthen runway slabs over Flushing Bay in Queens, NY. The strips were used to restore the ultimate strength capacity of the slabs and to stiffen them to prevent localized surface cracking and failure initiation from holes cut in the slabs for the installation of additional support piling.

The strengthening or repair of slabs is being increasingly considered to be of significant importance for the future, not just for purposes of strengthening and increase in load capacity, but also for the repair of deficient structures, such as those where local punching shear failures are seen in bridge decks between longitudinal girders. A recent example is related to the strengthening of a 103 m long, five-span bridge built over the California aqueduct in Northern California. The bridge was built in 1964 as a cast-in-place reinforced concrete T-girder bridge and due to a variety of reasons, both human error and deterioration, had undergone substantial deterioration in the form of longitudinal and transverse cracking of the deck leading to the existence of damage due to punching shear and the subsequent cracking of girders due to overall loss in stiffness. Due to the necessity of not obstructing the flow of water in the aqueduct, the need to ensure that construction material did not contaminate the water, and the need for rapidly effecting a repair without severely disrupting traffic, the use of conventional methods of repair was not deemed appropriate. Based on a detailed analysis of the bridge system, a rehabilitation scheme involving the use of composites to strengthen the slab in both the transverse and longitudinal directions was designed and installation was conducted using a set of scaffolds with sheeting in between to ensure that no debris entered the water.

Carbon/epoxy pultruded strips (51 mm wide with a thickness of 1.2 mm and having a longitudinal modulus and strength of 150 GPa and 3000 MPa, respectively), and stitch-bonded unidirectional carbon fabric impregnated in the field using an epoxy and wet lay-up (152 mm wide with an overall thickness using multiple fabric layers of 1.1 mm and having a longitudinal modulus and strength of 75 GPa and 930 MPa, respectively were used. The use of composites made it possible to complete the rehabilitation without having to remove utilities (gas pipelines) hanging from the deck, which would not have been possible if adhesively bonded or bolted steel plates or other conventional methods of strengthening had been used. Modal tests conducted both prior to, and after rehabilitation validated the strengthening scheme and it was shown that the most severely deteriorated deck slabs increased overall between 25–35% in stiffness, which more than compensated for the previous deterioration. In this case the use of composites enabled a rapid rehabilitation without disruption of both water and traffic flow, and was completed without need for removal and then re-installation of utilities.

A second typical example relates to the rehabilitation of a two-lane, 225 m long T-girder bridge built in 1969 carrying a large amount of interstate traffic between California and Nevada, including permit trucks that exceed the normal load designation. The bridge showed significant signs of deterioration due to cracking in the deck and local punching shear. Due to the high volume of traffic and its critical location there was a need to extend its service life to enable plans to be completed and funding secured for the construction of a new bridge. The over-riding requirement was that traffic not be delayed or disrupted during the rehabilitation. The constraints, including those of cost, again resulted in the selection of the externally bonded composite option as one of the most efficient. Rehabilitation was completed through the use both of prefabricated strips and fabric used, as appropriate, in different bays with the spacing and amount of material tailored to the requirements of each individual bay.

The rehabilitation operation was conducted without any change in traffic flow and was completed in a remote location (*vis-à-vis* facilities and heavy equipment) without recourse to equipment beyond manually placed scaffolds, and equipment related to fabrication and bonding of the composites (all of which could be transported in a small pickup). Based on the successful completion of the rehabilitation to a frame, the bridge was made capable of remaining in service for an extended period of time without any disruption in traffic patterns. Initial determination of service life based on methods of time dependent reliability show the effectiveness of this method. It is noted that initial studies of the effectiveness of the technique indicate that the rehabilitation costs can range between one-third and one-fifth that of a new bridge, depending on the level of rehabilitation required, while providing an extension of service life of at least 10–15 years.

Due to changing structural and/or functional requirements it is often necessary to introduce sectional openings (i.e. cut-outs) in existing slabs in buildings and industrial facilities. This can be due to the need for the installation of elevators, escalators, staircases, or even utility ducts (such as for heating and air conditioning) resulting in the creation of a cut-out which removes both the concrete and reinforcing steel. This can potentially cause a serious deficiency that has to be mitigated. Conventional methods of mitigation such as the addition of edge girders and columns at corners of the cut-outs, or the construction of load-bearing walls around the edges take up valuable floor space and in addition may not be aesthetically viable. Similar concerns are associated with the use of trusses to redistribute load. Although steel plates have successfully been used for flexural strengthening there are concerns related to corrosion at the steel/adhesive interface and at joints between plates as well as the difficulty associated with the handling of heavy plates in areas with restricted access/clearance. Also the addition of heavy plates to the tension face of slabs weakened through age and/or deterioration may not be desirable. The external bonding of FRP strips around the opening has been shown to be an effective mechanism for regain of capacity and redistribution of stress accruing from the placement of a through-thickness cut-out in concrete slabs. In addition to enabling strengthening of slab around the cut-outs the addition of the FRP strips reduces strain levels in the steel and results in the development of more desirable cracking pattern.

In addition to the uses described previously, FRP composites have also been used for the repair and strengthening of historical structures or those with local community value, due to the ease of application without causing change in the overall aesthetics and geometry of the structural components under consideration. In an interesting application in this area, three 65 m (215 ft) tall concrete smoke stacks of a cement plant in San Antonio, TX, were strengthened in 1998 using glass-fiber-reinforced polymer composites over the full height of the stacks to repair them from a structural aspect, while improving the aesthetics of a local landmark. The entire plant, with the smoke stacks as the signature elements, was renovated and converted into a shopping and entertainment complex. Similar approaches have been taken for the rehabilitation of historic buildings and bridges, enabling architectural details to be preserved while still ensuring safety and extended service-life.

5.6 Future trends

While it could reasonably be argued that composites have already found widespread acceptance in civil infrastructure applications faster than many other construction materials in the past, their full potential has still not been realized. While the lack of standards for the use of these materials on a routine basis is definitely one of the reasons for its slow acceptance, specifications and guidelines, as well as standards are being developed worldwide. The lack of long-term durability data is another obstacle, which will be overcome with time and as research in this area progresses. This data is critical, especially since in cases of rehabilitation composites are being used to extend the service-lives of already deteriorated structural systems, whereas in the case of new composite structures the initial cost is often higher than that of the incumbent formed of conventional materials and can only be justified based on more rapid installation without heavy equipment and lower life-time maintenance costs.

It must be noted that to date the materials used in civil infrastructure applications have not been tailored specifically for the environment and loading conditions prevalent. Thus, beyond the economic need for cheaper raw materials, there is a need for specially formulated resin systems that are not as sensitive to local conditions during cure, especially for those used in wet lay-up, as well as for tougher systems with higher resistance to moisture and alkalinity and higher glass transition temperatures. Although glass fibers are often the preferred form of reinforcement due to their higher levels of ultimate strain and greater impact resistance, their susceptibility to moisture, alkalinity and creep rupture makes their use in applications requiring long service life under sustained loads raises a level of concern.

The development of better sizings, leading to higher durability would be of great value. Carbon fibers, with their relative inertness to most environmental influences, and especially with the stiffness and strength attributes similar to the T700 grade, provide a good choice for rehabilitation and in new construction where stiffness is a primary concern. However, there is still need for the development of lower cost variations, and newer fibers, with higher modulus (such that the resulting composite modulus would approach that of structural steel) without the decrease in levels of strain capacity.

Besides the tremendous opportunities afforded in rehabilitation and new bridge structural systems, there are enormous opportunities, and challenges, in the development of systems capable of mitigating multiple threats. The unfortunate increase in terrorist activities has made it essential that critical structures be hardened/strengthened for blasts. Fiber reinforced composite jackets and overlays, similar to those used in seismic retrofit have already shown applicability and there needs to be further development of these technologies for the protection of bridges and buildings, both in the form of rehabilitation measures and through development of new composite components for critical areas. It is important that the inherent tailorability of composites be used to the full extent to ensure that multiple threats (such as these due to blasts, earthquakes, and hurricanes) are mitigated simultaneously without causing significant increases in size or mass of the structural elements.

With the increasing number of projects, especially in the area of rehabilitation, being completed there has been a marked increase in the level of comfort of civil engineers in using these materials. Although the durability of these materials can be significantly better than that of conventional materials it is emphasized that this is possible only through the selection of appropriate constituent materials, use of appropriate processes for manufacturing and fabrication, and incorporation of appropriate measures for environmental and damage protection. A schematic of needs from the constituent to the applications levels is shown in Fig. 5.7. It is important to emphasize that further research is needed at specific levels of materials, systems, and field implementation focusing on combined effects of stress and environmental exposure. The development of validated databases for materials sets aimed at application in a civil infrastructure environment and the achievement of high-quality products through control of processes along with the development of appropriate safety factors will result in the attainment of the potential of

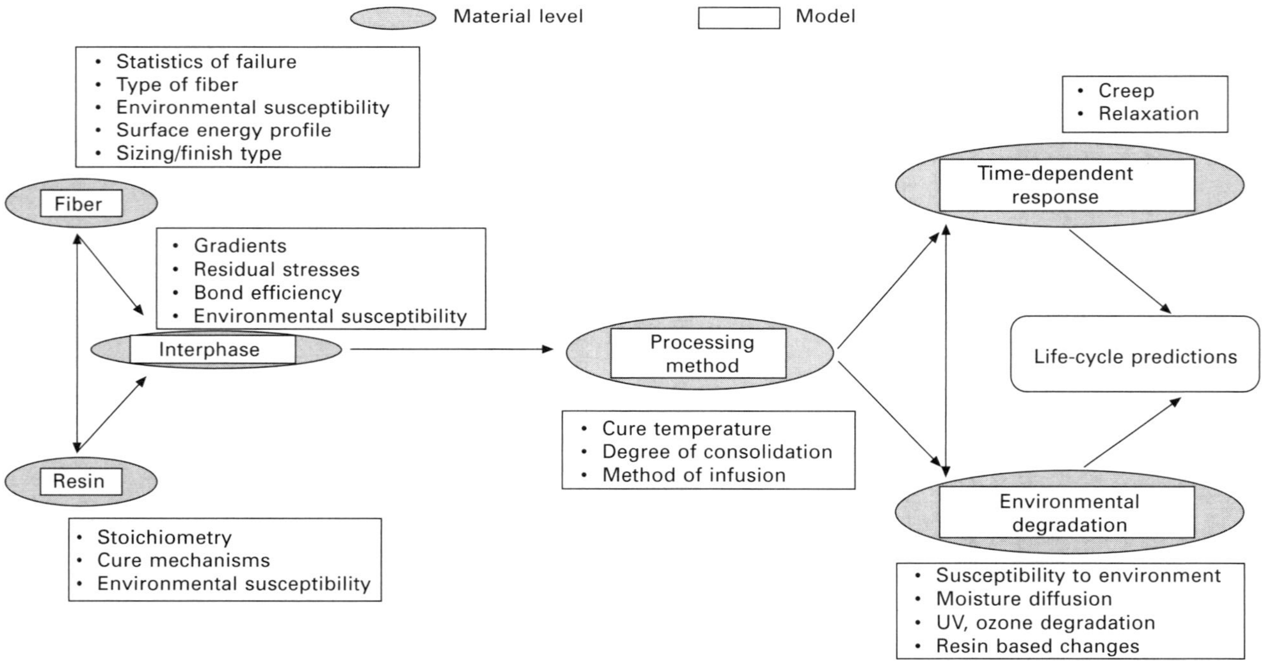

5.7 Schematic for overall evaluation of durability.

long life with low maintenance – without these the durability of FRP composites in civil infrastructure will remain both a myth and a mystery.

Civil infrastructure, with its tremendous challenges related to systems level costs (including that of long-term traffic disruption on the economy), service life, harsh and changing environments, and the need to blend aesthetics with functionality, offers opportunities hitherto not realized by FRP composite materials in other areas of application. It is perhaps in this area that the materials can fulfil their true potential as 'designer' materials while simultaneously becoming one of the palette of routinely used construction materials. It must be remembered that the ancient Egyptians realized the efficiency of composite materials in reinforcing mud bricks with straw – centuries later we have an opportunity to truly implement that concept.

5.7 References

Abanilla, M.A., Li, Y. and Karbhari, V.M. (2006) 'Durability Characterization of Wet Layup Graphite/Epoxy Composites Used in External Strengthening,' *Composites B*, 37(2/3), pp. 200–212.

Alkhrdaji, T., Nanni, A., Chen, G. and Barker, M. (1999), 'Upgrading the Transportation Infrastructure: Solid RC Decks Strengthened with FRP,' *Concrete International*, American Concrete Institute, Vol. 21, No. 10, October, pp. 37–41.

American Concrete Institute, (2002), '*Guide for the Design and Construction of Externally Bonded FRP Systems for Strengthening Concrete Structures*, ACI Committee 440, Report 440.2R-02, American Concrete Institute, Farmington Hills, MI.

Aradero, R., Lee, L. and Karbhari, V.M. (2005) 'Consideration of Material Variability in Reliability Analysis of FRP Strengthened Bridge Decks,' *Composite Structures* 70(4), pp. 430–443.

ASM Handbook Volume 21 Composites (2001), ASM International, Materials Park, OH.

Bacon, R. (1960), 'Growth, Structure and Properties of Graphite Whiskers,' *Journal of Applied Physics*, 31, pp. 283–290.

Bacon, R. and Smith W.H. (1965), *Proceedings of the 2nd Conference Ind. Carbon and Graphite, Soc. Chem. Ind*, pp. 203.

Bacon, R. and Tang, M.M. (1964), 'Carbonization of Cellulose Fibers, I,' *Carbon*, 2, pp. 211.

Blades, H. (1973), U.S. Patent 3, 767, 756.

Bresson, J. (1971), Nouvelles Rechershes et Applications Concernant l'Utilisation des Collages Dans les Structures, Béton Plaqué, Annales de l'Institute Technique dn Batiment et des Travaux Publics, No. 278, pp. 21–55.

Carolin, A. (2001), *Strengthening of Structures With CFRP*, Licentiate Thesis, Lulea University of Technology, Sweden.

Chang, K.K. (2001), 'Aramid Fibers,' in *ASM Handbook, Volume 21, Composites, Volume Chairs*: D.B. Miracle and S.L. Donaldson, ASM International, pp. 41–45.

Dussek, I.J. (1974), 'Strengthening of Bridge Beams and Similar Structures by Means of Epoxy-Resin-Bonded External Reinforcement,' *Transportation Research Record*, 785, pp. 21–24.

Edison, T.A. (1880), U.S. Patent 233, 898

El-Hacha, R., Wight, R.G. and Green, M.F. (2004) 'Flexural Behaviour of Concrete

Beams Strengthened with Prestressed CFRP Sheets subjected to Sustained Loading and Low Temperature,' *Canadian Journal of Civil Engineering,* April, Vol. 31, No. 2, pp. 239–252.

Eng, C. (2004), *The Use of MechanicalAnchorage in FRP Strengthening of Reinforced Concrete Beams*, Masters Thesis, Department of Structural Engineering, University of California San Diego.

Garden, H.N. and Hollaway, L.C. (1998), 'An Experimental Study of the Influence of Plate End Anchorage of Carbon Fibre Composite Plates Used to Strengthen Reinforced Concrete Beams,' *Composite Structures*, 42, pp. 177–188.

Green, M., Bisby, L., Beaudoin, Y. and Labossière, P. (2000), 'Effect of Climatic Conditions on the Bond Durability Between FRP Plate Reinforcements and Concrete,' *Canadian Journal of Civil Engineering*, 27(5), 949-959 (CJCE Special Issue on ISIS-Canada, October 2000).

Gupta, P.K. (1988), 'Glass Fibers for Composite Materials,' in *Fibre Reinforcements for Composite Materials*, editor A.R. Bunsell, Elsevier Publishers, pp.19–72.

Helbling, C., Abanilla, M., Lee, L. and Karbhari, V.M. (2005, in press) 'Issues of Variability and Durability Under Synergistic Exposure Conditions Related to Advanced Polymer Composites in the Civil Infrastructure,' *Composites A*.

Johnson, S.M. (1965), *Deterioration, Repair and Maintenance of Structures*, McGraw Hill Book Co., New York.

Jones, R.M. (1999), *Mechanics of Composite Materials*, 2nd edn, Taylor and Francis, Philadelphia, PA.

Kaiser, H. and Karbhari, V.M. (2003), 'Identification of Potential Defects in the Rehabilitation of Concrete Structures With FRP Composites,' *International Journal of Materials and Product Technology*, 19[6], 2003, pp. 498–520.

Kaiser, H., Navada, R. and Karbhari, V.M. (2003), 'A Fracture Mechanics Approach to Determination of the Effect of Defects in FRP Strengthening of Concrete,' *Proceedings of the 48th International SAMPE Symposium*, Long Beach, CA, pp. 1566–1580.

Karbhari, V.M. (1996), 'Issues in Joining of Composites to Concrete – Rehabilitation and Retrofit,' *Proceedings of the Composites '96 Manufacturing and Tooling Conference*, Anaheim, CA, Society of Manufacturing Engineers, pp. 345–363.

Karbhari, V.M. and Wilkins, D.J. (1993), 'Development of Composite Materials and Technology for Use in Bridge Structures,' presented at and published in the *Proceedings of the Symposium on Practical Solutions for Bridge Strengthening and Rehabilitation*, Ames, IA, pp. 211–219.

Karbhari, V.M. and Zhao, L. (1998), 'Issues Related to Composite Plating and Environmental Exposure Effects on Composite-Concrete Interface in External Strengthening,' *Composite Structures*, Vol. 40[3/4], pp. 293–304.

Karbhari, V.M., Chin, J.W. and Reynaud, D. (2001), *Gap Analysis for the Durability of Fiber Reinforced Polymer Composites in Civil Infrastructure,* American Society of Civil Engineers, Reston, VA.

Karbhari, V.M., Chin, J.W., Hunston, D., Benmokrane, B., Juska, T., Morgan, R., Lesko, J.J, Sorathia, U. and Reynaud, D. (2003a), 'Durability Gap Analysis for FRP Composites in Civil Infrastructure,' *ASCE Journal of Composites for Construction*, 7[3], pp. 238–247.

Karbhari, V.M., Zhang, J., Wu, L. and Reynaud, D. (2003b), 'Materials and Durability Characterization of Composite Wrap Systems Using NOL Ring Tests,' in *Composite Materials: Testing and Design Fourteenth Volume, ASTM STP 1436*, C. E. Bakis, ed., American Society for Testing and Materials, West Conshohocken, PA, pp. 151–165.

Kwolek, S.L. Morgan, P.W., Schaefgen, J.R. and Gulrich, L.W. (1977), *Macromolecules*, 10, pp. 1390–1396.

Lamanna, A.J., Bank, L.C. and Scott, D.W. (2004), 'Flexural Strengthening of Reinforced Concrete Beams by Mechanically Attaching Fiber-Reinforced Polymer Strips,' *ASCE Journal of Composites for Construction*, 8(3), pp. 203–210.

L'Hermite (1967), 'L'application des Colles et Resines dans la Construction,' *La Betong a Coffrage Portant*, Annales de L'institute Technique, No. 239.

Mallick, P.K. (1994), *Fiber-Reinforced Composites. Materials, Manufacturing and Design*, 2nd edn, M. Dekker, New York, NY.

McKenna, J.K. and Erki, M.A. (1994), 'Strengthening of Reinforced Concrete Flexural Members Using Externally Applied Steel Plates and Fibre Composite Sheets – A Survey,' *Canadian Journal of Civil Engineering*, Vol. 21(1), pp. 16–24.

Meier, U., Deuring, M., Meier, H. and Schwegler, G. (1992), 'Strengthening of Structures With CFRP Laminates, Research and Applications in Switzerland,' *Proceedings of Advanced Composite Materials in Bridges and Structures*, ACMBS 1, Canadian Society for Civil Engineering, pp. 243–251.

Mirmiran, A., Shahawy, M., Nanni, A. and Karbhari, V (2004), *Bonded Repair and Retrofit of Concrete Structures Using FRP Composites*, NCHRP Report 514, Transportation Research Board, Washington, DC.

Murthy, S., Myers, J.J. and Micelli, F. (2002), 'Environmental Effects on Concrete-FRP Bond under Various Degrees of Sustained Loading', Proceedings of the Second International Conference (CDCC 02) *Durability of Fiber Reinforced Polymer (FRP) Composites for Construction*, editors, Brahim Benmokrane, Ehab El-Salakawy, Montreal, Quebec, pp. 333–346.

Otani, S., Watanabe, S. Orion, H., Iikma, K. and Koitabashi, T. (1972), 'High Modulus Carbon Fibers from Pitch Materials,' *Bulletin of the Chemical Society of Japan*, Vol. 45, p. 3710.

Priestley, M.J.N., Seible, F. and Calvi, G.M. (1996), *Seismic Design and Retrofit of Bridges*, Wiley-Interscience, John Wiley & Sons, Inc., New York, NY.

Pritchard G. (2000), *Reinforced Plastics Durability*, Woodhead Publishing Limited, Cambridge, England.

Schuman, P.M. (2004), *Mechanical Anchorage for Shear Rehabilitation of Reinforced Concrete Structures With FRP: An Appropriate Design Method*, Doctoral Dissertation, Department of Structural Engineering, University of California San Diego.

Sharif, A., Al-Sulaimani, G.J., Basumbul, I.A., Baluch, M.H. and Ghaleb, B.N. (1994), 'Strengthening of Initially Loaded Reinforced Concrete Beams Using FRP Plates,' *ACI Structural Journal*, 91[2], pp. 160–168.

Shindo, A. (1961), Report No. 317, Government Industrial Research Institute, Osaka.

Spadea, G., Bencardino, F. and Swamy, R.N. (1998), 'Structural Behavior of Composite RC Beams With Externally Bonded CFRP,' *ASCE Journal of Composites for Construction*, 2(3), pp. 132–137.

Springer, G.S. (1981), *Environmental Effects on Composite Materials*, Vol. 1, Technomic Publishing Co., Lancaster, PA.

Taljsten, B. (1994), *Plate Bonding, Strengthening of Existing Concrete Structures With Epoxy Bonded Plates of Steel or Fibre Reinforced Plastics*, Doctoral Thesis, Lulea University of Technology.

Teng, J.G., Lam, L., Chan, W. and Wang, J. (2000), 'Retrofitting of Deficient RC Cantilever Slabs Using GFRP Strips,' *ASCE Journal of Composites for Construction*, 4(2), pp. 75–84.

TR-55 (2000), *Design Guidance for Strengthening Concrete Structures Using Fibre Composite Materials*, The Concrete Society, UK.

Watt, W., Phillips, L.N. and Johnson, W. (1986), *Engineer* London, 221.

Wilkins, D.J. and Karbhari, V.M. (1991), 'Concurrent Engineering for Composites,' *International Journal for Materials and Product Technology* 6 (3), pp. 257–268.

Zhang, J.S., Karbhari, V.M., Wu, L. and Reynaud, D. (2003), 'Field Exposure Based Durability Assessment of FRP Column Wrap Systems,' *Composites B*, Vol. 34(1), pp. 41–50.

6

Advanced engineered wood composites for use in civil engineering

H J DAGHER, University of Maine, USA

6.1 Introduction

Advanced engineered wood composites (AEWC) refer to a new class of materials resulting from the combination of wood with fiber reinforced polymers (FRP). These materials couple the advantages of wood which include high performance/cost and strength/weight ratios with the advantages of FRP which include high strength and stiffness and versatility.

FRPs are a versatile material system consisting of (i) synthetic fibers including glass, carbon, graphite and aramid in forms such as continuous rovings, chopped strands, mats, woven rovings or textiles, and (ii) a thermoset or thermoplastic polymer matrix which serves to bind the fibers together, transfer loads to the fibers, and protect them against environmental attack. The versatility of the constituent materials, the various fabrication techniques, and the ability to tailor the materials for the required application make FRP composites appealing as a reinforcement for wood.

Wood is a versatile structural material which can be modified into a variety of forms including lumber, veneers, flakes, strands or fibers. These in turn are combined with thermoset resins to form a variety of structural wood composites such as glulam beams, structural composite lumber (SCL), I-joists and a number of panel products. In North America, more wood products are used in construction than all other construction materials combined.

Combining two materials with compatible and complementary physical and mechanical properties can revolutionize construction techniques. In the mid 19th century, reinforcing concrete with steel significantly changed building and bridge construction throughout the world. The keys to the continued success of reinforced concrete are the high compressive strength and the low cost of concrete, the high tensile strength of steel which compensates for the low tensile strength of concrete, the bond strength between the two materials, and the compatibility between the thermal expansion properties of steel and concrete. Basic science and engineering research to improve the performance of reinforced and prestressed concrete is still ongoing today, more than a century after Joseph Monier of France made the first practical use of reinforced concrete.

6.1.1 Enabling advances in science and engineering

As the 21st century begins, many of the factors that contributed to the success of reinforced concrete are found in reinforcing wood with fiber reinforced polymers (FRPs). Lower grades of wood have high compressive strength and low cost; FRPs have high tensile strength which compensates for the lower tensile strength of (low-grade) wood; and FRPs are a very flexible class of materials which can be engineered to ensure compatibility with the properties of wood. Studies in the past 15 years by the University of Maine (UMaine) and others have shown the significant promise of combining wood and FRP. The UMaine studies have demonstrated, for example, that E-glass FRP reinforcement in the order of 3% can increase the strength of wood beams by over 100%.

Much of the published work on FRP reinforced wood has focused on short-term response of mostly reinforced rectangular glued laminated beams (glulam). A basic understanding of FRP-wood bond issues, optimum FRP composition for compatibility with wood, optimum FRP-wood structural member geometries and material properties, and modeling of long-term behavior are all fundamental science and engineering issues that must be addressed in any design. As in the development of reinforced and prestressed concrete, basic engineering and material science research are needed to unlock the full potential of a wide variety of FRP reinforced wood structural members, e.g., beams, columns, panels and connections.

The combination of wood and FRPs can lead to products which bridge the high-performance–high-cost FRP domain and the moderate-performance–low-cost wood domain. The resulting AEWC hybrids, with appropriate processing technologies, will yield a vast array of structural shapes with higher performance capacities and numerous applications. We expect AEWCs to compete not only with conventional engineered wood composites but with steel and concrete systems in structural applications.

6.1.2 AEWC significance

AEWC materials will have a significant impact on the way wood is used for construction in the 21st century. There are many engineering, economic and environmental reasons to combine FRP and wood. These are to:

1. Increase strength and stiffness.
2. Increase ductility which provides a safer failure mechanism.
3. Improve creep characteristics.

4. Reduce variability in mechanical properties which allows for higher design values.
5. Reduce volume effect in glulam girders.
6. Use of low-grade wood in construction.
7. Improve structural efficiency and reduce structural member size requirements and weight.
8. Improve wood connection properties.
9. Improve serviceability.
10. Reduce costs.
11. Reduce pressures on wood supply; because of environmental pressures and demand, the availability of quality wood needed in construction is shrinking. Reinforcing wood with FRP significantly reduces the size of the wood component and allows the use of lower quality wood. This matches the trends in resource availability of smaller diameter, lower quality wood.

6.2 Characteristics and properties

6.2.1 Lessons from the past: compatibility and durability

The idea of reinforcing wood is not new. Many studies on wood reinforcement have been performed in the past 40 years (Bulleit 1984). Often metallic reinforcement was used including steel bars, prestressed stranded cables, and stressed or unstressed bonded steel and aluminum plates (e.g. Boomsliter 1948; Mark 1961, 1963; Bohanan 1962; Sliker 1962; Peterson 1965; Lantos 1970; Krueger 1973, 1974a,b; Stern and Kumar 1973; Coleman and Hurst 1974; Hoyle 1975; Kobetz and Krueger 1976; Taylor *et al.* 1983; Dziuba 1985; Bulleit *et al.* 1989). While significant increases in strength and stiffness have been achieved, the problems encountered were generally related to incompatibilities between the wood and the reinforcing material. Wood beams reinforced with bonded aluminum sheets experienced metal-wood bond delamination with changes in moisture content of only a few percent (Sliker 1962). The differences in hygro-expansion and stiffness between the wood and reinforcing materials can lead to separation at the glue line, or tension failure in the wood near the glue line.

6.2.2 FRP versatility can overcome compatibility and durability problems

To improve durability, fiberglass has been used in a number of ways such as for beam reinforcement, as face material of wood-core sandwich panels, as external reinforcement for plywood, and in the form of prestressed strands (e.g. Wangaard 1964; Biblis 1965; Theakston 1965; America Plywood

Association 1972; Mitzner 1973; Boehme and Shultz 1974; Saucier and Holman 1975; Steinmetz 1977; Bulleit 1981; Spaun 1981; Davalos *et al.* 1992; Kimball 1995; and Dagher *et al.* 1995, 1996a, 1998, 1999). Unlike traditional steel and aluminum reinforcement, FRP reinforcement of wood composites can be successful because:

- The physical/mechanical/chemical properties of the FRP are very versatile. The FRP may be engineered to match and complement the orthotropic properties of wood; consequently, incompatibility problems between the wood and the reinforcing FRP are minimized.
- FRP materials (fibers/matrix) can be readily incorporated into many of the manufacturing processes currently used to produce structural wood composites.

Because AEWCs take advantage of the better physical and mechanical attributes of both wood and FRPs, they will result in performance/cost ratios that are superior to conventional structural wood composites. The cost/performance advantage of FRP-wood hybrids is expected to improve with time, because long-term trends indicate that the cost of wood is increasing due to rising population and demand, while FRP material costs are on the decline. FRP material suppliers are switching from low-volume, high-cost defense applications to high-volume, low-cost civil infrastructure markets. The composites industry estimates that the cost of FRP raw materials (reinforcements and resins) will increase at a rate less than inflation due to economies of scale which the industry is entering.

6.2.3 Examples of mechanical properties improvements

FRP reinforcement may be used to increase strength, stiffness, and ductility; reduce creep properties; and improve connection performance. For example, FRP-reinforced wood bending elements are obtained by reinforcing the tension and/or compression sides of solid or glued laminated (glulam) beams with fiber reinforced polymers (FRP). FRP tension reinforcement was observed to significantly increase girder bending strength (by over 100%) and to a lesser extent bending stiffness (by 10%–20%) (Dagher *et al.* 1998). Reinforcement also reduces variability in mechanical properties which allows for higher design values. As will be shown later, the volume effect in FRP-glulam beams gradually disappears with increasing tension reinforcement ratios, and the volume effect is eliminated at an E-glass tension reinforcement ratio of 2% (Lindyberg 2000). This result is particularly significant for long non-southern pine glulam girders where the volume effect may reduce bending strength by up to 30%. Tension reinforcement also increases ductility which provides for a safer failure mechanism. Ductility ratios near 3 to 1 have been calculated from beam laboratory test results (Lindyberg 2000; Dagher *et al.*

1998). With the increased strength and elimination of the volume effect, the smaller, high-performance FRP-glulam bridge girders may present a viable alternative to conventional glulams.

6.3 Applications

We will consider in further detail three applications of AEWC materials: FRP-glulam, FRP-reinforced wood-plastic composites (WPC), and FRP-reinforced oriented strand board panels for disaster-resistant housing.

6.3.1 FRP-glulams

FRP-glulams are obtained by reinforcing the tension and/or compression sides of glulam beams with fiber reinforced polymers (FRP). The FRP may be of any geometry but rectangular plates are most common. The FRP is used to increase the bending strength and stiffness of wood beams, much like steel reinforcement is used in concrete construction. Conventional glulam beams typically fail at knots or finger joints on the flexural tension side of the beam making tension reinforcement beneficial.

We start by describing the behavior of reinforced glulams through typical laboratory test results obtained at the University of Maine. Figure 6.1 compares the load-deflection performance of three glulam beams with tension reinforcement ratios ρ of 0% (control), 1%, and 3% E-glass-FRP (GFRP). The GFRP used has 120 ksi ultimate tensile strength (UTS) and a 6×10^6 psi modulus of elasticity (MOE). As in reinforced concrete, the tension reinforcement ratio ρ is defined as the cross-sectional area of the GFRP

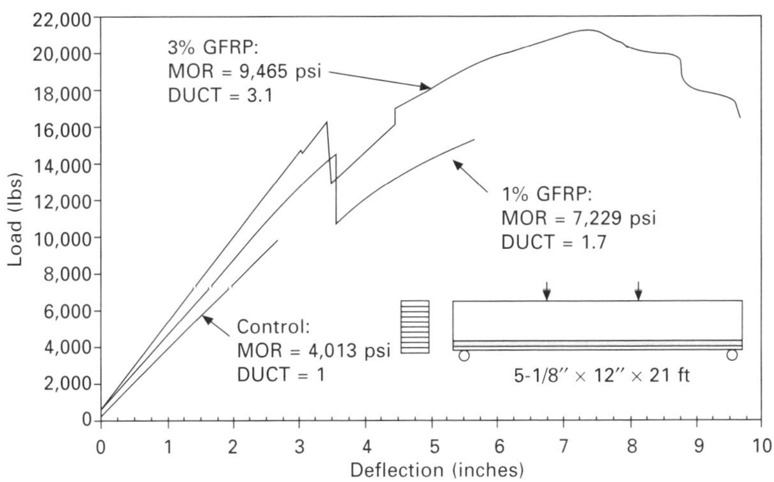

6.1 Behavior of FRP-glulam beams.

divided by the cross-sectional area of the wood above the GFRP. All three beams in Fig. 6.1 are 5-1/8″ × 12″, 21 ft long, loaded in four-point bending according to ASTM D 198 (2005b). The GFRP is placed immediately above the bottom wood lamination, herein called a bumper lam. The bumper lam serves to protect the FRP during shipping and handling and provides both UV and fire protection.

Figure 6.1 helps illustrate the behavior and benefits of tension-reinforced glulams. As the tension reinforcement ratio increases, the beam modulus of rupture (MOR) calculated using the cross-section properties also increases. The MOR is the extreme fiber stress that coincides with flexural failure. For the beam shown, the MOR more than doubles (4013 psi to 9465 psi) as the tension reinforcement ratio ρ is increased from 0% to 3%. The beam stiffness also increases with increased reinforcement, since the MOE of the GFRP is larger than the MOE of the wood. The modular ratio of GFRP/wood typically ranges from 3–6, depending on the wood properties and the fiber volume fraction of the GFRP. The beam bending stiffness also increases with increasing tension reinforcement, as may be seen by comparing the slopes of the three load-deflection curves. Gains in beam stiffness may be calculated using transformed section analysis; these gains are less than the corresponding gains in bending strength. For the beams in Fig. 6.3, gains in stiffness are approximately 15% with a reinforcement ratio of 3%.

Figure 6.1 also shows that the beam failure mode becomes significantly more ductile with increased tension reinforcement. At $\rho = 3\%$, the deflection at failure is nearly 9.75 inches (or span/26), compared with 2.75 inches for the control beam. The corresponding ductility ratio with 3% reinforcement is 3:1, compared to 1:0 with no reinforcement. The ductility ratio is defined here as the deflection at failure divided by the deflection at the end of the linear-elastic range. The increased ductility is due to a change in the failure mode from a brittle tension failure at a knot or a finger joint for conventional glulams to a compression failure of the wood for $\rho = 3\%$.

The drops visible within the load-deflection curves for $\rho = 1\%$ and 3% are due to the failure of the wood bumper lam and the displacement control used in the laboratory test protocol. After the bumper lam fails, the beams with $\rho = 1\%$ and 3% appear to lose then regain capacity, indicating that the bumper lam failure represents a serviceability limit state, rather than a strength limit state. For the beam with $\rho = 3\%$, the peak applied load is approximately 30% higher than the load corresponding with bumper lam failure. For improved structural efficiency, it is therefore preferable to find alternatives to the wood bumper lam.

Design approach and mechanics modeling

Here we describe a modeling and design approach for FRP-glulams embodied into a computer program called ReLAM for Reinforced LAMinated Beams (Lindyberg 2000). ReLAM is a numerical mechanics model which uses a moment-curvature (M-Φ) approach combined with Monte Carlo simulation to analyze a user specified glulam beam layup in four-point bending (see Fig. 6.2). The general approach of a ReLAM analysis includes:

1. The user defines a beam layup, consisting of in-grade lamstock and FRP laminations (see Fig. 6.2). The flatwise, long span bending modulus of elasticity (E), ultimate tensile stress (UTS), ultimate compressive stress (UCS), and the downslope of the bilinear compressive stress-strain curve (m) are taken from in-grade testing, and are defined as random variables with a specific distribution (e.g. normal, lognormal, 3-parameter Weibull) and parameters (e.g. mean, standard deviation, etc.).
2. Monte Carlo simulation is used to generate pseudo-random observations of the mechanical properties of each lamination.
3. A deterministic M-Φ analysis is performed using the simulated beam properties to obtain the bending stiffness and strength.
4. Steps 2 and 3 are repeated for the user-specified number of simulations.
5. Probability density functions (PDF) of beam bending strength and stiffness are obtained. The beam MOE, 5th percentile modulus of rupture ($MOR_{5\%}$) and the allowable bending stress (F_b) are calculated.

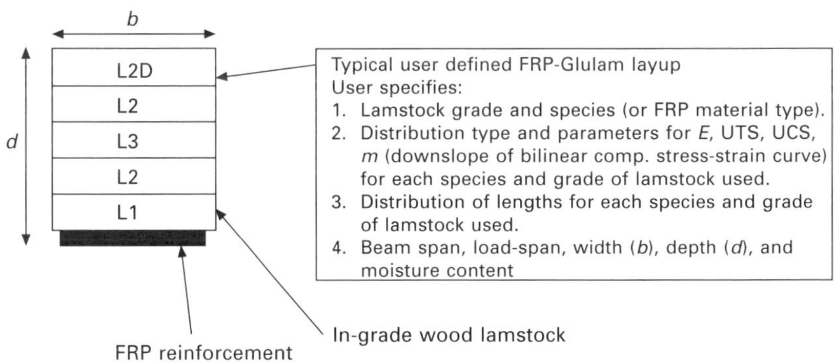

6.2 FRP-glulam cross-section analyzed by ReLAM.

Constitutive properties of wood and FRP laminations

A bilinear constitutive relationship for wood is used (see Fig. 6.3). The flatwise-bending modulus of elasticity E of the laminations is used to approximate the modulus of elasticity in both tension and compression. In tension, wood is assumed linear-elastic from zero load to the ultimate tensile stress (UTS). In compression, wood is assumed linear-elastic from zero load until the ultimate compressive stress (UCS). After UCS, stress decreases linearly to the ultimate compression strain (ε_{cult}) with a slope of E'. UCS,

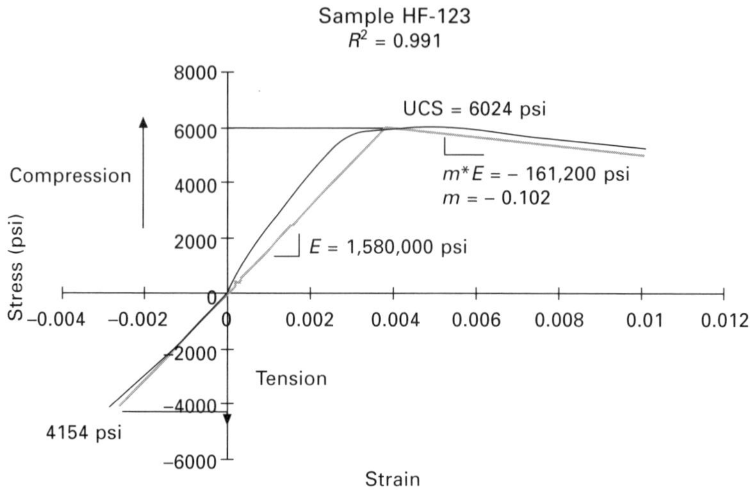

6.3 Typical bilinear constitutive relationship for wood used in ReLAM.

UTS, E and m are correlated random variables generated using statistical properties obtained from in-grade test data. The ReLAM input variable m is calculated as follows:

$$m = \frac{E'}{E} \tag{6.1}$$

The E-glass FRP constitutive relationship in tension is assumed linear-elastic-brittle. The ultimate tensile strength and modulus of elasticity are uncorrelated random variables generated using statistical properties obtained from test data.

Moment-curvature analysis

Once the constitutive relationships for each beam lamination are established, a moment-curvature (M-Φ) analysis is conducted for the beam cross-section. Figure 6.4 shows the cross-section of a beam with typical strain and stress distributions, and the neutral axis a distance Y below the top of the beam. Strain compatibility is given by:

$$\varepsilon(y) = \varepsilon_c - \frac{\varepsilon_c}{Y} * y \tag{6.2}$$

Where ε_c is the strain at the extreme compression fiber of the beam. The stress $\sigma(y)$ through the depth of the section is calculated, and stresses are checked at the top and bottom of each tension lamination. A tension lamination is considered to have failed when the tensile stress at the lamination midheight exceeds the lamination UTS.

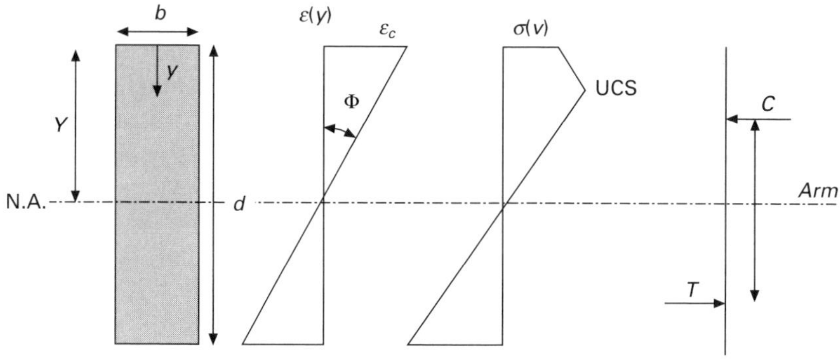

6.4 Stress and strain distribution, and horizontal forces in a rectangular cross-section.

In the compression region of the beam, stresses are checked at the top and bottom of each lamination and at the yield point (see Fig. 6.3) if the stress in a lamination exceeds the lamination UCS. The location of the yield point in the wood laminations is found using the strain distribution through the lamination thickness. Equilibrium is expressed by:

$$\Sigma F_x = 0 \Rightarrow \int_{\text{depth}} \sigma(y)\, dA = 0 \qquad 6.3$$

$$M_{\text{applied}} = M_{\text{internal}} = C \text{ (or } T\text{)} * Arm = \int_{\text{depth}} -y * \sigma(y)\, dA \qquad 6.4$$

Figure 6.5 shows the elevation of a unit length of beam subjected to a bending moment *M*. Under the applied moment, the ends of the segment rotate an angle θ, causing an extreme fiber strain equal to $2X$. The curvature (Φ) is thus given by:

$$\Phi = 2 * \Theta = \frac{2 * x}{Y} = \frac{\varepsilon_c}{Y} \qquad 6.5$$

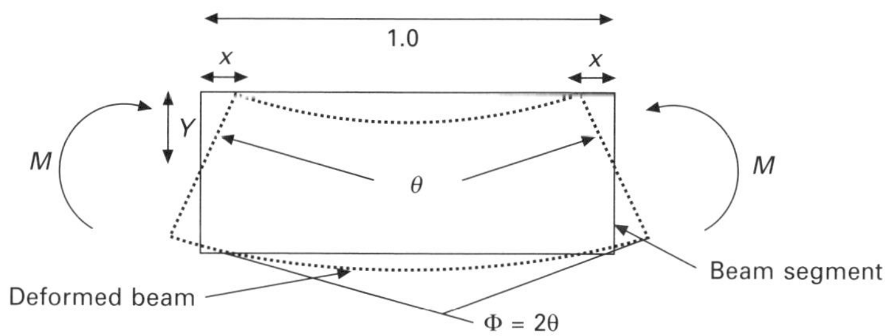

6.5 Curvature (Φ) of a unit length of beam.

For a selected value of compressive extreme fiber strain ε_c, the neutral axis location (Y) is obtained by solving a set of nonlinear equations using Newton's method (Stein, 1987). The equations include strain compatibility (eqn 6.2), the constitutive relationship (Fig. 6.3), and equilibrium (eqns 6.3 and 6.4). In this manner, the M-Φ relationship for an FRP-glulam cross-section is obtained from zero load to beam failure. In ReLAM, beam failure is defined by either one of two conditions: (i) The beam no longer carries load: $M = \int -y * \sigma(y)dA = 0$, or (ii) a specified limit strain (ε_{cult}) is reached at the extreme compression fiber.

For any value of the applied load, the beam deflection is obtained as follows. The moment diagram and the M-Φ relationship are used to obtain the beam curvature diagram. Let y be the beam deflection and x the location along the length of the beam, curvature is expressed as:

$$\Phi(x) = \frac{d^2 y}{dx^2} \qquad 6.6$$

The corresponding beam deflection is therefore (West, 1989):

$$y(x) = \iint_{\text{span}} \Phi(x)dx \qquad 6.7$$

Monte Carlo simulation

Monte Carlo simulations are used to generate correlated random variables for E, UTS, UCS, and m to account for the variability in mechanical properties of the lamstock and FRP (Marx and Evans 1986, 1988; Green and Kretschmann 1991; Taylor and Bender 1991; Dagher *et al.* 1998). The correlated random material properties used in the M-Φ model are obtained as follows:

1. Using a computer-generated array of random uniform variates, generate an array of standard normal ($N[0,1]$) variates, denoted as $\{z\}_{n\times1}$.
2. Obtain the lower triangular matrix $[T]_{n\times n}$ from the correlation matrix $[C]_{n\times n}$:

$$[C]_{n\times n} = [T] * [T]^T \qquad 6.8$$

3. Linearly combine the uncorrelated standard normal random vector $\{z\}_{n\times1}$ with $[T]$ to obtain the correlated standard normal variates $\{X\}_{n\times1}$:

$$\{X\}_{n\times1} = [T]_{n\times n} * \{z\}_{n\times1} \qquad 6.9$$

4. Transform the correlated standard normal variates (R_N) to the observations (y') from $N[\lambda_I, \xi_I]$ using eqn 6.11, and from $LN[\lambda_i, \xi_i]$ using eqn 6.11 (ASTM D 5124-96, 2001):

$$y' = \mu + \sigma * R_N \qquad 6.10$$

$$y' = \exp(\lambda + R_N * \xi) \qquad 6.11$$

5. For observations from the 3-parameter Weibull distribution, the vector of correlated standard normal variates are transformed to correlated standard uniform variates (u_i) using the inverse transform method. From this, observations from the 3-parameter Weibull distribution are calculated using eqn 6.12 (ASTM D 5124-96 2001).

$$y' = \mu + \sigma * (-L_n(u_i))^{1/\eta} \qquad 6.12$$

Length effects, finger-joints, and the laminating effect

The M-Φ analysis focuses on a single cross-section. Therefore, there is a need to account for the lengthwise variability of lamstock UTS (herein called the length effect) as well as finger joints in the tension region of the beam. There is also a need to account for the laminating effect, which is defined as the 'strength increase of lamination lumber as a result of being bonded into a glulam beam' (Falk and Colling 1995).

While E, UTS, and UCS all vary along the length of a piece of lamstock, only UTS has an established length effect, that is, the tensile strength decreases with increasing specimen length (Showalter *et al.* 1987; Taylor and Bender 1991). The following approach was developed to address the phenomenon. Lamstock UTS is first converted into a 2 ft length (adjusted from the 7 ft clear distance between the tension grips) using the length adjustment formula in ASTM D 1990-00 (2002):

$$\text{UTS}(2) = \text{UTS}(1) * \left[\frac{L(1)}{L(2)}\right]^{0.14} \qquad 6.13$$

When the M-Φ analysis begins, lamstock UTS is adjusted from 2 ft to the length of the load-span (distance between the load points in the ASTM D 198 test) using eqn 6.13. After failure of the first wood lamination in tension above the FRP reinforcement (or the first wood lamination in tension if there is no FRP), the remaining laminations are adjusted back to the 2 ft UTS. The assumption is that the first tensile failure occurs somewhere between the load heads – in the area of maximum moment – and that subsequent tensile failures of the remaining wood laminations occur in the immediate vicinity of the first tensile failure.

To address finger joints, the finger joint location in the tension region of the cross-section is determined by generating random lengths of lumber (based on the density function of lumber lengths) and assembling the simulated beam while noting the locations of the end joints. When end joints in the simulated beam occur in the region of maximum moment (between the load-heads in four-point bending, see Fig. 6.1), an ultimate tensile strength for the finger joint – UTS(FJ) – is generated based on finger joint test data provided by the laminator. This is the same general procedure used by Govindarajoo (1989) and Hernandez *et al.* (1992). In ReLAM, UTS(FJ) is treated as an independent random variable with a normal distribution.

The laminating effect (λ) was defined by Falk and Colling (1995) using the following equation:

$$\lambda = \frac{f_{b,gl}}{\text{UTS}} \qquad 6.14$$

Where $f_{b,gl}$ is the tensile stress in the beam at the extreme fiber of the bottom tension lamination at failure calculated using the gross section properties, and UTS is the ultimate tensile stress of the lamstock. When evaluated at the mean for a matched sample of beam and lamstock tests (e.g. λ = mean $f_{b,gl}$/ mean UTS), λ increases with decreasing lamstock UTS (Falk and Colling 1995). Therefore, the laminating effect is estimated in ReLAM by multiplying the distribution of lumber UTS by the factor λ. Where λ is a function of the mean UTS at a standard length (7 ft). Using beam and lamstock test data from Foschi and Barrett (1980), and Marx and Moody (1981a,b), the relationship between λ and the mean UTS was determined. Because ReLAM checks lamination stresses at the midheight of each lamination, λ was calculated from these studies using the stress at the midheight of the bottom lamination at failure, instead of the stress at the extreme tensile fiber of the beam. Only beams consisting of L1 and L3 grade laminations were considered, as higher grades do not display a laminating effect for 2 in. nominal thickness lamstock (Falk and Colling 1995). All tested beam MORs were adjusted to standard size (5-1/8 in. × 12 in. × 21 ft) using the NDS volume adjustment factor. Two sets of beams from the Marx and Moody study (1981a) were eliminated from the data set because they consisted of only two laminations and were likely not subject to the full laminating effect. The following laminating effect is used in ReLAM:

$$\lambda = -0.00073 * \text{UTS}_{\text{mean}} + 1.41354 \qquad 6.15$$

Validation of ReLAM modeling approach

Three sets of beam test data were used to validate ReLAM:

1. UMaine-1996 test program: a series of 90 (5.125 in. × 12 in. × 21 ft) FRP-reinforced glulam beams, utilizing *E*-rated low/medium grade western species, in a study undertaken by the University of Maine AEWC Center, Willamette Industries, and APA-The Engineered Wood Association in 1996 (Dagher *et al.* 1998).

2. AITC-1990 test program to determine the glulam volume effect: a series of unreinforced Douglas fir glulams, with spans ranging from 9.5 ft to 64 ft, tested to determine the glulam volume effect (Moody *et al.* 1990).
3. FHWA-2002 study: a series of eighty-five (85)36 ft and 48 ft span beams tested at the University of Maine as part of a FHWA-funded project to develop AASHTO Standards and Specifications for FRP-glulam bridges.

UMaine-1996 test program

One hundred and two (102) 22 ft long, 5.125 in. wide × 12 in. deep beams were fabricated by Willamette Industries. Half were manufactured using *E*-rated western hemlock lamstock, and the other half consisted of *E*-rated Douglas fir. All lamstock was 14 ft in length. The test beams were specifically sized to have a volume factor (C_v) of 1.0. Three GFRP reinforcement ratios (0%, 1.1%, and 3.3% by volume) were used for each species. The GFRP was placed directly above the lowest wood lamination in the beam (herein called bumper-lam) as shown in Fig. 6.6. The thickness of the bumper-lam was adjusted so that the nominal depth of each beam was approximately 12 in. A random layup of a special *E*-rated L2/L3 lamstock grade was used in all beams.

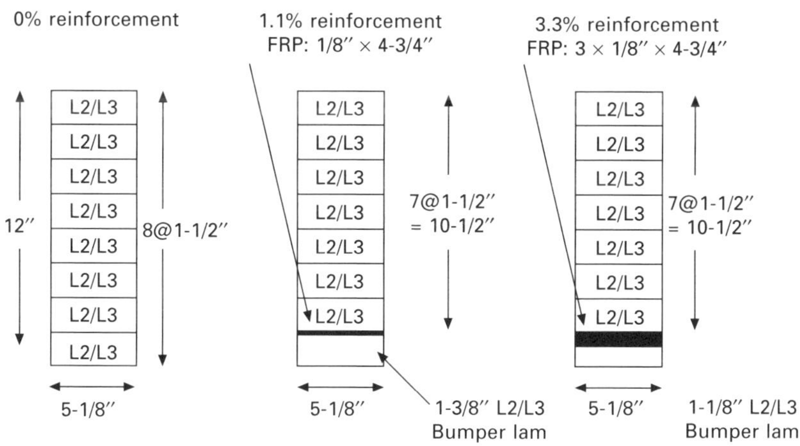

6.6 Cross-sections of control (0%), 1.1% FRP-reinforced, and 3.3% FRP-reinforced beams in the UMaine-1996 study.

Ninety (90) of the beams were tested under static four-point bending to failure, in accordance with ASTM D 198-05 (2005b). The 22 ft long beams were simply supported at a span of 21 ft (center-center of bearings) and subjected to vertical loads symmetrically applied at two points between the supports (see Fig. 6.1).

During manufacturing, lamstock was randomly sampled from the population used to fabricate the beams and set aside for testing to determine the correlated input variables (*E*, UTS, UCS, and *m*) used in ReLAM. The lamstock testing proceeded as follows: First, *E* was measured for each piece of lamstock using the flatwise center point bending test with a span of 150 in., following the definition in ASTM D 3737-04 (2004). Then, a 1 ft segment was cut from the initial piece for compression testing under ASTM D 198-05 (2005b), in order to measure UCS and *m* (see eqn 6.1). The remaining 13 ft segment was tested in tension under ASTM D 4761 (2005a) to measure UTS. The ReLAM input variables *E*, UTS, UCS, and *m* (adjusted to 12% moisture content) are listed in Table 6.1.

Table 6.1 UMaine-1996 lamstock mechanical properties (12% MC)

		Mean	St. dev.	Dist.
DF	*E*	1.54×10^6 psi	0.16×10^6 psi	Normal
	UTS(7 ft)	3097 psi	754 psi	Normal
	UTS(2 ft)*	3691 psi	899 psi	Normal
	UCS	6039 psi	842 psi	Normal
	m	0.25	0.22	Normal
	UTS(FJ)	5825psi	1002 psi	Normal
WH	*E*	1.40×10^6 psi	0.15×10^6 psi	Normal
	UTS(7 ft)	3588 psi	1141 psi	Normal
	UTS(2 ft)*	4276 psi	1360 psi	Normal
	UCS	5915 psi	756 psi	Normal
	m	0.19	0.34	Normal
	UTS(FJ)	5825psi	1001 psi	Normal

*UTS(2 ft) used as Relam input parameter.
**Laminating factor calculated from eqn 6.15, based on mean UTS(7ft).

Using the lamstock properties and correlations defined in Table 6.1, ReLAM was used to perform 2,000 beam simulations for each set of beams. ReLAM (Table 6.2) predicted the mean MOE within 1.0% over the 90 beam data set and the 5% lower exclusion limit $MOR_{5\%}$ within 1.3%.

AITC-1990 test program to determine the glulam volume effect

The test data used in this section were part of a study to determine the volume effect for glulam beams (Moody *et al.* 1990). The volume effect for glulam beams is defined as the decrease in the load-carrying capacity of the member as member size (volume) increases (AITC 1995b). This phenomenon is widely recognized to be driven by the occurrence of strength reducing

Table 6.2 ReLAM predictions versus test results (UMaine-1996 study)

Species	Reinf. ratio ρ	Sample size		Mean MOE ($\times 10^6$ psi)			$MOR_{5\%}$		
		Relam	Test	Relam	Test	Diff.	Relam	Test	% Diff.
WH	0%	2000	15	1.32	1.32	0%	3595	3240	+11%
WH	1.1%	2000	15	1.34	1.41	−5.0%	4290	4520	−5.1%
WH	3.3%	2000	15	1.48	1.51	−2.0%	5175	5150	+0.4%
DF	0%	2000	15	1.53	1.49	2.7%	3345	3490	−4.2%
DF	1.1%	2000	15	1.59	1.58	0.6%	4445	4245	+4.7%
DF	3.3%	2000	15	1.71	1.74	−1.7%	5950	5895	+0.9%
Total test beams:			90	Ave. diff:		−0.9%		Ave. diff:	+1.3%

defects (i.e. knots, pitch pockets, finger joints, etc.) in regions of high bending stress, particularly in the tensile region of the beam cross section. As opposed to the empirical AITC approach, the volume effect is calculated internally within ReLAM using basic mechanics principles. To validate the results, ReLAM was used to analyze the Douglas fir beam tests that were reported by Moody *et al.* (1990).

A total of 2000 ReLAM simulations were performed for each layup in Table 6.3. The depth, width, span, and load-span for each beam category was assumed deterministic, using the values listed in Table 6.3. The moisture content of all beams was assumed to be 12%. Table 6.4 compares the mean MOR and $MOR_{5\%}$ predicted by ReLAM with the test data published by Moody *et al.* (1990).

Table 6.3 AITC-1990 study beam test matrix (Moody *et al.*, 1990)

Beam series	Sample size	No. of laminations in layup	Nominal lumber size	Beam dimensions			
				Width (in.)	Depth (in.)	Span (ft)	Load span (ft)
IA	10	4	2 × 6	5.125	6	9.5	2
IB	10	8	2 × 6	5.125	12	19	4
IC	10	10	2 × 6	5.125	15	24	5
II	30	16	2 × 6	5.125	24	38	8
III	15	32	2 × 10	8.75	48	64	12

These results demonstrate that ReLAM was accurate in predicting both the mean MOR and the $MOR_{5\%}$ for the test beams. It should be noted that the majority (21 out of 30) of the 302-24 grade laminations used in the category II beams were described as average to high quality near the beam midspan, which possibly resulted in increased bending strength for the test beams

Table 6.4 ReLAM predictions vs. AITC-1990 unreinforced glulam test results

Beam series	Sample size		Mean MOR (psi)			COV of MOR (%)		$MOR_{5\%}$		
	Test	Relam	Test	Relam	Diff.	Test	Relam	Test	Relam	Diff.
IA	10	2000	9020*	9810	8.8%	16.1	30.0	5970*	5660	–5.2%
IB	10	2000	7110*	7070	–0.5%	17.0	24.0	4570*	4670	2.2%
IC	10	2000	6660*	6440	–3.3%	16.0	23.0	4420*	4450	0.7%
II	30	2000	6050	5320	–12%	14.5	17.0	4430	3970	–10.4%
III	15	2000	4850	4700	–3.1%	14.5	14.5	3640	3850	5.8%

* Adjusted to account for near maximum-sized defects in 302-24 tension laminations.

(Moody *et al.* 1990). Nonetheless, ReLAM maintained a high level of accuracy in predicting strength values over the entire range of beams, confirming that it accurately accounts for the volume effect in unreinforced glulam beams.

FHWA-2002 test program at umaine

The objective of the FHWA-2002 test program is to validate the ability of ReLAM to predict the volume effect for FRP-reinforced glulams. To this end, beams with 36 ft and 48 ft spans were tested in 4-point bending as shown in the test plan in Table 6.5.

When adequately reinforced in tension, glulam beam failure shifts from primarily tension, to compression yielding followed by tension/shear failure in the wood (Dagher *et al.* 1998). Therefore, the lamstock compression properties – primarily UCS – have more influence on the behavior of reinforced beams than the lamstock tension properties and the finger joints. Since UCS does not appear to be length dependent (ASTM D 1990-00 (2002)) nor finger joint dependent, ReLAM predicts that the volume effect gradually disappears with increasing tension reinforcement.

Following the same procedure used in the University of Maine 1996 study, lamstock was randomly sampled during beam fabrication and tested to determine E, UTS, UCS, and m for each grade used (L1 and L2), and each nominal lamstock dimension (2 × 6 and 2 × 8). At the time of writing, only the 2 × 6 lamstock testing had been completed (Table 6.6).

ReLAM was used to analyze the beams listed in Table 6.5. Since the 2 × 8 lamstock test data were not completed at the time of writing, the 2 × 6 lamstock properties were used as inputs for the 48 ft beams. Table 6.7 lists a comparison of ReLAM's analysis versus the beam test data, focusing on the mean MOE and the $MOR_{5\%}$.

While the FHWA-2002 lamstock and beam testing was not completed at the time of writing, the preliminary results indicate that ReLAM is accurate in the analysis of medium and long span FRP-glulams. Further, ReLAM's prediction of the eventual elimination of the volume effect with increased tension reinforcement appears to be confirmed (Table 6.7).

Summary and conclusions: FRP-glulams

A probabilistic mechanics-based numerical model for FRP-glulam beams that is currently being used to develop AASHTO bridge design specifications at the University of Maine was described. The model's 5th percentile modulus of rupture predictions were compared with those calculated from 250 laboratory beam test results. The model shows excellent agreement with the laboratory test results. The following conclusions are drawn from the laboratory testing and modeling work:

Table 6.5 FHWA-2002 test matrix for validation of ReLAM volume effect predictions

Beam series	Layup	Species	Grades		Reinf ratio ρ	Sample size	Width (in.)	Depth (in.)	Span (ft)
			Comp. zone	Core and tension zone					
1	A	Douglas-fir	L2/L3 (100%)		0%	6	5.125	21	36
2	A	Douglas-fir			0%	6	6.75	28.5	48
3	B	Douglas-fir	L1 (25%)	L2/L3	1.2%	30	5.125	21	36
4	B	Douglas-fir		(73.8%)	1.2%	21	6.75	28.5	48
5	C	Douglas-fir	L1 (25%)	L2/L3	2.4%	30	5.125	21	36
6	C	Douglas-fir		(72.6%)	2.4%	21	6.75	28.5	48

Table 6.6 FHWA-2002 lamstock mechanical properties for 2 × 6 Douglas-Fir (12% MC)

Grade	Property	Mean	St. dev.	Distribution
L1	*E*	1.90 Msi	0.19 Msi	Normal
	UTS (7 ft)	3640 psi	1225 psi	Normal
	UTS (2 ft)	4338 psi	1460 psi	Normal
	UCS	5497 psi	1110 psi	Normal
	m	0.197	0.148	Normal
	UTS(FJ)	5825 psi	1001 psi	Normal
L2/L3	*E*	1.58 Msi	0.22 Msi	Normal
	UTS (7 ft)	2346 psi	897 psi	Normal
	UTS (2 ft)	2796 psi	1069 psi	Normal
	UCS	4499 psi	1191 psi	Normal
	m	0.366	0.314	Normal
	UTS (FJ)	5825 psi	1001 psi	Normal

- GFRP tension reinforcement can significantly increases girder bending strength (by over 100%) and to a lesser extent bending stiffness (10–20%). For strength-controlled designs, reduction of structural member size requirements and weight are therefore possible. Reinforcement also reduces variability in mechanical properties which allows for higher design values.
- The volume effect in FRP-glulam beams gradually disappears with increasing tension reinforcement, and it is eliminated at an E-glass tension reinforcement ratio of approximately 2%. This result is particularly significant for long non-southern pine girders where the volume effect may reduce bending strength by up to 30%.
- Tension reinforcement increases ductility which provides for a safer failure mechanism. Ductility ratios near 3 to 1 have been calculated from beam laboratory test results.
- Reinforcement reduces pressures on wood supply and reduces the high-grade requirements for wood tension laminations. This matches the trends in resource availability of smaller diameter, lower quality wood.

The validated ReLAM numerical model is now being used to develop design tables and simplified design equations that will be included in a new AASHTO Guide Specifications for FRP-glulams. In addition to static structural design, the AASHTO Guide Specifications will include fatigue design criteria and material specifications.

Table 6.7 ReLAM predictions vs. FHWA-2002 test results

Beam series	Span (ft)	Reinf. ratio ρ	Sample size		Mean MOE ($\times 10^6$ psi)			$MOR_{5\%}$ (psi)		
			Relam	Test	Relam	Test	Diff.	Relam	Test (point estimate)	% Diff.
3	36	1.2%	2000	30	1.81	1.97	–8.1%	4760	5455	–12.9%
4	48	1.2%	2000	16*	1.82	1.90	–4.1%	5015	5445	–7.9%
5	36	2.4%	2000	30	1.93	2.1	–8.2%	5790	6160	–6.0%
6	48	2.4%	2000	9*	1.93	2.1	–8.2%	5810	6015	–3.4%
Total beams tested:				85	Average diff:		–7.2%	Average diff:		–7.6%

*Full test series of 21 beams not completed at time of writing

6.3.2 FRP-reinforced wood-plastic composites

In recent years, increased environmental concerns for forest protection coupled with environmental safety concerns and stringent regulations for wood treatment chemicals have created the impetus to find alternatives to pressure treated wood for the construction industry. Durability of wood is another concern, which arises from many factors such as fungal attacks, moisture intrusion, and mechanical degradation from creep. Research has shown that wood hybrid composites using thermoplastic polymers and wood fibers are capable of resisting both moisture and biological decay. These materials are commonly called wood plastic composites (WPC). As a result, there is a movement in both the industry and research facilities to find new ways of developing and utilizing wood plastic composites for the construction market. Although applications for WPC are many, the economic use of WPC for structural members such as beams and posts has not yet been realized.

Wood plastic composites such as extruded PVC-wood fiber lumber are suitable for marine applications because of their excellent moisture resistance and very low maintenance (Chetanachan *et al.* 2001). They can also be cut, nailed, and screwed just like wood using the same equipment used in wood construction (Patterson 2001). In this section, we describe a project where the research team at the University of Maine's Advanced Engineered Wood Composites Center (AEWC) used PVC wood plastic composite lumber to design a reinforced built-up section (RBS) beam that can be used for structural applications. Suitable adhesives were examined and shear block tests were performed to determine adhesive performance. A method of manufacturing the RBS beam was developed and the adhesive quantity along with clamping parameters was optimized to increase quick turn out (Dagher *et al.* 2003).

As with reinforced glulams described earlier, RBS design may be achieved using a nonlinear, layered moment-curvature model. The inherent low stiffness of thermoplastic polymeric materials dictates that reinforcement of the beams with high stiffness synthetic fibers is needed to achieve the desired structural strengths and meet deflection requirements.

Example: design and analysis of reinforced WPC box section

We consider a WPC beam with an open rectangular box cross section. The materials used to fabricate the beam are commercially available wood plastic composite (WPC) boards and adhesives identified as feasible and effective from preliminary bond tests.

Using conventional beam analysis, the mechanical properties listed in Table 6.8 for PVC WPC material, and the adhesive bond strengths listed in Table 6.3, different cross-sections were analyzed that consisted of $5\,\frac{1}{2}'' \times \frac{3}{4}''$ wood-plastic boards bonded with an appropriate adhesive. The final section

Table 6.8 Mechanical properties of certainteed boardwalk PVC-WPC

Mechanical property	ASTM test method	No. specimens	Mean	COV
Ultimate tensile stress (UTS)	D 638	20	1970 psi	3.6%
Tensile modulus of elasticity (MOE_T)	D 638		0.250 Msi	6.0%
Ultimate compressive stress (UCS)	D 198	6	2470 psi	3.2%
Compressive modulus of elasticity (MOE_C)	D 198		0.250 Msi	6.9%
Shear strength	D 905	11	1220 psi	9%

selected is shown in Fig. 6.7. The shear stresses at the bonded interfaces, shear stresses in the wood plastic composite, and flexural tensile stresses in the wood plastic composite are of particular concern for these box sections. Assuming four-point bending, different span lengths were analyzed to determine whether bond shear failure or wood plastic shear failure governed beam strength for each cross-section. Based on selected material properties and experimentally determined bond strengths, bond line failure should not control the strength of a well-constructed box section. Further, the computed factors of safety against bond and shear failure indicate that it may be feasible to increase flexural strength through the use of fiber reinforced composites.

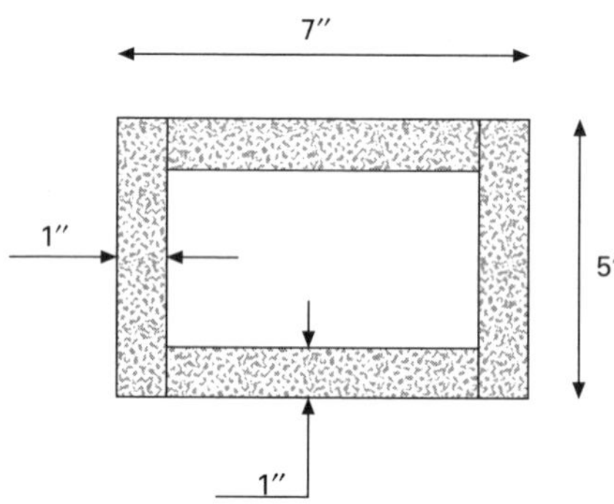

6.7 Unreinforced WPC box beam cross-section.

Mechanical properties of certainteed PVC WPC

Commercially available PVC wood plastic composite specimens were obtained and tested in tension and compression. Tension testing was performed following ASTM D 638-03 (2003), *Standard Test Method for Tensile Properties of Plastics.* Ultimate tensile stress (UTS) was calculated for each specimen by dividing the maximum tensile load by the cross-section area at the point of tensile failure. All specimens failed within the necked-down region. Tensile modulus of elasticity (MOE_T) was calculated using the initial tensile stress-strain slope. The mean and coefficient of variation (COV) for UTS and MOE_T are listed in Table 6.8.

Compression testing was performed using the short-column compression test, described in Section 12 of ASTM D 198-85a(2005b), *Standard Test Methods of Static Tests of Lumber in Structural Sizes.* This test requires that specimens have a slenderness ratio (l/r = length to least radius of gyration) of 17 or less. In order to achieve the $l/r < 17$, three boards (each 1″ × 5″ × 12″) were glued together to form a 3″ thick × 5″ wide × 12″ long specimen. Ultimate compressive stress (UCS) was calculated for each specimen by dividing the maximum compression load by the cross-sectional area of the specimen. The compressive MOE (MOE_C) was calculated using the initial compression stress-strain slope. Table 6.8 lists the mean and COV for UCS and MOE_C. The mechanical properties listed in Table 6.8 were used as inputs in preliminary modeling and RBS development.

Shear strength was measured by fabricating compression shear specimens and testing them in accordance with ASTM D 905-03 (2003), *Standard Test Methods for Strength Properties of Adhesive Bonds in Shear by Compression Loading.*

Manufacturing issues

The potential of assembling PVC wood plastic composites into larger cross-sectional elements was investigated through the use of solvent bonding techniques. For developing the bonding protocols, commercially available PVC wood plastic deck board was examined with several adhesives. To produce built-up beam sections, the PVC lumber had to be sawn and/or planed which exposes a different surface compared to the surface in contact with the extrusion die. Therefore, bonding of sawn surfaces and die surfaces were examined and it was found that sawn surfaces produced a better bond.

Shear block tests were conducted following ASTM D 905-03 (2003) in order to evaluate bond strengths for the PVC WPC to itself (WPC-WPC bonding) as well as to an E-glass fiber-reinforced polymer (GFRP) laminate reinforcements (WPC-GFRP). The WPC-WPC specimens typically failed along a line approximately 1/8 in. away from the bond line. The failure in these cases propagated deep into the blocks instead of failing along a straight plane. The WPC-GFRP specimens failed along a relatively straight plane. The failure could be approximated at 95% WPC material failure. The bond strength was high (935–1156 psi) and relatively close to the cohesive shear strength of the WPC (1200 psi) in all cases.

Reinforced box beam flexural tests

To study the effects of reinforcement on the box beams, 15 WPC box beams – each 5 in. deep × 7 in. wide × 10 ft long – were manufactured with cross-sections as shown in Figs 6.7 and 6.8. These are in addition to 13 control box beams (unreinforced). Full length GFRP reinforcement was then bonded to the WPC sections in the form of preconsolidated laminates – each 0.033 in. thick by 7 in. wide – on the tension and/or compression side using cold press techniques. Four reinforcement ratios (ρ^1) were used (see Fig. 6.8 and Table 6.9).

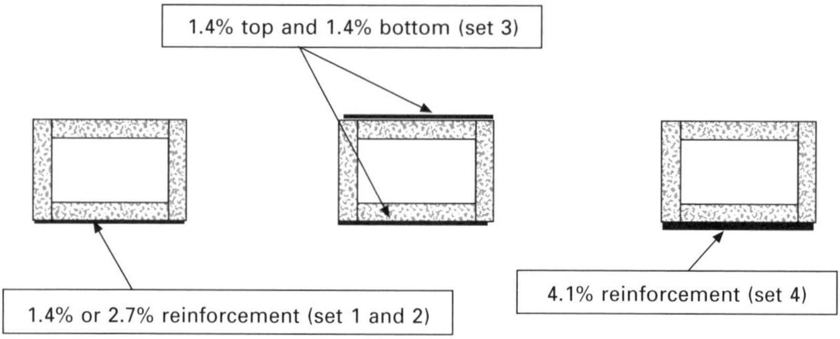

6.8 Reinforced WPC box beam cross-sections.

Table 6.9 Reinforced box beam test matrix

Beam set	$\rho_{tension}$	$\rho_{compression}$	Sample size	Number of layers 0.033″ thick GFRP laminate
1	1.4%	0%	4	1 layer in tension
2	2.7%	0%	4	2 layers in tension
3	1.4%	1.4%	4	1 layer in tension, 1 layer in compression
4	4.1%	0%	3	3 layers in tension

Both the reinforced and control (unreinforced) box beams were tested in 1/3rd-point bending, in order to determine the apparent MOE, MOR, and reinforcement bond behavior. The 1/3rd-point flexure test was conducted using a 55 kip servo hydraulic mechanical tester under constant displacement rate as required by ASTM D 198-05 (2005b).

The elastic (initial tangent) apparent MOE was calculated according to ASTM D 198-05 (2005b) flexure testing method. The MOE was calculated from the initial linear slope of the load vs. midpoint deflection curves. Since

[1] ρ = area of GFRP/area of WPC.

the WPC behaved nonlinearly, the tangent to the initial linear slope between 10% and 50% of the maximum load was used to calculate the apparent MOE, using the general formula published in ASTM D 198-05 (2005b),

$$MOE = \frac{PL^3}{56.35 * \Delta * I} \qquad 6.6$$

where I is the moment of inertia of the hollow rectangular cross section, L is the beam span (9 ft), and P/Δ is the initial linear slope of the load deflection test data.

Of the 15 reinforced WPC beams tested, only three failed prior to reaching a midspan deflection of 10 in. If the beam did not fail prior to 10 in. deflection, testing was stopped, the load at 10 in. midspan deflection was recorded as the maximum load (P_{max}), and this value was used in comparisons of beam flexural strength. For beams that failed prior to 10 in. midspan deflection, the modulus of rupture (MOR) was calculated based on the load at failure (P_{fail}) using the general formula published in ASTM D 198-05 (2005b):

$$MOR = \frac{P_{fail} * L}{6 * S} \qquad 6.7$$

where S is the section modulus of the hollow, reinforced WPC section.

Table 6.10 provides the bending stiffness (MOE) and bending strength between the unreinforced control beams, as well as each of the four sets of reinforced WPC beams. This table shows that GFRP reinforcement of up to 4.1% by volume in tension increases MOE by over 100% and bending strength by over 70%. Reinforcing in both tension at 1.4% and compression at 1.4% by volume yielded over a 100% increase in MOE and a 79% increase in bending strength. The results demonstrate that it is feasible to manufacture WPC structural beams and that the flexural properties of these beams can be substantially improved with the addition of GFRP reinforcement.

Table 6.10 Comparison of reinforced beam test results with control beam test results

Beam set	$\rho_{tension}$	$\rho_{compression}$	Sample size	Mean MOE (Msi)	% increase over control	Mean P_{max} or P_{fail} (lbs)	% increase over control
Control	0%	0%	13	0.240	N/A	3976	
1	1.4%	0%	4	0.356	48.3%	6031	51.7%
2	2.7%	0%	4	0.436	81.7%	6500	63.5%
3	1.4%	1.4%	4	0.495	106%	6806	71.2%
4	4.1%	0%	3	0.491	105%	7104*	78.7%

* Excludes specimen that failed prematurely due to GFRP delamination.

Mechanics modeling of both unreinforced and reinforced WPC sections

Using the compression and tension material level test data for the WPC (see, for example, Table 6.8), the WPC beams tested may be modeled using a moment-curvature approach as described earlier for reinforced glulam beams. Using a layered moment-curvature analysis combined with Monte Carlo simulation, a computer model called WPC-LAM was developed to predict the population distribution of bending strength and stiffness for a user specified reinforced WPC beam layup. For analysis of both the unreinforced and reinforced WPC sections, an elastic plastic approximation of the WPC stress-strain relationship was used in both tension and compression (see Fig. 6.9). The input parameters for WPC-LAM are as follows:

- Modulus of elasticity (MOE): based on the test data listed in Table 6.8, tension MOE (MOE_T) was assumed to be equal to compression MOE (MOE_C). The mean and COV for MOE_C listed in Table 6.8 was used in the WPC-LAM analysis.
- Ultimate tensile stress (UTS): the mean and COV of UTS listed in Table 6.8 was used in the WPC-LAM analysis.
- Ultimate compressive stress (UCS): the mean and COV of UCS listed in Table 6.8 was used in the WPC-LAM analysis.

All beams were analyzed in 1/3rd-point bending, with a span of 9 ft and a load span of 3 ft. Table 6.11 compares the results of the WPC-LAM analysis of the control (unreinforced) beams with the test data.

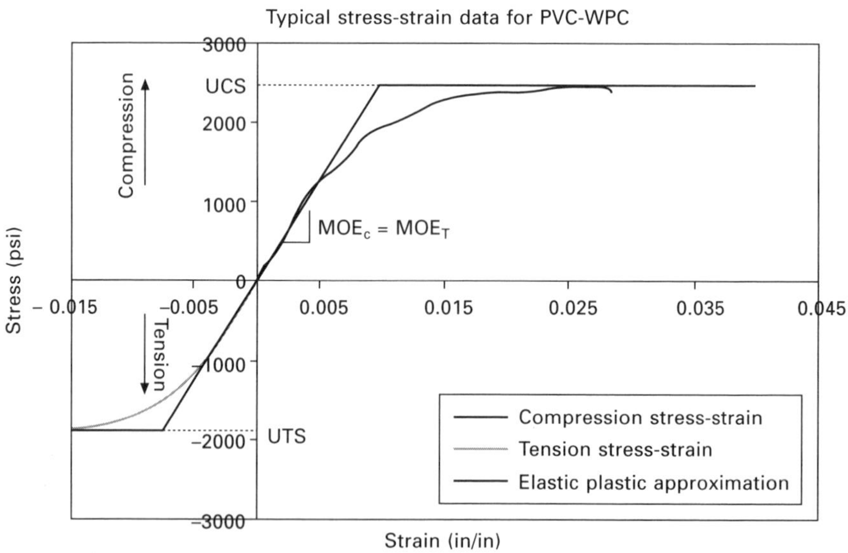

6.9 Stress-strain properties of PVC-WPC and elastic-plastic approximation used in WPC-LAM.

Table 6.11 Comparison of WPC-LAM analysis to control (unreinforced) beam test results

	MOE (Msi)		P_{fail} (lbs)		Deflection at P_{fail} (in.)	MOR (psi)	
	Mean	COV	Mean	COV	Mean	Mean	COV
Test	0.240	4.9%	3976	4.6%	6.49	2975	6.0%
WPC-LAM	0.248	4.2%	3960	3.2%	5.85	2962	3.0%
% difference	+3.3%		–0.4%		–9.9%	–0.4%	

Table 6.12 Comparison of WPC-LAM analysis to set 2 beam test results ($\rho_{tension}$ = 2.7%, $\rho_{compression}$ = 0%)

	MOE (Msi)		P_{max} (lbs) at 10″ deflection		Deflection at P_{max} (in.)
	Mean	COV	Mean	COV	Mean
Test	0.436	2.6%	6500	5.8%	10
WPC-LAM	0.420	3%	6250	5%	10
% difference	–3.7%		–3.8%		0%

The reinforced beams were also analyzed using WPC-LAM, assuming that the GFRP reinforcement was linear-elastic until failure. Table 6.12 compares the results of the WPC-LAM analysis of the Set 2 tension-reinforced beams with the test data. Figure 6.10 compares the WPC-LAM load deflection

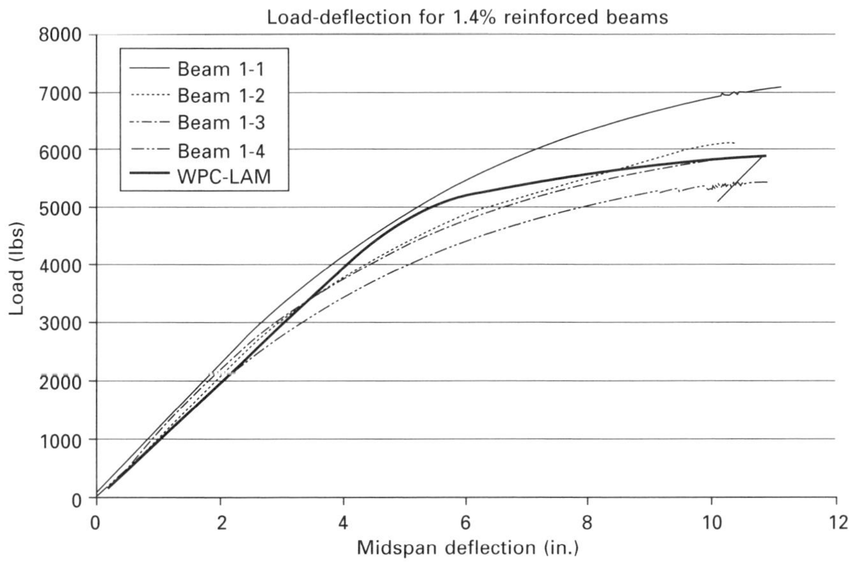

6.10 Load-deflection curves for four reinforced box beams with 1.4% tension reinforcement, and WPC-LAM predictions.

curve with the test load deflection curves for the Set 1 tension-reinforced beams.

RBS: Summary and conclusions

An example of the design and modeling of a reinforced built-up wood plastic composite section (RBS) suitable for use in structural applications was described. This included developing bonding protocols for wood plastic composite boards so that they can be assembled into a built-up section, developing a nonlinear numerical model to predict the behavior of RBS beams in flexure and confirming the model predictions through production and testing of RBS beams. Notable conclusions are:

- Structural analysis and testing of PVC-WPC box beams indicate that a rectangular box section is practical from both a structural and manufacturing viewpoint, and that it is feasible to improve the strength and stiffness of the wood plastic composite box section using advanced composites.
- Adhesive screening tests indicate that solvent bonding can produce WPC-WPC adhesive strengths approaching 1000 psi, or nearly 80% of the strength of the WPC base material.
- Shear block tests show that synthetic fabrics such as glass fiber laminates can be bonded to PVC wood plastic composites using epoxy adhesives. Bond line shear strengths of approximately 1000 psi can be achieved using this system.
- The unreinforced box beams had an average MOE of 0.241 msi, and a mean MOR of 2700 psi. The box beams with a reinforcement ratio of 1.4% GFRP on top and 1.4% GFRP on bottom had a 100% increase in MOE and approximately 80% increase in MOR. The reinforcement ratio is the cross-sectional area of the GFRP laminate divided by the cross-sectional area of the WPC
- In addition to increasing strength, tension reinforcement adds significant ductility to a WPC beam
- A computer program – called WPC-LAM – that incorporates layered, moment-curvature analysis and Monte-Carlo simulation was used both for the unreinforced and reinforced WPC beam test data. The model predicts the nonlinear load-deflection curve to failure, as well as the MOE and MOR of the beams, and the 5% lower exclusion limit (LEL) of the strength. The 5% LEL is typically used as the basis to calculate design values for wood-based composite beams.
- For the unreinforced beams, WPC-LAM predicted the mean MOE and mean MOR within 3.3% and 0.4% of the test data, respectively. For the reinforced beams, WPC-LAM predicted mean MOE within 8%, and was

extremely accurate in predicting the nonlinear load-deflection behavior of the beams from initial load through failure. With WPC lam, the tools are available to optimize reinforced WPC section shapes for various structural applications.

Finally, creep and creep-rupture issues, durability, and excessive creep deflections are major concerns for WPC beams used in structural applications. Ongoing work at the University of Maine shows that FRP reinforcement significantly reduces creep deflections in WPC beams and may ultimately provide a cost effective solution for the creep issues of WPC sections.

6.3.3 FRP-reinforced sheathing panels

According to recent United States Census Bureau data, the in-place value of housing constructed during 1998 was $189 billion, which represents 28% of the total value of all 1998 building construction (http://www.census.gov/pub/const). The vast majority of new homes are built using conventional, stick built lumber construction. As such, increasing the disaster resistance of wood framed residential construction is a critical research need.

A sufficiently strong, ductile and energy dissipating lateral force resisting system (LFRS) is crucial for adequate building performance in extreme wind or seismic events. In housing construction, the LFRS typically consists of interconnected vertical and horizontal diaphragms. In stick built construction these diaphragms are generally fabricated from oriented strand board (OSB) or plywood sheathing panels which are nailed to lumber framing. OSB use continues to increase at the expense of plywood sheathing, and today OSB has taken a significant portion of the plywood market for sheathing and flooring applications.

It is well known that the performance of wood-framed shear walls (vertical diaphragms) with panelized sheathing is highly dependent on the nail spacing at the panel edges. This section summarizes research efforts into the development and structural testing of an advanced OSB (AOSB) panel that have been conducted at the University of Maine. The AOSB panel is selectively reinforced with fiber reinforced polymer composites (FRPs) at the panel edges to increase the lateral resistance of conventional wood-framed shear walls. The AOSB panel is a sandwich panel consisting of thin outer sheets of OSB with an FRP composite sandwiched between OSB panels at the edges. The reinforcement has been designed specifically to (see Fig. 6.11):

1. Reduce nail edge tear failures around the sheathing panel perimeter. Shear wall failures are often precipitated by inadequate panel perimeter nail edge distance or a larger than allowed perimeter nail spacing. This leads to edge splitting of the sheathing panel around the nails, or 'edge tear', preventing the full shear capacity of the panel from being developed.

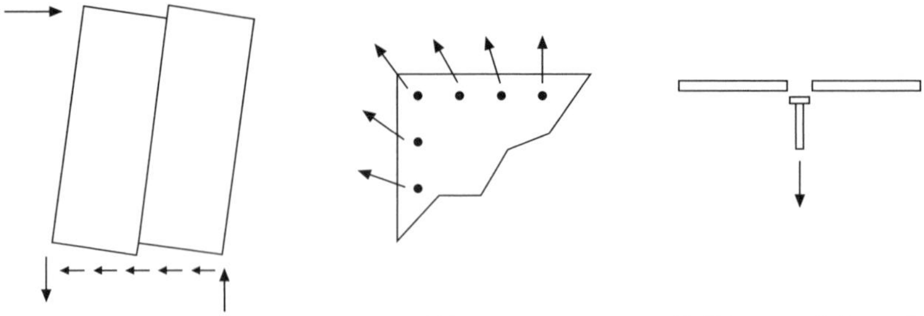

6.11 Connection failure modes in OSB sheathing panels.

2. Reduce nail head pull-through failures. Nail heads pulling through the sheathing panels have been observed in laterally loaded shear walls as well as roof diaphragms subject to negative pressure from high winds.
3. Improve energy dissipation capacity under load cycling. The tests of Shenton *et al.* (1998) showed that enlarging of nail holes due to localized crushing of panel fibers reduces the energy dissipation capabilities of shear walls after the initial load cycle.
4. Meet or exceed the weather exposure requirements of conventional OSB. Sheathing panels used in shear wall and horizontal diaphragm applications are not typically classified as exterior exposure, however they may be exposed to inclement weather during construction. Specific attention has been paid to the strength and water resistance of the adhesive bond between the OSB and FRP, to ensure this bond performs as well or better than the internal flake to flake bond of the OSB itself.

The AOSB panel directly addresses commonly observed deficiencies in the performance of conventional wood framed shear walls. It will allow significant advance in wood shear wall technologies to:

- Improve durability and thereby reduce costs. The AOSB panel is expected to reduce structural and nonstructural damage under hurricane and seismic events. This will result in lower insurance and reconstruction costs and lower overall cost of home ownership.
- Reduce the risk of life, injury, and property destruction from natural hazards. The AOSB technology will allow stronger and more energy absorbing shear walls.

Design and optimization of reinforcing

A key to the success of AOSB is selecting the reinforcing materials and associated process parameters to meet panel strength and durability criteria. A variety of resin/fiber systems were fabricated and their effectiveness tested via monotonic tests of single-nail sheathing-to-framing connections per ASTM D 1761-88 (2000) (see Fig. 6.5). Panel durability when exposed to moisture and internal shear strength were evaluated by testing both wet and dry specimens in compression shear per ASTM D 1037-99 (1999). Based on these test results, a panel consisting of two sheets of OSB sandwiching a narrow internal layer of FRP was developed to have the best combination of durability, ductility (defined as energy absorption capacity), strength, and ease of fabrication.

1. Single-connection static capacity increased by 39%. Figure 6.12 shows the smoothed average connector load displacement response of the AOSB specimens and the unreinforced 11.1 mm control specimens. The use of AOSB changes the predominant connection failure mode from edge tear to nail pullout, increasing capacity by 39%.
2. Single-connection energy-absorption increased by 351%. Cyclic connection tests were also performed with the same connection test rig. The loading protocol was developed using the recommendations of the CUREE wood-framed housing project for the control specimens. Both the conventional OSB and AOSB specimens were subjected to the same loading protocol to allow easy comparison of damage accumulation. The AOSB specimens were found to lose stiffness less quickly under load cycling and maintained significant strength and energy absorption capacity after being subjected to the full cyclic loading protocol.
3. 8 ft × 8 ft shear wall energy absorption increased by up to 73%. Following the material selection and single-nail connection tests, monotonic tests

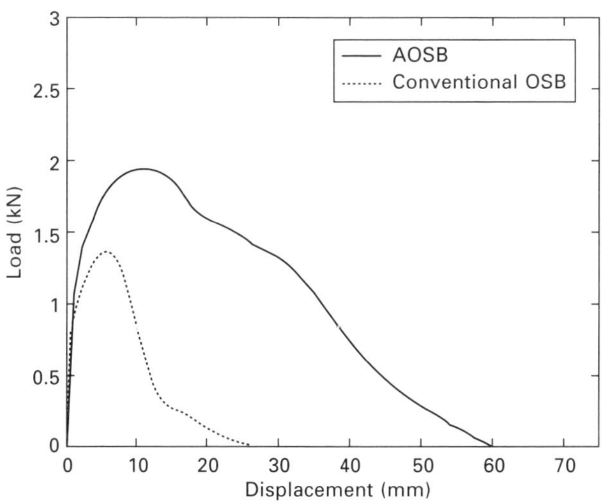

6.12 Monotonic load-deflection for individual nail in single shear: OSB-to-stud and AOSB-to-stud.

of full-size two-panel shear walls were performed. A total of 12 wall tests were conducted: three control specimens and three AOSB specimens with 102 mm edge nailing, and three control specimens and three AOSB specimens with 152 mm edge nailing. The AOSB specimens with 102 mm nail spacing showed an average increase in peak strength of 24%, but the difference in strength between the AOSB and control specimens with 152 mm perimeter nail spacing was less than 1%. However, for both nail spacings, nail edge tear and pull-through failures were much less common for AOSB, and wall ductility (energy absorption capacity) was increased.

Cyclic tests were performed on 12 nominally identical additional shear wall specimens. Separate load protocols were developed based on the CUREE protocols for both the AOSB and control walls. The hysteresis loops of the AOSB specimens exhibited significantly less pinching and loss of capacity relative to the monotonic response. Overall, the AOSB specimens exhibited less damage due to load cycling, and absorbed an average of 52% more energy than the control specimens with a 152 mm nail spacing and 73% more energy with a 102 mm nail spacing.

6.4 Conclusions

While the benefits of AEWC materials are many, research is needed to develop the underlying science and processing parameters which will allow the full benefits of AEWC materials to be realized. Specific research objectives include:

- developing FRP material systems that are compatible with wood, adhesive systems, and wood preservatives;
- establishing microstructure-property-performance relationships;
- developing durability and long-term performance models;
- Developing and verifying models of structural element behavior under service and ultimate loads;
- Simultaneously optimizing structural systems, structural element shapes and AEWC material composition; and
- establishing reliability-based design (RBD) criteria.

AEWC hybrids are unique in that two different classes of materials, FRP and wood, are used together; thus the principles governing the short- and long-term structural behavior differ in many ways from those involving only one of the two materials. Wood exhibits large changes, relative to FRPs, in strength, stiffness, dimensions, and creep properties with changes in ambient relative humidity. Accordingly, the FRP in the hybrid system will experience long-term stresses and strains that would not be present in the FRP alone. The short-term behavior of AEWC hybrids is also different from that of either the FRP or wood. While the bending failure of a wood beam is typically brittle, the corresponding failure of a wood beam properly reinforced on the tension side with FRPs is ductile (Dagher *et al.* 1995, 1996b, 1998).

Although there is interest in the reinforcement of wood with FRP systems, there needs to be a systematic investigation of the controlling mechanisms and interactions between materials necessary to develop an optimized, durable hybrid composite. The keys to the successful and safe design of AEWC are:

- designing a new class of FRP reinforcing materials that are compatible with the hygro-expansion and visco-elastic characteristics of wood;
- developing and maintaining over time the interface (bond) between the two materials needed to ensure full composite action (a core component of this research will focus on the mechanisms of composite and bond durability. The general field of degradation science is important for a wide range of material systems, as recognized by its designation as one of four strategic areas within the NSF Civil Infrastructure Systems program.)
- developing a basic understanding of the short- and long-term behavior of AEWC structural elements including performance over the full range of loading, ultimate strength, ductility, creep, fatigue, and moisture/temperature/UV cycling.

6.5 References

American Institute of Timber Construction (AITC). 1995. Use of a volume effect factor in design of glued laminated timber of softwood species. AITC Technical Note No. 17. AITC. Vancouver, WA.

American Plywood Association. 1972. Basic panel properties of plywood overlaid with fiberglass-reinforced plastic. American. Plywood Assoc. Research. Report. No. 119, Part 1, Madison, WI.

American Society for Testing and Materials (ASTM). 1999. *Standard test methods for evaluating properties of wood-base fiber and particle panel materials.* ASTM D 1037-99. ASTM, Philadelphia, PA.

American Society for Testing and Materials (ASTM) 2000. *Standard test methods for mechanical fasteners in wood.* ASTM D 1761-88 (2000)e1. ASTM, Philadelphia, PA.

American Society for Testing and Materials (ASTM) 2001. *Standard practice for testing and use of a random number generator in lumber and wood products simulation.* ASTM D 5124-96. Philadelphia, PA.

American Society for Testing and Materials (ASTM) 2002. *Standard practice for establishing allowable properties for visually-graded dimension lumber from in-grade tests of full-size specimens.* ASTM D 1990-00 (2002)e1. Philadelphia, PA.

American Society for Testing and Materials (ASTM) 2003. *Standard test method for strength properties of adhesive bonds in shear by compression loading.* ASTM D 905-03. ASTM. Philadelphia, PA.

American Society for Testing and Materials (ASTM) 2003. *Standard test method for tensile properties of plastics.* ASTM D 638-03. ASTM. Philadelphia, PA.

American Society for Testing and Materials (ASTM) 2004. *Standard Practice for*

Establishing Allowable Properties for Structural Glued Laminated Timber (Glulam). ASTM D 3737 04. Philadelphia, PA.

American Society for Testing and Materials (ASTM) 2005a. *Standard Test Methods for Mechanical Properties of Lumber and Wood-Base Structural Material.* ASTM D 4761-05. ASTM, Philadelphia, PA.

American Society for Testing and Materials (ASTM) 2005b. Standard test methods of static tests of lumber in structural sizes. ASTM D 198-05a. Philadelphia, PA.

Biblis EJ. 1965. Analysis of wood-fiberglass composite beams within and beyond the elastic region. *Forest Products J.* 15(2): 81–88.

Boehme C, Schultz U. 1974. *Load bearing behavior of GFRP sandwich.* Holz Roh-Werkst, 32(7): 250–256 (in German).

Bohanan B. 1962. Prestressed wood members. *Forest Products J.* 12(12): 596–602.

Boomsliter GP. 1948. Timber beams reinforced with spiral drive dowels. *West Virginia Univ. Bulletin*, October, Morgantown, WV.

Bulleit WM. 1981. Models for fiberglass reinforced particleboard in flexure. *Proc. Fifteenth Washington State University International Symposium on Particleboard.* Pullman, WA.

Bulleit WM. 1984. Reinforcement of wood materials: a review. *Wood and Fiber Sci.* 16(3): 391–397.

Bulleit WM, Sandberg LB, Woods GJ. 1989. Steel-reinforced glued laminated timber. *J. Struct. Engrg.*, ASCE 115(2): 433–444.

Chetanachan W, Sookkho D, Sutthitavil, W, Chantasatrasamy, N and Sinsermsuksakul, R. 2001. PVC: A new look in construction. *Journal of Vinyl & Additive Technology* 7(3): 134–138.

Coleman GE and Hurst HT. 1974. Timber structures reinforced with light gage steel. *Forest Products J.* 24(7): 45–53.

Dagher HJ, Lindyberg RF. 1999. *FRP-reinforced wood in bridge applications.* Paper presented at the Transportation Research Board Conference, Jan 13, Washington, DC.

Dagher HJ, Kimball T, Abdel-Magid B, Shaler S. 1995. *Behavior of FRP-Reinforced Glulam Beams.* Report to the United States Dept. of Agriculture and the National Science Foundation, Civil Engineering Dept., Univ. of Maine.

Dagher HJ, Kimball T, Abdel-Magid B, Shaler S. 1996a. Effect of FRP reinforcement on low-grade eastern hemlock glulams. *Proc., 1996 National Conference on Wood Transportation Structures*, Madison, WI.

Dagher HJ, Breton J, Abdel-Magid B, and Shaler S. (1996b). *Bond strength, material failure, cyclic delam for FRP-DF, FRP-WWH, and FRP-FRP*, Report AEWC 96-1, Advanced Engineered Wood Composites Center, Univ. of Maine, Orono, Maine, pp 21.

Dagher HJ, Abdel-Magid B, Lindyberg RF, Poulin J, Shaler SM. 1998. *Static bending test results of douglas fir and western hemlock FRP-reinforced glulam beams.* Advanced Engineered Wood Composites Center Report No. 98-4. University of Maine, Orono.

Dagher HJ, Iqbal MA, Lindyberg R, Gardner D, Davids W. 2003. *Advanced wood composites materials and section mechanics reinforced section design*, Report for the Office of Naval Research, Advanced Engineered Wood Composite Center. University of Maine. Orono, ME.

Davalos JF, Barbero E, Munipalle U. 1992. Glued-laminated timber beams reinforced with E-glass/polyester pultruded composites. *Proc. 10th structures Congress, ASCE*, San Antonio, TX, 47–50.

Dziuba T. 1985. The ultimate strength of wooden beams with tension reinforcement. *Holzforschung und Holzverwertung*, 37(6): 115–119 (in German).

Falk RH, Colling F. 1995. Laminating effects in glued-laminated timber beams. *Journal of Structural Engineering*, ASCE. 121(12): 1857–1863.

Foschi RO, Barrett JD. 1980. Glued laminated beam strength: a model. *Journal of Structural Engineering*, ASCE. 106(ST8): 1735–1754.

Govindarajoo R. 1989. *Simulation modeling and analyses of straight horizontally laminated timber beams*. Ph.D. Thesis. Purdue University.

Green DW, Kretschmann DE. 1991. Lumber property relationships for engineering design standards. *Wood and Fiber Sci.* 23(3): 436–456.

Hernandez R, Bender DA, Richburg BA, Kline KS. 1992. Probabilistic modeling of glued-laminated timber beams. *Wood and Fiber Sci.* 24(3): 294–306.

Hoyle RJ. 1975. Steel-reinforced wood beam design. *Forest Products J.* 25(4): 17–23.

Kimball T. 1995. *Feasibility of glulam beams reinforced with FRP sheets*. M.S. Thesis, Department of Civil & Environmental Engineering, University of Maine, Orono, ME.

Kobetz RW and Krueger GP. 1976. Ultimate strength design of reinforced timber: Biaxial stress failure criteria. *Wood Sci.* 8(4): 252–262.

Krueger GP. 1973. Ultimate strength design of reinforced timber. *Wood Sci.* 6(2):175–186.

Krueger GP, Eddy FM. 1974a. Ultimate strength design of reinforced timber: Moment-rotation characteristics. *Wood Sci.* 6(4): 330–344.

Krueger GP, Sandberg LB. 1974b. Ultimate strength design of reinforced timber: Evaluation of design parameters. *Wood Sci.* 6(4): 316–330.

Lantos G. 1970. The flexural behavior of steel reinforced laminated timber beams. *Wood Sci.* 2(3): 136–143.

Lin TY, Burns. NH. 1981. *Design of Prestressed Concrete Structures.* Third edition. Wiley & Sons. New York.

Lindyberg R. 2000. ReLAM: *A Probabilistic Nonlinear Computer Model for the Analysis of Reinforced Glulam Beams in Bending.* Ph.D. Thesis. University of Maine. Orono, ME.

Mark R. 1961. Wood-aluminum beams within and beyond the elastic range. Part I: Rectangular sections. *Forest Products J.* 11(10): 477–484.

Mark R. 1963. Wood-aluminum beams within and beyond the elastic range. Part I: Trapezoidal sections. *Forest Products J.* 13(11): 508–516.

Marx CM, Evans JW. 1986. Tensile strength of AITC 302-24 grade tension laminations. *Forest Products J.* 36(1): 13–19.

Marx CM, Evans JW. 1988. Tensile strength of laminating grades of lumber. *Forest Products J.* 38(7/8): 6–14.

Marx CM, Moody RC. 1981a. *Bending strength of shallow glued-laminated beams of a uniform grade.* USDA Forest Service. FPL 380.

Marx CM, Moody RC. 1981b. *Strength and stiffness of small glued-laminated beams with different qualities of tension laminations.* USDA Forest Service. FPL 381.

Mitzner RC. 1973. Durability and maintenance of plywood overlaid with fiberglass reinforced plastic. *Am. Plywood Assoc. Res. Report No. 119, Part 3*, Amer. Plywood Assoc., Madison, WI.

Moody RC, Falk RH, Williamson T. 1990. Strength of glulam beams – volume effects. *Proceedings, 1990 International Timber Engineering Conference*, Vol. 1. Tokyo, Japan. 176–182.

Patterson J. 2001. New opportunities with wood-flour-foamed PVC. *Journal of Vinyl & Additive Technology* 7(3): 138–141.

Peterson J. 1965. Wood beams prestressed with bonded tension elements. *J. of Struct. Engrg.*, ASCE. 91(1): 103–119.

Saucier JR, Holman JA. 1975. Structural particleboard reinforced with glass fiber – Progress in its development. *Forest Products J.* 25(9): 69–72.

Shenton III, HW, Dinehart DW and Elliott TE. (1998). Stiffness and energy degradation of woodframe shear walls, *Canadian Journal of Civil Engineering*, 25(3).

Showalter KL, Woeste FE, Bendtsen BA. 1987. *Effects of length on tensile strength in structural lumber*. USDA Forest Service. FPL-RP-482.

Sliker A. 1962. Reinforced wood laminated beams. *Forest Products J.* 12(1): 91–96.

Spaun FD. 1981. Reinforcement of wood with fiberglass. *Forest Prod. J.* 31(4): 26–33.

Stein, S.K. 1987. *Calculus and Analytic Geometry* (4th edn). McGraw-Hill Book Co. New York.

Steinmetz PE. 1977. *Resin systems and glass reinforcements to improve dry-formed hardboard.* USDA Forest Service, Forest Products Laboratory Report, FPL 284.

Stern EG, Kumar VK. 1973. Flitch beams. *Forest Products J.* 23(5): 40–47.

Taylor SE, Bender DA. 1991. Stochastic model for localized tensile strength and modulus of elasticity in lumber. *Wood and Fiber Sci.* 23(4): 501–519.

Taylor RJ, Batchelor B, Dalen KV. 1983. *Prestressed wood bridges*. Struct. Res. Report SRR-83-01, Ministry of Transp. and Communication, Downsview, Ontario, Canada.

Theakston FH. 1965. A feasibility study for strengthening timber beams with fiberglass. *Canadian Agricultural Eng.* Jan: 17–19.

Wangaard F. 1964. Elastic deflection of wood-fiberglass composite beams. *Forest Prod. J.* 14(6): 256–260.

West H.H. 1989. *Analysis of Structures* (2nd edn). Wiley and Sons. New York.

7
Sustainable materials for the built environment

J HARRISON, TecEco Pty Ltd, Australia

7.1 Introduction

For the last several billion years nature has nurtured the planet evolving complex eco-systems (the biosphere) that conserve and recycle energy and materials. Much is to be learned from the study of mature climax eco-systems which demonstrate complex web-like integration, conserving energy, all of which is ultimately derived from the sun as it flows through the system and materials which are constantly recycled, the waste from one natural metabolism being the input of another.

Plants and animals live together in mutually interdependent ways that become increasingly specialized, developing mechanisms for regulation preventing overgrowth or dominance. The wonder of nature is the complexity of this interdependence and how extremely efficient it is at recycling the nutrients essential to life.

Occasional aberrations are triggered climatically, by fire, cataclysmic events or in some other way. Species are wiped out (as were the dinosaurs) and new species or groups of species appear to take over. In the past, however, the biosphere has always come back to a balance characterized by complexity and integration but this can take thousands and sometimes millions of years.

Along came humans. In the last hundred years or so, and for the first time in geological history, we have become masters of the destiny of the unique blue-green planet we live on. We are like no other cataclysmic event, yet we are responsible for many of the changes affecting the geosphere-biosphere from salinity, deforestation and pollution to the global carbon dioxide balance. Climate change is the most visible result – storms, droughts, floods and the like are rising in frequency and severity and the consensus is that we are to blame.

In nature climax communities are stable. Our presence on the planet is non-climax, non-stable and non-sustainable. Driven by our intelligence, greed and arguably, cheap fossil fuel energy, and like a new predator before which no living thing can stand, we are taking over. We are, however, the agents of our own eventual doom. We are choking and poisoning ourselves – slowly. It's a bit like the old agar dish experiment from biology at school – remember how it works? You put a speck of bacteria in the middle of a food supply loaded in agar jelly and behold – the bacteria grows in a ring to the edge of the dish, dying in the middle as the poisons it has released in its world (the dish) kill it. There is no escape. We are so ignorant that few understand the flows, checks and balances vital to the maintenance of the biosphere as a whole. There will be a natural correction. Will it involve our own extinction? Should we wait to find out?

First the village smithy, then James Watt and the steam engine followed by oil, abundant energy and thousands if not millions of innovations later and we have a tiger by the tail called the techno-process. It is bigger than we are, more ubiquitous, far reaching and in its name some five or six hundred billion tonnes of matter are moved about the planet to create the twenty or thirty billion tonnes of new materials we actually use every year. The tremendous appetite of the techno-process is irreversibly changing the planet. The earth that nurtures us has limits that we have now most certainly exceeded.

Resources are supplied by the geosphere-biosphere one way or another and are not infinite. Needs change and the things we make out of materials wear out. Eventually everything is thrown away. All this activity has an impact on the planet and it seems vital earth systems are unable to cope and are rapidly going out of balance. It appears impossible for humans to correct the problem on such a large scale. What options does that leave us? Kyoto is a symbolic start but that is all. What does nature teach us?

Economics could perhaps be defined as the set of common behaviors that string together our interactions for survival. It is about the application of resources to needs. Economics drives both the techno-process and nature – both are based on survival but there are some fundamental differences between our techno-process and natural systems. Nature uniquely embraces integration and balance, seen as desirable by economists but unfortunately missing from the techno-process. Economists should study ecology for a few clues about where we are going wrong.

The cataclysmic event in our evolution has been the development of machines energized by fossil fuels. The resulting techno-process is simple, linear, non-integrated and arguably non-climax. Linear systems cannot be balanced because they cannot possibly contribute as much as they take. Climax eco-systems in the biosphere are, on the other hand, characterized by complex integration and balance.

Efficiency is important for profit which drives the allocation of resources. Unfortunately, we seem to understand efficiency only in a linear sense, not an integrated one. The greatest proponent of efficiency, Henry Ford, developed a linear production line to which resources were delivered. There was no concern for resource issues beyond the factory gate – that was somebody else's problem or nature would provide. Enterprise-based efficiency espoused by Henry Ford neglects the value of the natural capital or the planet as a whole.

Climax ecologies are characterized by extremely efficient systems in which all processes are integrated. For example, a leaf is technically designed to minimize water loss and maximize photosynthetic production. When the leaf falls to the ground it is eaten by bugs, grubs and bacteria and eventually it provides nutrients for the trees above it in what is a highly efficient process that retains embodied energy from the sun and recycles materials indefinitely.

Liquid and gaseous fossil fuels are now running out and the techno-process cannot continue the way it has in the past. The planet is in crisis. It is time for change, so the total net consumption of energy and materials is much less. Can the intelligence of the computer chip provide the connections to close the loops in our linear techno-process, can we invent new materials that do not have such an impact on the planet. Can we live in harmony with the planet? These are the big questions.

The only driving force humans answer to on a large scale is economics but, like a mirror, economics is really only a measurable reflection of how we really are, how we think and act. Economics is the driver of the techno-process. Technology however defines what moves through it and how. In this simple understanding lies the clue. Maybe we can redefine materials so economics drives more sustainable processes. Can we re-invent our physical world? In my view we are going to have to if we want to survive. Natural climax eco-systems involve conservation of energy and materials, integration and thus recycling and provide the example as to how this could be done. How can we mimic nature and yet still obey the rules of economics?

Can we harness economic forces to bring about change in what is a linear, substantially unintegrated techno-process[1] and develop a more sustainable regional industrial ecology which conserves materials and energy in the system as a whole, to a desirable extent that complements nature and that is highly integrated with much more recycling and reuse.

Technology, which primitively used, created the industrial revolution and the linear techno-process, can be turned to the greater cause of producing an industrial ecology which is integrated, efficient and which has minimal impact on the planet. The need is obvious. Technology is the means. Linked with the will we may yet reduce our footprint on the planet until it is hardly noticeable. As part of this a paradigm shift in the technology underlying materials is required.

The opportunity for change is greatest in the built environment which, after all, encompasses the greatest materials flows on the planet with the largest take and waste impacts. Taking into account infrastructure, governments

[1] Take, manipulate use and waste.

are uniquely the largest constructors on the planet and have the opportunity, if not the responsibility, to encourage and implement new technologies that will make the difference. Architects, engineers and specifiers are uniquely positioned to take advantage of the changes that are occurring and drag the rest of the supply chain into delivering sustainability.

7.1.1 Major themes

1. Sustainability is a goal not a benchmark.
2. There is a techno-process describing our activities on the planet of taking resources, manipulating molecules, making new 'things', using these 'things' and eventually throwing them 'away'.
   A special case of the techno-process is the extraction and burning of fossil fuels.
3. Materials are everything between the take and waste.
4. By re-engineering materials we can reduce the volume and impact of the take and waste.
5. The built environment offers enormous opportunities for re-engineering materials and improving sustainability.
6. The only way to make any of the above happen on a global scale is economically.

7.1.2 Theme statement

The technology paradigm defines what is or is not a resource in an economic system that drives materials flows through the techno-process. By harnessing basic human psychology through cultural change to achieve greater demand for sustainable outcomes delivered by evolving and changing techno-processes that sustainably deliver cost effective solutions, economics will define more sustainable resource flows.

7.2 The current situation

Global population, consumption per capita and our footprint on the planet is exploding. The impacts are affecting every indicator of the health of the planet. The world population passed 6 billion in 1999 and at the current rate of growth the world will have 7 billion people soon after the year 2010 (Fig. 7.1). The overwhelming share of world population growth is taking place in developing countries (95.2% in 1990–2000, an estimated 97.6% in 2000–10 and 98.4% in 2010–20) where the population has more than doubled in 35 years, growing from 1.89 billion in 1955 to 4.13 billion in 1990 (UNCHS, 2004a).

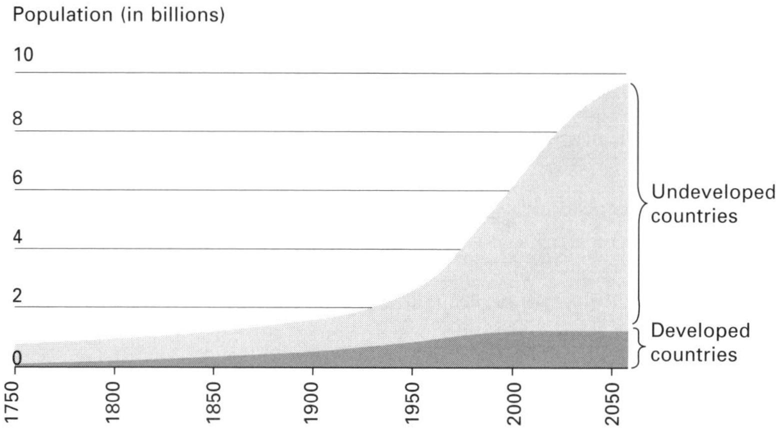

7.1 World population growth 1750–2050. Population in billions (after UN, 2002).

> Significant proportions of population increases in the developing countries have been and will be absorbed by urban areas (71.8% in 1990–2000, an estimated 83.4% in 2000–10 and 93.4 in 2010–20). Urban settlements in the developing countries are, at present, growing five times as fast as those in the developed countries. Cities in the developing countries are already faced by enormous backlogs in shelter, infrastructure and services and confronted with increasingly overcrowded transportation systems, insufficient water supply, deteriorating sanitation and environmental pollution. (UNCHS, 2004a)

The World Resource Institute recently published a document titled *The Weight of Nations* which demonstrates that over the last 15 years per capita consumption of resources to produce materials and the per capita production of wastes are increasing in spite of increasing efficiencies and a slight de-coupling of GDP per capita from energy use per capita and material use per capita. It follows that gross consumption and gross waste production are also increasing (WRI, 2004).

In October 2004, the World Wildlife Fund released *The Living Planet Report 2004* which confirms that ecological overshoot has become a reality: humanity is now consuming over 20% more natural resources than the Earth can produce (WWF, 2004) (Fig. 7.2). The WWF report exemplifies the relevance of the ecological footprint tool for decision-makers and defines humanity's challenge for the 21st century as learning to live within the means of our planet. According to the report 'deficit spending' leads to the destruction of ecological assets, on which our economy depends, such as depleted groundwater, collapsing fisheries, CO_2 accumulation in the atmosphere, and deforestation. 'Human beings already use over half the world's accessible

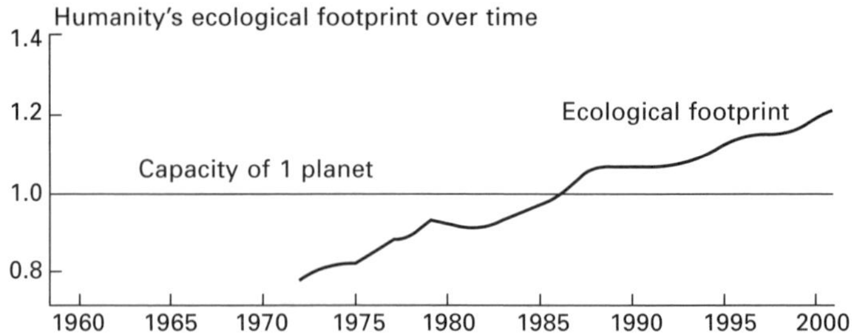

7.2 The WWF's world ecological footprint index (WWF, 2004).

surface freshwater, have transformed one-third to one-half of its land surface, fix more nitrogen than do all natural systems on land, and appropriate more than two-fifths of the planet's entire land-based primary biological productivity' (Vitousek, *et al.*, 1997). 'The doubling of these burdens with rising population will displace many of the millions of other species, undermining the very web of life.' (Hawken, *et al.*, 2000.)

World systems are interconnected and under stress. Daily we hear about salinity, water, weather, fishery, waste, global warming and other issues. Most news about the health of the planet is unfortunately bad. Much of it is documented in a recent report on the International Geosphere-Biosphere website (Steffen and Tyson, 2001).[2] Of particular concern and therefore the most studied is the problem of CO_2 in the atmosphere. The level of CO_2 in the air from the burning of fossil fuels is rising too rapidly for natural processes to absorb and has risen from 280 parts per million in pre-industrial times to around 380 parts per million in 2004. Between 2001 and 2002 the rise was from 371.02 to 373.10 parts per million, an increase of 2.08 over the year. Then it rose again in 2003 to 375.64 parts per million, an annual increase of 2.54 – the first time the rise was over two parts per million in two years.

Just as concerning is the fact that methane concentrations have also risen by 145% over the same period and oxygen levels have fallen. Before the industrial revolution gaseous chlorines did not exist in the atmosphere. By 1996 there were 2731 parts per trillion, most of these produced in the 20th century (WRI, 1998) (Fig. 7.3). The well documented result of the increase in CO_2 in the atmosphere has been global warming and climate change accompanied by devastating sea level rises (Fig. 7.4).

We move around some 500 (Hawken *et al.*, 2000) or 600 billion tonnes of material every year and use a trifling 50 billion tonnes.[3] Most of the rest is

[2] IGBP SCIENCE No. 4 downloadable from the International Geosphere – Biosphere website.

[3] My own very rough estimate.

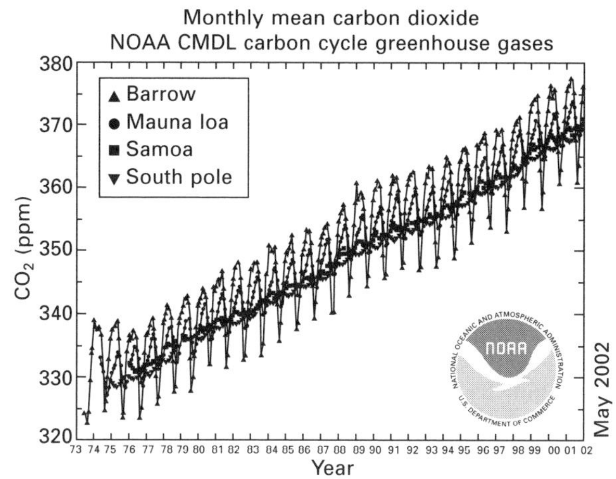

Atmospheric carbon dioxide mixing ratios determined from the continuous monitoring programs at the 4 NOAA CMDL baseline observatories. Principal investigator: Dr. Pieter Tans, NOAA CMDL Carbon Cycle Greenhouse Gases, Boulder, Colorado, (303) 497-6678 (ptans@cmdl.noaa.gov,http://www.emdl.noaa.gov/ccgg).

7.3 Global CO_2 balance (CMDL, 2003).

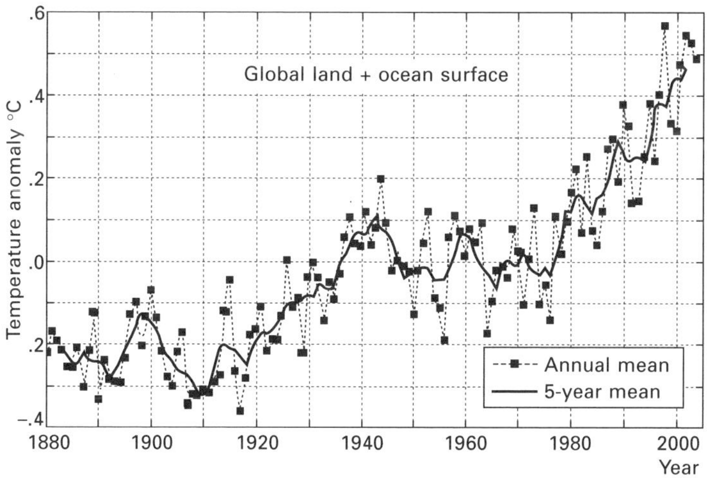

7.4 Global temperature anomaly (GISS, 2005).

waste. Even that which is used is eventually wasted. Fortunately much of this waste is biodegradable, however, there are still several hundred kilos of non-biodegradable waste per person on the planet to safely dispose of. The production and disposal of waste is also a major global issue. Simply put, there is ample evidence that increases in consumption per person and population growth have compounded to unsustainable levels.

7.3 Sustainability

Definitions as to what the words sustainable, sustainability and so on mean abound. The concept of sustainability was brought to prominence by the then Prime Minister of Norway, Mrs Gro Harlem Brundtland in the World Commission for Sustainable Development which defined sustainable development as 'development that meets the needs of the present without compromising the ability of future generations to meet their own needs' (Bruntland, 1987). 'Living within the means of the planet.' Value for impact. 'The concept of striving for sustainable development is not so much a "destination" at the end of a road but like life, more a "manner of travelling".' (CICA, 2002). We can say, however, we are succeeding when our footprint is reduced, when trends are reversed such as the level of CO_2 in the atmosphere and when water quality, sanitation and other benchmark measures improve.

Sustainable materials and processes are all relative. Something that is sustainable in one context may not be in another; what is sustainable today may not be tomorrow and so on. We can, however, always compare our efforts towards sustainability to nature. Natural processes[4] are usually as a whole balanced, integrated and efficient, all major features of sustainable processes. Some aspects of sustainability are easier to understand than others. One thing is clear however. To live more sustainably we must reduce the impact of our activities on the earth's systems by what I call the techno-process described on page 284.

7.4 The Earth's natural systems

The planet has well-connected natural systems. Carbon dioxide emitted in one country is rapidly mixed throughout the atmosphere; pollutants released into the ocean in one location are transported all over. Local and regional emissions create global environmental problems. There is in effect a 'global commons.' Earth system science involves:

1. Exploring interactions among the major components of the earth's system. These are:
   - the biosphere – living organisms and the part of the earth and its atmosphere in which living organisms exist or that is capable of supporting life
   - the atmosphere – earth's vital blanket of gases
   - the geosphere – the solid earth including the continental and oceanic crust as well as the various layers of the earth's interior
   - the hydro-sphere – earth's circulating water systems
   - the energy system – the energy that powers all earth's systems.

[4] See www.biomimicry.org

2. Distinguishing natural from human-induced causes of change.
3. Understanding and predicting the consequences of change.

As we learn about how the earth works from our study of biology, geology, hydrology, metrology and so on we have recognized patterns that help us understand how our planet works. Some of the most important patterns are:

- networks – how life and planetary systems are connected to each other
- cycles – how matter gets used over and over again
- flows – how matter and energy move from place to place
- balance – how change is regulated.

It is difficult completely to understand let alone quantify the complex flows and balances that go on around the planet. We do know, however, that pollution from wastes of various kinds has affected atmospheric composition, land cover, marine ecosystems, coastal zones, freshwater systems, global biological diversity and many other global systems.

Due to our pervading interference the constituent components of matter which are molecules are no longer produced or used in equilibrium. Complex carbon based molecules put together by living matter over many millions of years are being used as if there was no tomorrow.

7.4.1 Carbon and oxygen flows

Rising atmospheric carbon dioxide concentration and its potential impact on future climate is an issue of enormous global significance. Also of concern is the depletion of oxygen in the atmosphere. Both carbon and oxygen are present in a combined form as CO_2 because of their importance in reduction and oxidization processes. Carbon dioxide is currently around 380 ppm whilst oxygen as O_2 is just over 20%. Nitrogen as N_2 is ubiquitous at around 78%.

The carbon cycle

Even though oxygen in the atmosphere is declining there is a lot more of it than carbon so of greatest concern at the present time is the balance of the carbon cycle which is a complex series of processes through which many of the carbon atoms in existence rotate (Fig. 7.5). Carbon is a major courier of energy. Plants capture energy from the sun using the DNA driven photosynthesis process as a result of which carbon is fixed in their structures. Animals consume food produced in this way releasing the stored up energy and CO_2. We also burn what was once living to obtain energy and release CO_2.

Living plants and the geological remains of living plants and animals are major reservoirs of carbon. While there is considerable debate about the magnitude and location of these sinks, there is strong evidence that they are very significant. The carbon cycle is out of balance and as concentrations in

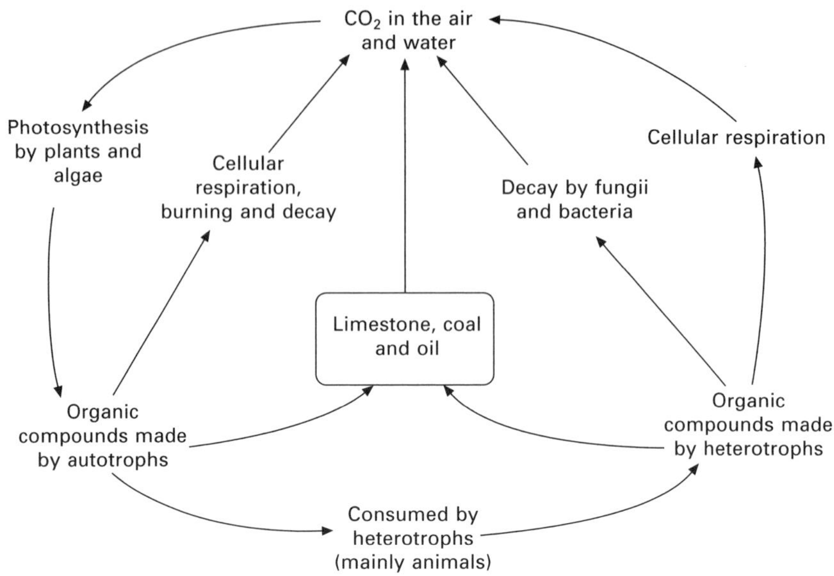

7.5 The carbon cycle (adapted from Kimball (2004).

the atmosphere increase the earth is undergoing 'global warming' and climatic disruption. Agriculture and many cities are threatened by climate change and a predicted rise in sea level. Unless reversed, climate change will, in time, impact severely on the world as we know it. We have the power to reverse climate change. Do we have the sense?

Large quantities of carbon dioxide are released by burning fossil fuels – according to the Wood's Hole Institute (Haughton, 2004) 'The concentration of CO_2 in the atmosphere has already increased by about 30% since the start of the industrial revolution sometime around the middle of the 19th century and will continue to increase unless societies choose to change their ways.' We are also destroying the living carbon sink by manipulating land surfaces for water management, settlement, food and energy production. Fig. 7.4 shows, there has been a strong warming trend over the past 30 years, a trend that has been shown to be due primarily to increasing greenhouse gases in the atmosphere. El Niños, in which warm water spreads over the tropical Pacific Ocean, are one major cause of fluctuations about the long-term trend. There was a very strong El Niño in 1998, while a weak El Niño existed for the last several months of 2002. The fact that 2002 is almost as warm as the unusual warmth of 1998 is confirmation that the underlying global warming trend is continuing (GISS, 2002)

According to the Wood's Hole Institute (Haughton, 2004) 'Most of the increase in atmospheric CO_2 concentrations came from and will continue to come from the use of fossil fuels (coal, oil, and natural gas) for energy, but about 25% of the increase over the last 150 years came from changes in land

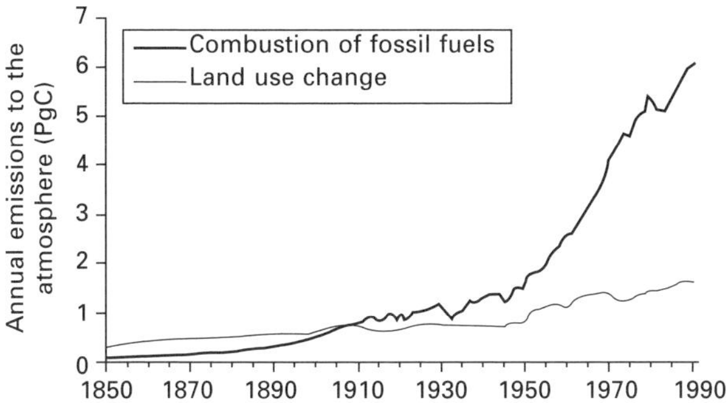

7.6 The clearing of forests and the cultivation of soils for food production (after Haughton, (2004).

use, for example, the clearing of forests and the cultivation of soils for food production' (Fig. 7.6).

Numbers in the carbon cycle

Summary totals presented by the Wood's Hole Institute (Haughton, 2004) for carbon in the carbon cycle during the decade ending 2000 (in billion metric tonnes or petograms) are as shown in Table 7.1. In general agreement with other research there appears to be a discrepancy which is believed to be because there have been greater than measurable amounts taken up by living plant sinks and possibly the ocean. We may also be increasing the overall size of our atmosphere (Fig. 7.7).

Table 7.1

Atmospheric increase	=	Emissions from Fossil fuels	+	Net emissions from changes in land use	–	Oceanic uptake	–	Missing carbon sink
3.2(±0.2)	=	6.3(±0.4)	+	2.2(±0.8)	–	2.4(±0.7)	–	2.9(±1.2)
Converting* to tonnes CO_2 this is:								
Atmospheric increase	=	Emissions from fossil fuels	+	Net emissions from changes in land use	–	Oceanic uptake	–	Missing carbon sink
11.73 (±0.73)	=	23.086 (±1.47)	+	8.062 (±2.93)	–	8.795 (±2.57)	–	10.627(±4.40)

*By dividing by the atomic weight of carbon (12.01) and multiplying by the molecular weight of CO_2 (44.01).

A description of all the other earth systems including, for example, the hydrosphere, of which freshwater systems are a part and in relation to which water quality, salinity and supply are major issues or sea fisheries, which are in major decline, is beyond the scope of this chapter.

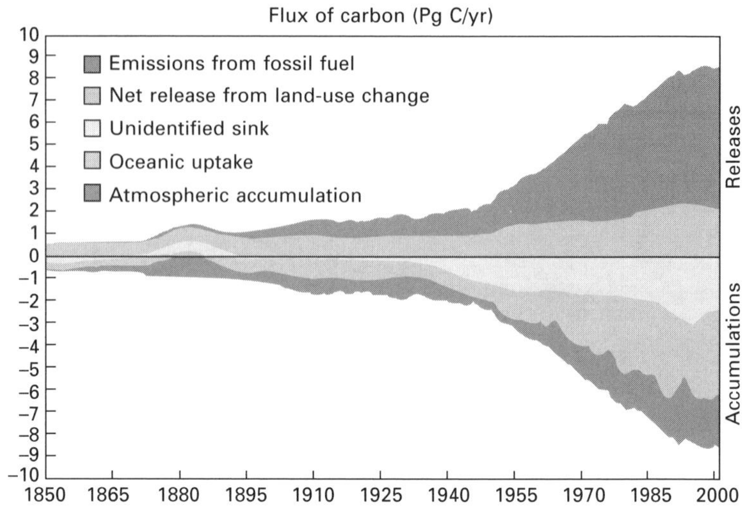

7.7 Sequestration and emission of carbon dioxide (Haughton, 2004).

Oxygen depletion?

Oxygen is ubiquitous at around 47% by weight of crustal earth. Like carbon it is substantially present in a combined form because of their importance in reduction and oxidization processes. There is a lot of it – some 600 moles above every square metre and 250 moles below in the oceans. The stoichiometry of the reactions

C (organic, inorganic, etc.) + $O_2 \rightarrow CO_2$ and $H_2 + O_2 \rightarrow H_2O$

suggest that with the burning of fossil fuels or hydrogen the level of oxygen in the atmosphere should be declining.

According to recent studies such as that conducted by Ray Langenfelds from CSIRO Atmospheric Research this is the case. Langenfelds showed a 0.03% reduction in oxygen levels in just 20 years using air samples taken from Cape Grim in remote Western Tasmania (Langenfields, 2003). The level as at the date of the study was 20.92%, down from 20.95% 20 years earlier. According to scientist Richard Matear from CSIRO Marine, a recent series of experiments showed a decline in the level of oxygen at between 500 and 1500 metres in samples taken from the great southern ocean taken from the Aurora Australis between Tasmania and Antartica (Matear, 2002). The decline at about 0.0015% (or about 3 ppm out of 209,500 ppm) per year is very slight and much less than the stoichiometry of the above equation predicts and this is thought to be because of biological catch up (Fig. 7.8).

The anonymous illustration in Fig. 7.8 shows the oxygen concentration fluctuating by about 30 ppm throughout the year (relative to nitrogen gas) and an annual depletion of 3–4 ppm superimposed upon the annual cycle.

The figure given in most textbooks for the level of oxygen in the air is

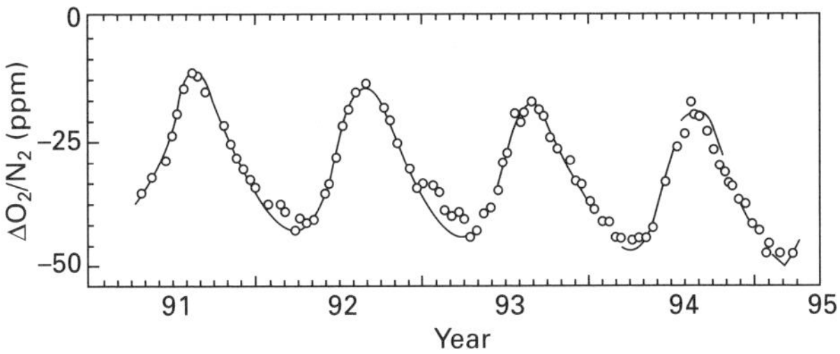

7.8 Oxygen depletion (unofficially released data from a northern hemisphere sampling station – no references available!).

20.95%, however, studies supporting this were conducted prior to 1965. In pristine and isolated places the figure today is more like 20.92% as found by the CSIRO in remote western Tasmania. In more populated areas where land has been cleared the average figure has fallen to more like 19% or 20%. In some cities (where the majority of the world's population live) the level is even lower due to consumption, higher levels of CO_2 and pollution and figures as low as 12–18% have been recorded. As 7% is considered starvation level and some 90% of the world's population live in cities, this is concerning.

What is the optimal level of oxygen? Field geologists have found that the evidence from small pockets of air trapped in fossilized amber that the level of oxygen some 80 million years ago was somewhat higher at around 30%; was that an optimal level and has it been falling since then? More research obviously needs to be done in this area and made public. It is also important when considering emissions and sequestration to not only consider the affects on CO_2 emissions but also on O_2 depletion as well.

TecEco, my company, has uniquely proposed changing the carbon cycle by using carbon as a major component of the built environment. The modifications proposed by TecEco include a kiln technology that allows the capture of CO_2 for geological sequestration and a way of using CO_2 in concretes, a major component of construction (see Table 7.2 on page 320).

7.4.2 Biomimicry

The term biomimicry was popularized by the book of the same name written by Janine Benyus (1997) and is a method of solving problems that uses natural processes and systems as a source of knowledge and inspiration. It involves nature as model, measure and mentor. The theory behind biomimicry is that natural processes and systems have evolved over several billion years through a process of research and development commonly referred to as evolution. A recurring theme in natural systems is the cyclical flow of matter in such a way that there is no waste. This concept can be represented by the simple formula: waste = food indicating that anything coming to the end of its useful life becomes food for something else.

Currently the world's carbon cycle is out of balance because the waste of fossil fuel burning (CO_2) is not being utilized as a fuel/food for other processes. TecEco have addressed this issue with their eco-cements by using carbon as an element of the built environment and for wastes by utilizing them for their physical property rather than chemical composition.

7.5 The impact of current technology

According to the American Association for the Advancement of Science Population and Environment Atlas (AAAS, 2004)

> Consumption and technology impact on the environment by way of two major types of human activity.
>
> First, we use resources. We occupy or pre-empt the use of space, and so modify or remove entirely the habitats of many wild species. We extract or take resources – growing food, catching fish, mining minerals, pumping groundwater or oil. This affects the stock of resources available for humans and for other species in the future.
>
> Second, we dump wastes – not just those that consumers throw away, but all the waste solids, liquids and gases that are generated from raw material to final product. These affect the state of land, groundwater, rivers, oceans, atmosphere and climate.

These activities are essentially the take and waste impacts of what I shall call the techno-process[5] which describes our primitive and unconnected industrial metabolism and the linear flow of matter through our economy.

$$\text{Take} \rightarrow \text{Manipulate} \rightarrow \text{Make} \rightarrow \text{Use} \rightarrow \text{Waste}$$

The techno-process is a mostly enterprise-based linear process of extracting resources, manipulating them as required (modifying substances), making something with them, using what is made and then throwing away what no longer has utility.[6] The impact of the techno-process on the planet is significant. Resources are not unlimited and the planet does not have an infinite capacity to reabsorb wastes. Given the diverse nature and huge scale of the techno-process and the many linkages of this process to the geosphere-biosphere, any solution will involve doing something in all ways possible at all stages of the techno-process from take to waste and involve a large number of people, professions and industries.

To allow us to live more equitably and sustainably within the means of the planet and possibly even for our own long-term survival take and waste

---

[5] Also variously called the materials supply or materials cycle, industrial metabolism, etc.

[6] Utility is an economic term for value to the user.

impacts need to be reduced and preferably eliminated to what is renewable and preferably biodegradable. Embracing philosophical catchwords are Reducing, Re-using, Recycling and Recovering.

> Zero waste is a philosophy and a design principle for the 21st Century; it is not simply about putting an end to landfilling. Aiming for zero waste is not an end-of-pipe solution. That is why it heralds fundamental change. Aiming for zero waste means designing products and packaging with reuse and recycling in mind. It means ending subsidies for wasting. It means closing the gap between landfill prices and their true costs. It means making manufacturers take responsibility for the entire lifecycle of their products and packaging. Zero waste efforts, just like recycling efforts before, will change the face of solid waste management in the future. Instead of managing wastes, we will manage resources and strive to eliminate waste. (GRRN, 2000)

There is no doubt that in the past technology has been a major player and at the root of much of the problem. It is time for science to balance the ledger in the form of new sources of energy and new materials with a much lower impact that solve issues such as global warming, climate change and waste because they change the molecular flows that are harming the planet (Fig. 7.9).

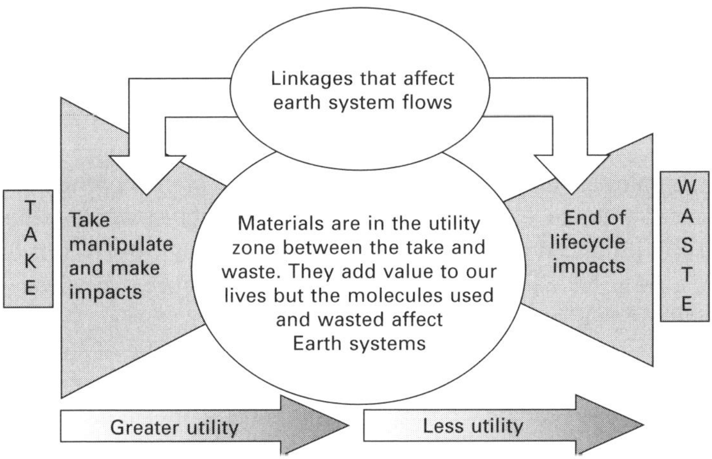

7.9 Materials and the take and waste impacts of the techno-process.

The techno-process is powered by fossil fuel and reversing damaging underlying molecular flows associated with it, such as the accumulation of CO_2, will best be achieved using non-fossil-based energy such as solar or energy directly derived from solar sources such as wind or wave. The techno-process has delivered a plethora of unnatural chemicals and materials with impacts at every step of the process. There is waste in the process and at the end of it. It is therefore fundamental that we think about the materials we use and the molecules they are made of.

7.5.1 The importance of materials

Materials are everything between the take and waste. They are fundamental to the techno-process and determine many properties including weight, embodied energies, fuel related and chemical emissions, lifetime energies, user comfort and health, use of recycled wastes, durability, recyclability and the properties of wastes returned to the geosphere-biosphere. Resources extracted from the geosphere-biosphere can be classified broadly into several non-exclusive types. They are short or longer use, biologically or geologically derived, surface or sub-surface, etc., but most of these classifications are less relevant to sustainability than the extent to which resources are renewable or not renewable.

The techno-process is very inefficient in that large quantities of renewable and non-renewable resources are extracted to produce small quantities of materials which themselves are used to produce even smaller volumes of things actually used, many of which do not retain utility[7] very long before they are in turn thrown away. There is no such place as away.

> The global flow of matter, some 500 billion tons per year, most of it wasted, is largely invisible. Yet obtaining, moving, using, and disposing of it is steadily undermining the health of the planet, which is showing ever greater signs of stress, even of biological breakdown. (Hawken *et al.*, 2000)

Examples abound. One study found that around 93% of materials used in production do not end up in saleable products but in waste, while 80% of products produced are discarded after a single use (Von Weizsäcker *et al.*, 1997). Some excellent examples of this wasteful process are given by David Schaller (Schaller, 2004) for organic materials and foods.

Mining is a major culprit as it involves the extraction of a non-renewable resource, and the processing of significant volumes of material for a small quantity of end product as most substances being mined are present only in small percentages, the rest being discarded as tailings. Consider a one-litre tin of white paint that will cover 16 m^2 of wall. During its manufacture 360 g of synthetic rutile, 95 g of chlorine, 118 g of petroleum coke, 235 g of oxygen, 350 g of nitrogen, 60 g of carbon, 60 g of lime, 9 litres of water and 20 MJ of energy were used and 28 kg of rock, sand and clay were moved (Herbertson and Green, 2003)

---

[7] Value to us.

Another example is the high proportion of new materials delivered to construction sites that are wasted. Some studies show that between a half and three-fourths of the materials used in our industrial economy are generated and treated as waste before ever entering the economy. They are not seen or treated as commodities and are not valued as such (Schaller, 2004). These materials contain undesirable concentrations of substances dangerous to the environment and many are wasted throughout the techno-process. Some when wasted have dangerous impacts. To reduce impact on linkages to the geosphere-biosphere it is essential we reduce the throughput and impact of the techno-process. The incentive is that economically we would be better off.

7.5.2 Impacts

In recent decades many environmental indicators have moved outside the range in which they have varied for the past half a million years. We are altering our life support system and potentially pushing the planet into a far less productive state.

> The world faces significant environmental problems: shortages of clean and accessible freshwater, degradation of terrestrial and aquatic ecosystems, increases in soil erosion, loss of biodiversity, changes in the chemistry of the atmosphere, declines in fisheries, and the possibility of significant changes in climate. These changes are occurring over and above the stresses imposed by the natural variability of a dynamic planet and are intersecting with the effects of past and existing patterns of conflict, poverty, disease, and malnutrition. (Steffen and Tyson, 2001)

Take effects of the techo-process

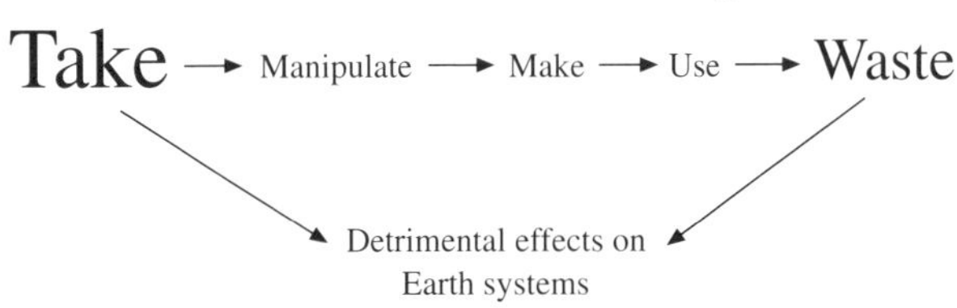

In the past the main cause of concern was that resources would not be sufficient to sustain the human race let alone the techno-process (even if it was not called that) ‘Frequent warnings were issued that we faced massive famines, or that we would “run out” of essential fuels and minerals’ (AAAS, 2004).

The first and most well known to predict such gloom and doom was Thomas Malthus who wrote numerous papers on the subject including *An Essay on the Principle of Population* (Malthus, 1798) predicting widespread food shortages with increases in population. Neo-Malthusians believe that increases in population will eventually degrade environmental resources leading to devastation. They may yet be right. Until the last few decades there was little evidence of population growth associated with resource consumption being limited by the supply of resources mainly due to the input of technology and market forces which improved discovery, extraction and production. When a particular mineral, for example, looked as if it might be becoming scarce market forces drove the price up encouraging further exploration and the discovery of new deposits. This process of new discovery has recently come under considerable pressure. Oil production and reserves for example are now thought to be near climax and set to fall (see Fig. 7.11 on page 292).

Renewable resources such as water, fish stocks, even the air we breathe, are today of much greater concern because they are now understood to be much more fragile and influenced strongly by overuse and pollution. Of major concern is the problem of wastes created in producing and consuming resources. Wastes are the output of the techno-process; some are insidious poisons that are distributed in the global commons as discussed above, others are not so dangerous to life but their sheer volume presents problems. In the mid-1990s, for example, countries belonging to the OECD produced 1.5 billion tons of industrial waste and 579 million tons of municipal waste – an annual total of almost two tons of waste for every person (AAAS, 2004). OECD countries produce much more waste per capita than underdeveloped countries in which by necessity the peoples tend to live more holistically and in harmony with nature.

Waste effects of the techno-process

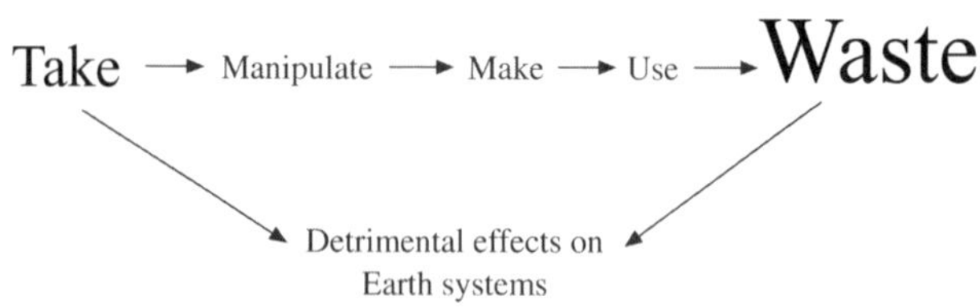

A high proportion of wastes is from construction and demolition activities. A waste audit in South Australia in 1998 found

> 53.1% by weight of the waste South Australians send to landfill is generated from construction and demolition activity (including waste fill). Waste materials in this category include concrete, bricks, tiles, steel, glass, metal, wood, asphalt and plastics. Domestic waste comprised 27.5% by weight of waste sent to landfill and commercial and industrial waste comprised 15.7% by weight. Around three-quarters of commercial and industrial waste is comprised of garbage, food/kitchen waste, cardboard, paper, wood, and plastic bags/film. (Anon, 2004)

Many of these wastes are potentially recyclable. According to Maria Atkinson of the Green Building Council of Australia the figure nationally of waste going to landfill from construction and deconstruction activities (predominantly from refurbishment activity) was around 40% (Atkinson, 2003).

The figures for building materials in waste streams vary around the globe and one of the problems is that the method in which audits are conducted also varies making it hard to obtain comparative statistics. The flow of unwanted or waste materials is affecting our planet. The solid wastes that are not incinerated generally go to landfill and pollute water courses and the local area. Liquid and gaseous pollutants are more insidious and spread invisibly in the global commons.

Landfill

Landfill is the technical term for filling large holes in the ground with waste. These holes may be specially excavated for the purpose, may be old quarries, mine shafts and even railway cuttings. More recently, mountains or islands made of waste have been created.

Apart from wasting what are potentially resources, landfill sites produce climate changing gases such as methane which is some 25–27 times more powerful than CO_2 as a greenhouse gas but remains in the atmosphere only for about ten years and so loses its greenhouse effect quickly compared to CO_2 which remains in the atmosphere significantly longer. The current atmospheric concentration of methane is 1.8 ppm or $25 \times 1.8 = 45$ ppm CO_2 equivalent. This is 12% of CO_2 concentration and its growing 2.5 times as fast (University of Oregon, 2004). The current concentration of CO_2 on the other hand is around 370 ppm.

Landfill can cause ill health in the area, lead to the contamination of land, underground water, streams and coastal waters and gives rise to various nuisances including increased traffic, noise, odours, smoke, dust, litter and pests. 'The real cost of dumping is not borne by the producer of the waste or the disposer, but by the people whose health and property values are destroyed when wastes migrate onto their property and by the taxpayers who clean it up' (Sanjour, 2001).

Recently the headline in *The Mercury*, a local Tasmanian newspaper, said 'Call for Landfill Cancer Enquiry' (2004) The paper went on to say that

> The site of an old Hobart landfill should be tested for groundwater contamination after a number of cancers have been discovered in the area over 30 years. ... The opposition first pushed for testing of the old tip last June and renewed its call this month after the release of the State of the Environment Report. This said that 100 of Tasmania's 176 landfill sites may have contaminated ground water. ... The potential damage to an ecosystem or to a community's health can be significant and the costs of

cleaning up the site can be high. ... A Department of Mineral Resources study in 2002 examined 10 tip sites and found potentially dangerous groundwater contamination in six.

According to the EPA

Resources that simply become waste are not available for future generations and extraction and harvesting of additional resources can have long-term environmental impacts. Even as we implement protective waste management programs, toxic chemicals can still find their way into the environment throughout the life cycle of materials in extraction, production, transportation, use, and reuse. Persistent, bio accumulative and toxic chemicals released into the environment can present long-term risks to human health and the environment, even when released in small quantities. (EPA, 2003)

7.5.3 Combined impacts

Most techno-processes have combined impacts at both the take and waste end. Consider some of the larger materials flows.

Resource use impacts

There is an explosion of substances through the techno-process many of which have damaging linkages with the geosphere-biosphere associated with them. Examples abound of the use of resources we extract from the planet that result in unsustainable linkages. Mentioned above was the example of a modern paint. Global warming is the biggest problem and the major contribution from fossil fuel burning, cement production and gas flaring during the period 1950 to 1995 is shown in Fig. 7.10.

The growth of energy consumption is closely correlated with increases in gross national product which is a measure of economic development. The current consumption patterns of fossil fuels, as well as contributing to emissions, is not sustainable and neither is the production of cement (see 'Cement production' on page 292) The emissions from the burning of fossil fuels and production of cement are shown in Fig. 7.10.

Fossil fuels

Fossil fuels are at the root of many of our problems, they powered the industrial revolution which attracted millions from self-sufficiency in sustainable country and village economies to cities with factories that were superficially more efficient but not in a holistic or ecological sense. Fossil fuels are not only a causal factor in the development of the environmentally damaging techno-process (Fig. 7.11) but it can also be seen that burning

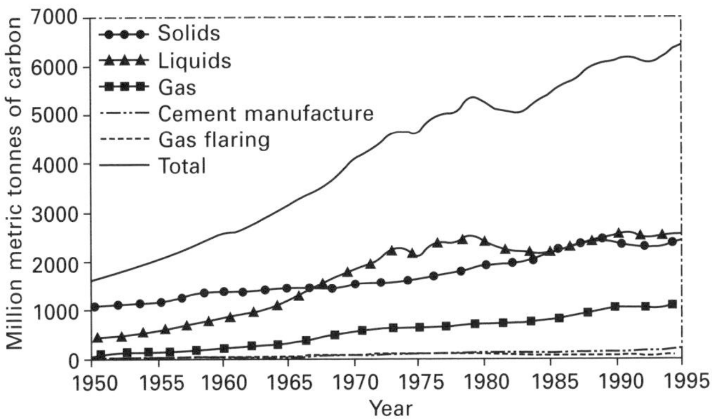

7.10 Global CO_2 emissions from fossil fuel burning, cement production, and gas flaring for 1950–1995 (Marland, *et al.*, 1999).

them to produce energy is responsible for the greater proportion of anthropogenic[8] emissions.

There is a strong need to kick the fossil fuel habit, however, this is unlikely to happen unless alternative sources of energy become more economical. This may be sooner than we think as 'just under half of the world's total endowment of oil and gas has been extracted already, and that output will begin to decline within the next five years, pushing prices up sharply' (Campbell, 2002). Most geologists, however, concur that thirty rather than five years is more likely (Fig. 7.11).

Even if we kick the fossil fuel habit before it kicks us, it will take centuries to bring the carbon balance back down to the levels of the 1950s. Abatement and sequestration on a massive scale are essential. 'Complementary to traditional areas of energy research, such as improving energy efficiency or shifting to renewable or nuclear energy sources, carbon sequestration will allow continued use of fossil energy, buying decades of time needed for transitioning into less carbon-intensive and more energy-efficient methods for generating energy in the future' (NETL, 2004). It is encouraging that there is a slow shift to renewable energy as Fig. 7.12 shows.

Australia is one of the biggest energy consumers per capita in the world and has tripled consumption in the past 20 years with 94% of our current energy consumption from fossil fuels (*The Sunday Tasmanian* 2001). As a consequence Australia contributes approximately 1.5% of the world total anthropogenic greenhouse gases. This, however, represents 33.3 tonnes per capita emission for 1990, the highest of all the OECD countries, including the USA, which records the next highest emission of 23.0 tonnes per capita (Durie, 1998).

[8] Anthropogenic – human sourced.

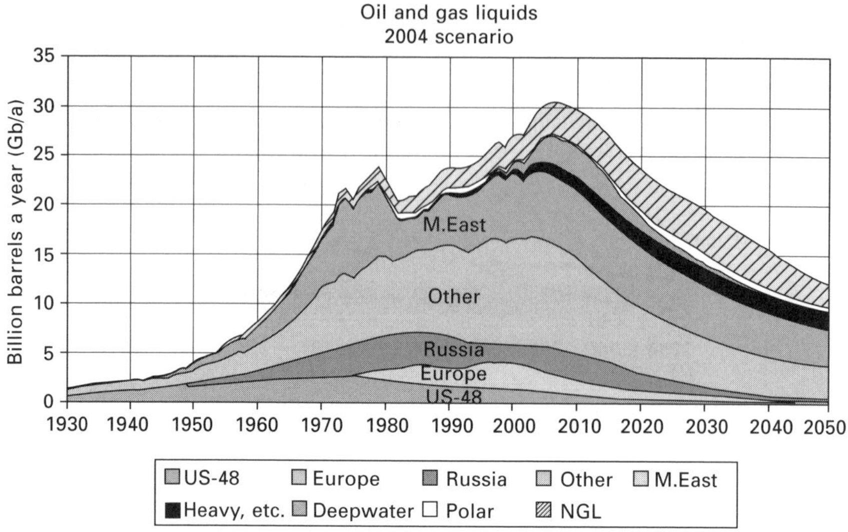

7.11 Oil and gas liquids 2004 scenario – updated by Colin J. Campbell, 15 April 2004, for the ASPO web site (Campbell, 2004).

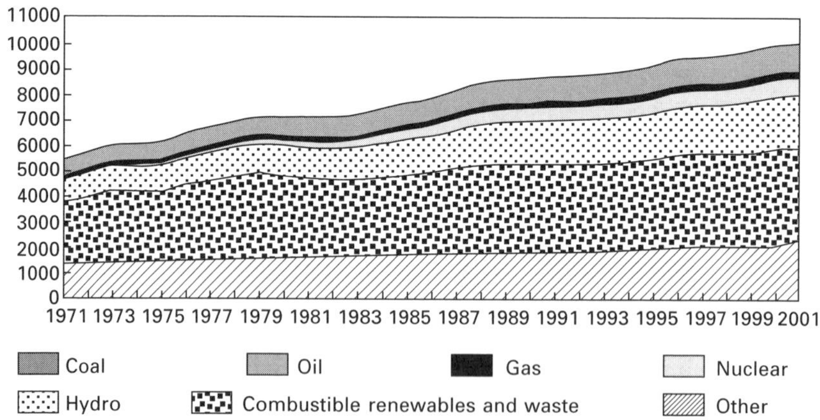

7.12 Evolution from 1971 to 2001 of world total primary energy supply (IEA, 2003).

Cement production

Another major source of CO_2 emissions is the production of cement. In 2003 the world produced 1.86 billion tonnes of cement (Van Oss, 2004) – about a quarter of a tonne for every man, woman and child on Earth. Production is now probably well over 2 billion tonnes. Over 13 billion tonnes of concrete are the result which is over 2 tonnes per person per annum on the planet. Contrary to lay understanding, Portland cement concretes have low embodied energies and relatively high thermal capacity compared to other building materials such as aluminium and steel and are therefore relatively environmentally friendly (Fig. 7.13). However, concrete, based mainly on Portland cement clinker, is the most widely used material on Earth. As of 2004 some 2.00 billion tonnes of Portland cement (OPC) were produced globally (USGS, 2004) (Fig. 7.14), enough to produce over 7 cubic km of concrete per year or over two tonnes or one cubic metre per person on the planet.

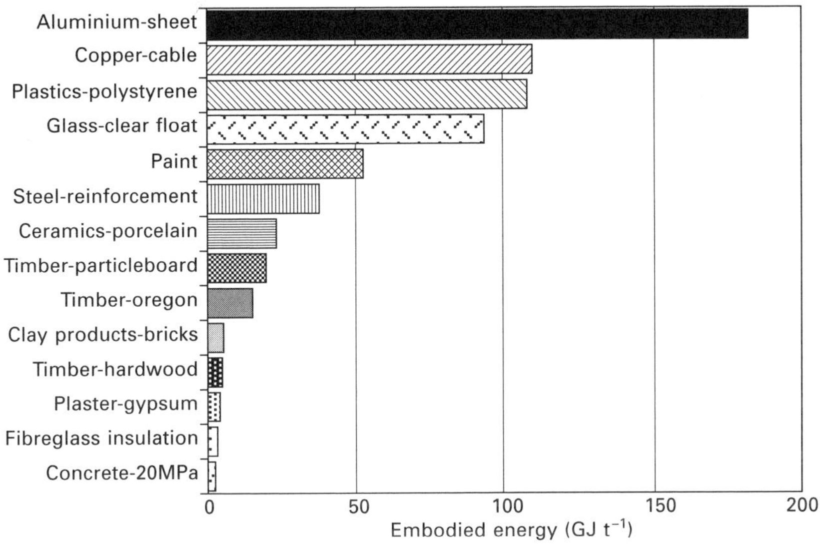

7.13 Embodied energy of building materials (Tucker, 2000).

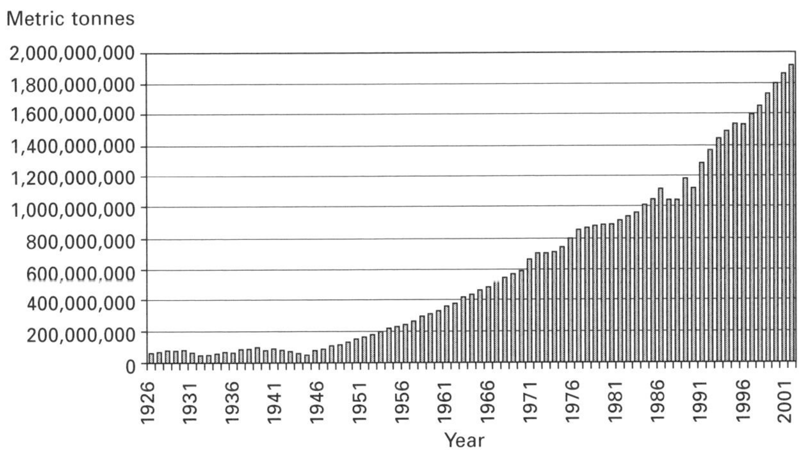

7.14 Cement production = carbon dioxide emissions from cement production 1926–2002 (Van Oss *et al.*, 2003).

As a consequence of the huge volume of Portland cement manufactured, considerable energy is consumed (Fig. 7.15) resulting in CO_2 emissions. CO_2 is also released chemically from the calcination of limestone used in the manufacturing process. Various figures are given in the literature for the intensity of carbon emission with Portland cement production and these range from 0.74 tonnes CO_2/tonne cement (Hendriks *et al.*, 2002) to as high as 1.24 tonne determined by researchers at the Oak Ridge National Laboratories (Wilson, 1993) and 1.30 tonne (Tucker, 2002). The figure of one tonne of carbon dioxide for every tonne of Portland cement manufactured (Pearce, 1997) given by *New Scientist Magazine* is generally accepted.

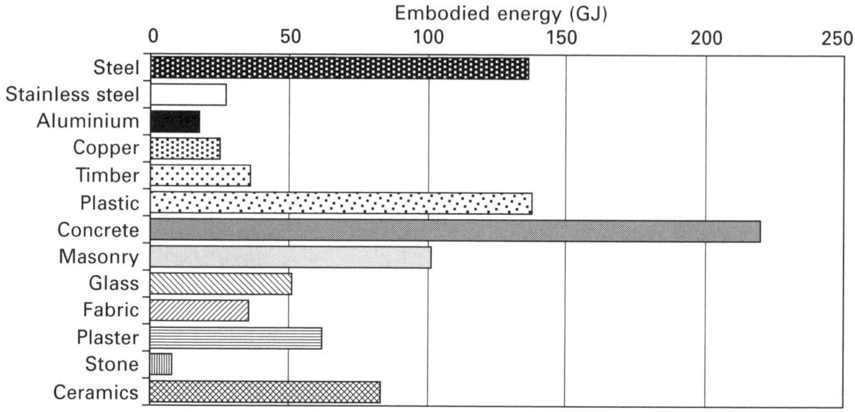

7.15 Embodied energy in buildings (Tucker, 2000).

Because of the huge volume used, Portland cement concrete is the biggest single contributor to embodied energy in most buildings. As a consequence Portland cement concretes account for more embodied energy than any other material in the construction sector (Tucker, 2000). The manufacture of OPC is one the biggest single contributors to the greenhouse effect after the burning of fossil fuels, accounting for between 5% (Hendriks, *et al.*, 2002) and 10% (Pearce, 1997) of global anthropogenic CO_2 emissions. Global production of cement is likely to increase significantly over the coming decades as:

- global population grows
- GDP grows[9]
- urban development continues
- industrialization increases.

Associated with such huge usage and growing demand is the enormous potential for improvement in properties and sustainability. New cements that

[9] Especially in China, India and other so-called underdeveloped countries.

may solve this problem are being developed by TecEco Pty Ltd based on blending magnesium oxide with other hydraulic cements. Magnesium is a small lightweight atom and its carbonates contain proportionally a lot of CO_2. TecEco have also designed a kiln that combines calcining and grinding in a closed system whereby CO_2 can be captured and geologically sequestered during manufacture. The carbonation that then occurs in TecEco eco-cement concretes used in the built environment equates to sequestration on a massive scale. As Fred Pearce reported in *New Scientist Magazine* (Pearce, 2002), 'There is a way to make our city streets as green as the Amazon rainforest'.

Biosphere impacts

What distinguishes our planet from any other we have yet discovered is that there is life, comprising carbon-based molecules which have evolved in a delicate balance with the rest of the atoms and molecules that make up planet earth. Living matter is different from dead matter in that it contains genetic coding and has the ability to join and recombine atoms and molecules from the environment to build new replicates of itself for the future. This delicate balance is characterized by the flow of substances from the dead world to living matter and in reverse and has gone on for billions of years. Since the dawn of history and in particular the industrial revolution, our ability and willingness to manipulate everything around us has, however, wreaked havoc with living life forms in the biosphere.

Seventy percent of Earth's coral reefs will cease to exist in the next 40 years if the current warming trend continues. Every year, around one thousand (1000) species are driven to extinction. The loss of tropical forests is especially alarming as they are the habitat of over 50% of the world's species of plants and animals. The list of impacts is a long one. There is no doubt that we are seriously disturbing the 'web of life' and evidence is mounting of worsening future problems unless we can reduce the impact of the take and waste associated with the techno-process.

Many of the transformed atoms and molecules produced as part of our techno-process activities are thrown 'away', entering the global commons where they negatively affect our wellbeing and that of other living organisms on the planet. There is no such place as 'away'. Persistent chemicals are not confined to the area in which they are disposed of. Gases like Chlorofluoro hydrocarbons (CFCs) move around in the atmosphere and are now found all over the globe. CFCs were developed during the 1930s and found widespread application after World War II. They are halogenated hydrocarbons, mostly trichlorofluoromethane and dichlorodifluoromethane and were used extensively as aerosol-spray propellants, refrigerants, solvents, and foam-blowing agents because they were thought to be non-toxic. They are non-flammable and readily converted from a liquid to a gas and vice versa.

CFCs were, however, found to pose a serious environmental threat because in the stratosphere they broke down under ultraviolet light releasing chlorine which destroyed ozone. A hole over Antarctica developed in the ozone layer which shields living organisms on earth from the harmful effects of the sun's ultraviolet radiation.

> Because of a growing concern over stratospheric ozone depletion and its attendant dangers, a ban was imposed on the use of CFCs in aerosol-spray dispensers in the late 1970s by the United States, Canada, and the Scandinavian countries. In 1990, 93 nations agreed to end production of ozone-depleting chemicals by the end of the century, and in 1992 most of those same countries agreed to end their production of CFCs by 1996. (c-f-c.com, 2004)

Pollution by heavy metals is ubiquitous in the global commons and of particular concern. The term 'heavy metal' refers to metals that are relatively high in density and toxic or poisonous even at low concentrations. Examples include mercury (Hg), cadmium (Cd), arsenic (As), chromium (Cr), thallium (Tl), and lead (Pb). They are particularly dangerous when released as wastes because they tend to bio-accumulate up the food chain. Marine organisms can, for example, consume a particularly dangerous form of mercury called methylmercury. When fish eat these organisms, the methylmercury is not excreted, but retained in bodily tissues. The older the fish and the more organisms further down the food chain they have consumed, the greater the amount of methylmercury in their tissues. The accumulated methylmercury is concentrated as it is passed up the food chain and any organism at the top, such as us, faces a serious risk of mercury poisoning by eating such fish.

Heavy metals can enter water supplies from industrial and consumer waste, or even from acidic rain breaking down soils and releasing heavy metals into streams, lakes, rivers, and groundwater. We are constantly being alerted to the fact that living organisms in the far reaches of the globe contain significant traces of organic and metallic pollutants and that the deepest marine sediments, remotest glaciers and icecaps are contaminated. The list of contaminants is frightening and long. Of even greater concern are radioactive wastes or accidental releases such as occurred at Chernobyl.

7.6 Managing change

> If you want to truly understand something, try to change it. – Kurt Lewin

> He who rejects change is the architect of decay. The only human institution which rejects progress is the cemetery. – Harold Wilson

Change is inevitable! In the case of the environment, desirable or we will go the way of the dinosaurs. Change is the solution.

7.6.1 The need for change

That earth systems are in trouble is evidenced by a variety of symptoms including global warming, pollution, salinity, loss of biodiversity and many others on a long list. The most frightening to our own survival is loss of oxygen in the atmosphere and increase in carbon dioxide causing global warming and significant climate change. Climate change is in the news all the time. As Mark Twain said, 'Everybody is talking about the weather but nobody does anything about it.' Britain, Germany and one or two other countries are the only ones that appear to have grasped the significance of the problem.

According to the UK Hadley Centre, most of the world's forests will begin to turn from sinks to sources – dying off and emitting carbon – by around 2040 (Gelbspan, 2004). Recently Robin Batterham, Australia's chief scientist, said he supported the Federal Government's decision not to ratify the Kyoto Protocol on climate change because the reductions it set were not high enough. 'I'm talking about enormous reductions – 80 per cent by the end of the century,' Dr Batterham said. 'Fifty per cent by 2050, I think, is realistic' (Batterham, 2004) and in the UK chief scientist Sir David King recently said that climate change is the biggest problem that civilization has had to face in 5000 years (Brown, 2004).

The scientific community and conservationists are now rigorously debating source reduction as the only available main option for ultimately reducing the greenhouse effect. Even the USGS *Commodities Annual 2000* has this to say on the cement pages (USGS, 2000). 'There continued to be concern over the environmental impact of cement manufacture, particularly the emissions of carbon dioxide and cement kiln dust (CKD).' The recent opinion of Rajendra Pachauri, chairman of the United Nations' Intergovernmental Panel on Climate Change is that Kyoto looks like it is a treaty but goes nowhere near far enough (Doyle, 2004). In a recent book Meyer Hillman has proposed carbon rationing as the only viable means to keep the carbon dioxide concentration in the atmosphere below 450 ppm (Hillman and Fawcett, 2004). Others believe that abatement and sequestration on a massive scale are essential 'because we will not be able to free ourselves from the fossil fuel habit' (NETL, 2004).

It is important in all this discussion to discern what can realistically be achieved and what cannot. As in dieting, whereby weight loss is achieved most effectively by both exercise and reduction in the consumption of food, reductions in atmospheric carbon will best be achieved by action in all possible ways. Politically, sequestration is the most acceptable as it does not involve one person, state or country attempting to regulate the consumption of carbon by another, for which force will ultimately be required. Who or what is to blame? Are there just too many of us or shall we blame technology? Is technology the culprit or a potential saviour? Or are we guilty through our naivety in using all that consumerism post the industrial revolution has delivered us. Is there any way out of this dilemma? Can the industrial juggernaut we have unleashed deliver more sustainable outcomes?

Village and tribal economies of yesteryear and still prevalent in some third world countries were actually very sustainable. Everybody had a role and all were interested in the ability of the surrounding area to sustainably continue to produce. One of my first supporters and an early shareholder, Bob Johannes, a marine scientist and Pew fellow, wrote a classic textbook about island ecologies/economies centred on marine life in a lagoon (Johannes, 1992). For centuries the people on the island lived in harmony with nature. They managed to live well without fridges and washing machines and all the other trappings of modern life. Do we have to dematerialise to quite such an extent to live more sustainably or can we preserve some of the trappings of modern living we are addicted to? I cannot answer all these questions but I do think we can live more sustainably without affecting our lifestyles too much and this chapter is about how. The bad news is that it would help if we could curb our population growth and dematerialize/deenergize to some extent!

The bottom line is that population, along with consumption and technology, are factors in waste and pollution. It is not population growth *per se* that is the culprit however; it is the increase in flows through the techno-process previously defined. The consumption of resources and level of production of wastes or pollutants impact on the environment. The severity of impact is a function of the number of people, the amount each person consumes, and the amount of waste created in the whole process from extraction, production and packaging to the consumer and his or her dustbin or sewage outlet.

Several efforts have been made to put numbers on the relative share of responsibility for the rising impacts and pollution. Environmentalist Barry Commoner studied examples from the United States between 1946 and 1968. Population growth accounted for only 14 to 18% of the increase in synthetic organic pesticides, in nitrogen oxides and in tetraethyl lead from vehicles. It was responsible for only 7% of the increase in non-returnable beer bottles and a mere 3% of the increase in phosphorus from detergents. In almost every case, technology was the dominant factor. A later study by Commoner of nitrates, cars and electricity in 65 developing countries came to similar conclusions (Commoner, 1972).

On the other hand, studies of changes in air pollutants (SO_2, nitrogen oxides, smoke and CO_2) in countries of the Organisation for Economic Co-operation and Development (OECD) between 1970 and 1988 showed technology as a downward pressure in all four cases – mainly through increased energy efficiency in the case of CO_2 and nitrogen oxides, and through cleaner technology in the case of SO_2 and smoke. Population growth was responsible for a quarter of the upward pressure on emissions, while consumption was responsible for three-quarters (Harrison, 1992).

These studies substantially miss the point however. The answer is that we are to blame. It is true to say that fewer of us would produce less of a problem. It is also true to say that if we did not have the tools, devices and processes that technological advances have provided we would not have made quite such a mess of the planet. So where to from this realization? First understand human nature. We are unlikely to want to consume less or give up our cars and appliances. We are not going to dematerialize/de-energize and do without modern conveniences, transport, etc. On the contrary, third world countries are rushing into materialization/energization at an alarming rate.

The answer is that we must move technology into new and more sustainable paradigms that have a much lighter impact on the planet. A materials revolution particularly in the built environment that may well solve the problem is the most acceptable way. Let's take up the challenge and get on with it!

7.6.2 Getting over barriers

Learned helplessness

Is it is possible that why we seem so unable to move forward in relation to global warming and other environmental issues is a learned helplessness (Seligman, 1990) in relation to what is an absurd consensus? Leith Sharp, of Harvard University (Sharp, 2003) says quite rightly

> People are conditioned to conform to group perceptions and to doubt and withhold their individual perceptions if they are in conflict with the shared reality of those around them. This has enormous significance when considering how people are currently responding to the demise of the planetary systems that support human life. The degree of inaction around this profoundly life threatening situation can perhaps best be explained by viewing our state as a massive 'absurd consensus' that is the product of our social conditioning which has enforced our subservience to, and blind confidence in, shared societal constructs of reality.

What, then, is it that has conditioned so many of us to learn helplessness in the face of the absurd notion that the state of the planet and problems such as global warming are something we cannot do anything about?

Societies we live in today extol the virtues of the self. According to Professor Seligman (1990) 'in the past quarter century, events occurred that so weakened our commitment to larger entities as to leave us almost naked before the ordinary assaults of life.' We have lost hope in the capacity in society to cure basic human ills and so in many ways have shifted our commitment to the self. Others call the age of rampant individualism we live in the 'me' age. Phillip Sutton, director, Policy and Strategy, Green Innovations Inc. says

> We have not taken steps to mobilise people, we need to think very carefully about how the processes of imagination might be catalysed and how the mobilisation of people might be accomplished. What is blocking these processes, what is or might drive them forward? How can we trigger effective action? (Sutton, 2002)

The media have a vital role in conditioning us all to believe that global warming and environmental problems are not inevitable and that we should and can do something about it.

The role of the media

Public understanding of climate change and other environmental problems is very important and conditioned by the media. People need convincing that they can do something about global environmental problems and that they must get on with it. The mass media plays a central role in the formation of opinions and images about climate change. The coverage by the press, radio and television of climate change and other environmental issues and how readers, listeners and viewers receive and process the information delivered to them is very important.

Social responses to climate change and other environmental issues are not exclusively determined by scientific knowledge. On the contrary, the media has an interpretative role. Using social actors in complex communication processes the cultural meaning of such environmental phenomena is shaped. Public perceptions of social risk evolve as a result of diverse patterns of interpretation from experts, policymakers, journalists and citizens. These risk constructs are regulated by culturally specific decision-making processes. The analysis of societies' ability to adjust to global climate change and how scientific knowledge is integrated in social and individual sense-making about environmental issues relies on knowledge about these interpretative processes.

The goodwill of the media is essential. Players in the media must take seriously their important role of conditioning people to action on sustainability. By exalting the achievements of our environmental entrepreneurs, instilling a philosophy of 'can do' and extolling us all to work for sustainability the media can help lead us away from threats to our very survival.

The role of educationalists

Our teachers have an important role as well. We must learn from the outset the value of, and how to relate to, the environment and a 'can do, can fix' attitude.

Education not only provides the scientific and technical skills required, it

> also provides the motivation, justification, and social support for pursuing and applying them. Education increases the capacities of people to transform their visions of society into operational realities. It is for this reason that education is the primary agent of transformation towards sustainable development. It is also for this reason that society must be deeply concerned that much of the education presently on offer falls far short of what is required. Improving the quality and coverage of education and reorienting its goals to recognize the importance of sustainable development must be among society's highest priorities'. (UNESCO, 2004)

Imperfect markets

Markets exist for the exchange of goods and ideas, but is the long-term view required for sustainability likely to cause changes in the way they operate? According to Fiona Wain from Environment Business Australia (Wain, 2004b)

> There is a school of thought which promotes the concept that human intellect will invent new technologies to avert disaster. But human ingenuity relies on market demand, and markets fail the system completely when they cannot respond quickly to threats that are *perceived* as slow and insidious.
>
> In order for technology to provide solutions to environmental challenges two things must happen:
>
> - The market must receive signals that relate to current challenges, the need for change, and the value of innovation.
> - The market must then pull that innovation through by championing its value.
>
> This in turn attracts investment and creates the longer term framework for amortisation and profit which in their turn encourage further R&D and benchmarking.

Fiona Wain (2004b) is right in so far as new technologies will help us avert disaster. This is because to some extent sustainability and resource efficiency are the same thing, I do not however believe her pessimistic view of what markets can deliver is justified. Due to the imperfections of markets including their inability to in part recognize the value of sustainability, see across technical paradigms or toward the longer term, there is a role for governments to intervene or to at least set trends.

Economies of scale

Economies of scale were discovered by the likes of Henry Ford and remain as a difficult barrier for any new technology or player in the industry to overcome. Arguably economies of scale are as large a barrier as the formula based standards that support the status quo. To nurture new technologies, a level playing field and incentives are required. As it is the role of governments to govern for the common good providing such business conditions is their prerogative (see Appendix on page 340). Even though governments through policy can introduce change that brings about economies of scale it is important that building technologies that seek sustainability are also eventually fundamentally economically viable. Given long-run economies of scale, the new calcium-magnesium cements from TecEco are more economic as well as better for the environment.

Other economic barriers to innovation and market entry

Economic barriers to innovation and market entry include the high cost of manufacturing plant and equipment, monopolistic and oligopolistic behaviours, transport costs, etc.

Barriers in the construction industry

The construction industry has inherent characteristics that restrict the rate of adoption of new technologies required to meet the challenges of the future and achieve greater sustainability. These include:

- conservativism, industry dogmatism and culture
- formula based standards
- expensive manufacturing infrastructure
- low margin product
- economic barriers to innovation and market entry.

Conservatism, dogmatism and culture

Buildings and construction methods have not changed all that much since the times of the Romans and in the past when they have changed they have not changed quickly. New technologies are usually slow to be adopted and this conservatism is entrenched in modern-day time by a culture of litigation and fault finding inducing a fear of change. Engineers also tend to think within a standards framework and not beyond. There is a saying in the industry that nobody gets fired for specifying bricks and mortar. In regional areas things tend to get done in certain ways and there are definite differences from country to country. Sustainability in construction does not mean loss of cultural influence and dogmatism should never stand in the way of common sense.

Formula or prescription-based standards

If the construction industry is to play a major role then it must shake off the shackles of the past and innovate. To do this it will have to abandon the formula-based approach to standards which grew out of the industrial environment of the early twentieth century. Standards are important for the protection of people, but if prescription based, stifle creativity. Why the industry is so bound to prescription-based standards from the past must be questioned. Performance-based standards make much more sense. National governments can force change by legislating for performance-based standards. If they did, innovation would return to reward us with much more rapid change towards sustainability.

Expensive manufacturing infrastructure

Infrastructure is expensive and particularly expensive in the business of manufacturing for construction. Plant for brick making, timber and plywood products, cement and so on costs a lot of money and the consequence is a cost barrier for change.

Low margin product

Cement, bricks and many other basic materials used in construction tend to be standardized, with little distinction between one brand and another. The result is price competition and low margins.

7.6.3 The economics of change towards sustainability

Our understanding of the flows and interactions in the global commons is very inadequate. The widely held view is that sustainable management strategies are complex to devise and politically difficult to introduce. It is important in all this discussion to discern what can realistically be achieved and what cannot. As in dieting, whereby weight loss is achieved most affectively by both exercise and reduction in the consumption of food, reductions in atmospheric carbon and other reductions in our impact will best be achieved by action in all possible ways. In relation to global warming political sequestration is the most acceptable as it does not involve one person, state or country attempting to regulate the consumption of carbon by another, for which force will ultimately be required.

What if sequestration and waste utilization on a massive scale was economic? Underlying economic processes are human behaviors. What this mean is that if harnessed, human behavior would deliver greater sustainability. How then can we achieve this? Economics is both a servant and a master. Economic forces have been behind most wars and most people are slaves to economics, yet we can be the masters. Working for sustainability, market orientated forces will make all the difference as people would demand much more sustainable products. The supply curve would shift to meet what was demanded and eventually a new more sustainable equilibrium would become established (see Fig. 7.16). The challenge is how to move the supply and demand of resources towards more sustainable outcomes.

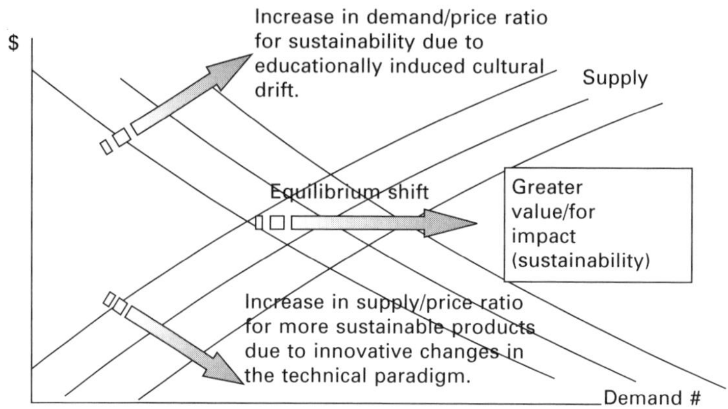

7.16 Achieving sustainability as an economic process.

A systems approach would define more sustainable outcomes as being of greater value for less impact. Herbertson and Sutton in *Green Processing* define value as meeting human needs economically and impacts as being harmful consequences for natural systems (see Fig. 7.17).

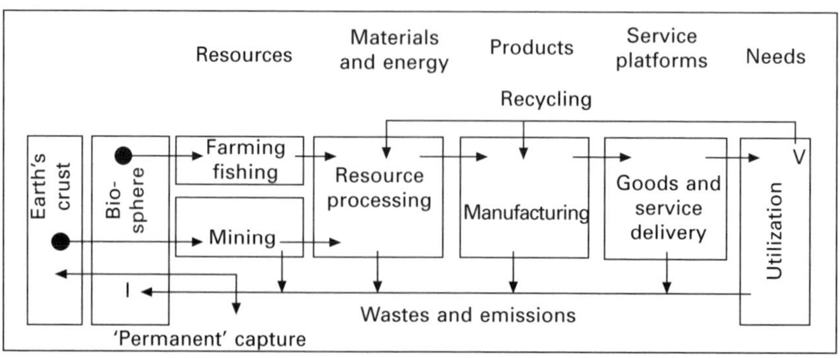

7.17 A systems perspective of value for impact (V/I) (Herbertson and Sutton, 2002).

To move towards economically driven sustainability we need to:

- Get over the idea that we are helpless and cannot do anything about the problem and understand that sustainability is good business (see 'Learned helplessness' on page 299 and section 7.10).
- Cultural change (as individuals demand sustainable products, as communities stimulate and harness human behaviours which underlie the demand for sustainable products) can be achieved through cultural change push by the media, educationalists, governments and other leaders (see below).
- Deliver more sustainable technologies through innovation (see below).

7.6.4 Drivers for change

Apart from the need for sustainability there are other irresistible forces of inexorable magnitude that will force change.

Cultural change

My experience so far is that it will be younger people who move the sustainability agenda forward. The greatest opportunity is in cities where architects and specifiers are leading the way. Many new developments are happening. I am, for example, involved with the Low Carbon Network and they have built some very innovative new sustainable structures. The media and educators are working to help spread the word and induce cultural change. The media publicize our environmental problems because sensationalism sells and maybe because of a trace of social responsibility. Educators teach about them because it is the right thing to do.

The challenge is to get sustainability happening in mainstream construction as that is where change will have the greatest impact. Purchasers are now aware that more sustainable buildings save heating and cooling costs and insulation is the norm. Multistorey buildings with a fraction of the lifetime energy costs are now routinely being constructed in many countries because they sell better. In time the materials buildings are made of will also change with more sustainable outcomes.

Paradigm shifts in technology

Small improvements in the technical deliverance of more sustainable products, materials and processes are important but it is the paradigm shifts in technology that will deliver results. In the past, improvements in technology brought us health and living standards that resulted in the explosion in population that has taken place in the last century. Along with increases in population and living standards came pollution at ever increasing levels. Fresh water is rare in some countries and even the air we breathe is often contaminated with the fallout from human activities. Technology brought us acid rain, pollution and the greenhouse gas problem and some scientists such as myself now believe that only technology can extricate us from it.

> The recognition, throughout the European Union and indeed the rest of the world, that sustainable growth should be a controlling process for guiding the development of technology through the millennium and beyond, is now well established.
>
> The more difficult problem is defining the content and thrust of the effort necessary to convert the recognition into practice and then demonstrating, through the best practice, the techniques and technology appropriate to reinforce the value of sustainability. The value can and should be measured by both environmental as well as financial gains and there is no more logical start point than Targeted Research Action on Waste Minimisation and Recycling. (TRAWMAR) (MIRO, 2002)

Given the scale of the problem, improvements in current technologies will not be enough. Paradigm shifts in technologies that redefine materials, hence the resources and molecular flows that underlie their movement through the economy are essential. 'By enabling us to make productive use of particular raw materials, technology determines what constitutes a physical resource' (Pilzer, 1990). The technological base of world economies will have to change strongly towards sustainability for there to be a significant reduction in anthropogenic global greenhouse gas emissions, wastes and the other environmental problems.

New materials with low embodied energies and emissions that deliver more than just strength or durability are urgently required. Many of these will be composites combining properties previously considered mutually exclusive such as thermal capacity and insulating ability. To survive as a species we must find ways of sustainably creating the urban environment. According to Dr Joe Herbertson (2004), 'Taking a fundamental, systems approach is the key to major innovation; moving beyond current best practice to sustainable solutions.'

I systematically worked on the problem of emissions and waste utilization in concretes and that is how I developed TecEco, tec and eco-cements and a kiln technology that captures CO_2, combines grinding and calcining and uses solar or solar derived power for efficiency. I saw the big picture long before I worked out the detail. The TecEco kiln, tec and eco-cement technologies are examples of benchmark paradigm shifts in technology. They demonstrate that it is possible to think laterally and deliver technologies that are more sustainable in the context of our greater understanding of earth systems.

Robotics as a driver

Challenging the traditional construction paradigm and driven by potentially lower costs is robotics. In the USA and elsewhere in the world researchers are looking at using robots to literally print buildings. It is all quite simple from a software, computer hardware and mechanical engineering point of view. The problem is in developing new construction materials with the right flow characteristics[10] so they can be squeezed out like toothpaste, yet retain their shape until hardened.

Once new materials suitable for the way robots work have been developed economics will drive their acceptance for construction. The one material fits all purposes approach will increasingly have only limited relevance. Concretes, for example, will need to evolve from being just a high strength grey material to a smorgasbord of composites that can be squeezed out of a variety of nozzles for use by a robotic workforce for the varying requirements of a structure (Fig. 7.18). Various materials from structural containing fibers for reinforcement to void filling and insulating will be required and like a color inkjet printer will be selected as specified by the design.

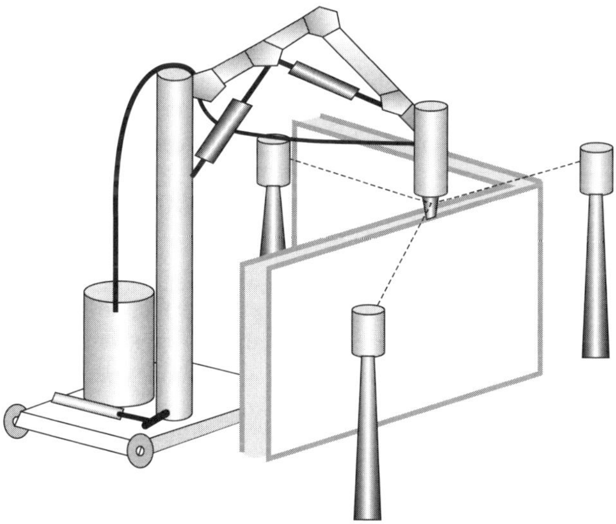

7.18 Robotic construction of cementitious composite walling.

Accuracy will be far greater than currently possible, wonderful architectural shapes as yet unthought of will be used and fibers will provide reinforcing. Walls will most likely have a low strength foamed insulating cementitious material between the faces making services easy to add at a later time. Conduits could also be provided by design.

[10] Bingham plastic rheology.

The use of robots in construction will reduce the waste of new construction materials immensely. Just as an inkjet printer uses only the right amount of ink, only the exact amount of material will be used. The introduction of robots to construction will also mean more wastes can be utilized for building materials. More self-hardening materials will be required, not fewer, as well as mineral binders, like TecEco's new cements, that have the right flow characteristics and obvious advantage of being able to utilize a large quantity of wastes.

7.6.5 The process of change

Four important threads are discussed:

1. Reducing, Re-using, Recycling and Recovering.
2. Re-engineering the materials we use.
3. Changing the molecular flows using non-fossil fuel energy.
4. Realizing that Sustainability is Good Business Sense.

Consider the techno-process. It has an input and output rate and volume dimension. It also involves the flow of matter. Everything between the take and waste is a material of one sort or another. The take and waste of the techno-process both have impacts and linkages to the wider geosphere-biosphere. Most noticeable are at the macro level. Most dangerous may well be at the molecular level. We are producing too much CO_2, too many mobile heavy metals, too many compounds containing halogen carbon double bonds and so on.

The techno-process is uniquely driven by human behavior. All sorts of people are involved in a value chain and each and every one of them is thinking about earning a living so the solution must also be economic (see section 7.6.3) otherwise it just will not happen. The techno-process globally is huge and diverse. There is therefore no simple solution because it is so pervasive and diverse problems associated with it must be tackled in every way possible. One set of approaches is to, step by step, change the techno-process until it becomes a climax industrial ecology characterized by:

- using more renewable and fewer non-renewable resources
- slowing down the rate of the process
- slowing down the volume of the process
- integrating the process beyond the enterprise level (implying much more recycling)
- reusing and remanufacturing
- wasting only what is biodegradable or can be reabsorbed by natural processes.

These changes are embraced by the popular catchwords of Reducing, Re-using, Recycling and Recovering (see section 7.7).

Another way that is connected to the other two ways is to change the materials we use. Materials are everything between the take and waste. Fundamentally changing what materials are composed of and how they are made will change what we take and what we waste. Changing the take and waste will change the molecular flows that underlie the techno-process that are of concern such as the production of CO_2 and other greenhouse gases. The idea that we can solve the problem by re-engineering materials is a major theme of this document (see section 7.8).

A third way is to reduce the impact of the molecular flows (too many CFCs too much CO_2, heavy metals in the global commons, etc.) underlying the flow of materials through the techno-process by reversing them, using processes driven by free and abundant solar or solar derived energy rather than fossil fuels.

The TecEco process for saving the planet involves such a bold step and because it is economic has a good chance of success (see section 7.9). The solution is thus a multi-dimensional and holistic re-tuning of the techno-process. The main weight of our footprint on the planet is in cities. Cities therefore offer huge opportunities for this process. Industrial ecology, implying integration across enterprises so that the waste output of one kind of activity can be resource input for another, is most easily achieved in cities. Cities make possible the treatment of industrial wastes because of the economies of scale and the agglomeration economies of having many similar industries together (Gismondi and Rees, 2004). It follows that sustainability is good business sense (see section 7.6.3). TecEco, my company, has uniquely proposed building cities of carbonate based eco-cement concretes as a way of sequestering massive amounts of carbon and using a wide range of wastes for their physical properties. More information on the exciting new development is available from the website at www.tececo.com.

7.7 Reducing the environmental impact of technology

It is essential that the human race, with all the power it has over the environment, moves rapidly to reduce the impact of the techno-sphere on the geosphere-biosphere before it is too late. This will not happen because it is the right thing to do. William McDonough and Michael Braungart, in an approach inspired by design and nature, make the case that an industrial system that 'takes, makes and wastes' can become a creator of goods and services that generate ecological as well as social and economic value (McDonough and Braungart, 2002). In this section it is argued that the impact of the techno-process can be reduced by introducing paradigm shifts in technology, particularly those relating to materials. It is important that these technologies are also fundamentally economic otherwise they are inefficient and not viable in the long run.

Economic viability attracts investment, and insufficient investment has in the past found its way into sustainability. Natural capital is undervalued. Consider the techno-process:

Take → Manipulate → Make → Use → Waste

It has an input and output rate and volume dimension. If the impacts of the techno-process on the geosphere-biosphere are to be reduced then the rate and volume of flows through the process need to be reduced.

A large proportion of what passes through the techno-process resides in it as the materials with which we build our techno-world (the techno-sphere). The rate and volume of the flow to produce or waste these materials can be reduced by:

- reusing the materials even if in a different form
- reducing the input to produce the same use or 'utility' of the materials
- recycling so that fewer resources are required to be extracted.

By reducing, re-using and recycling the process becomes:

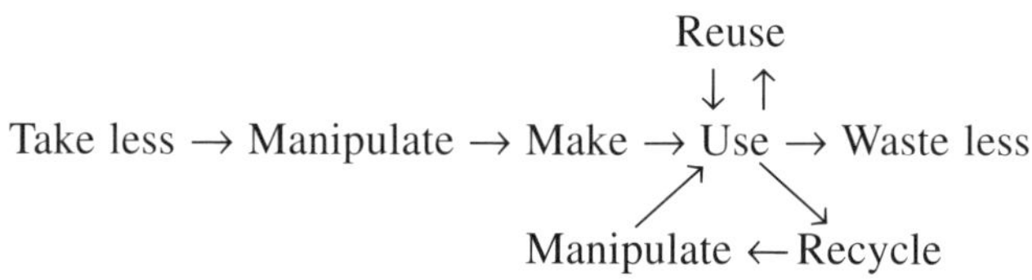

And more desirably:

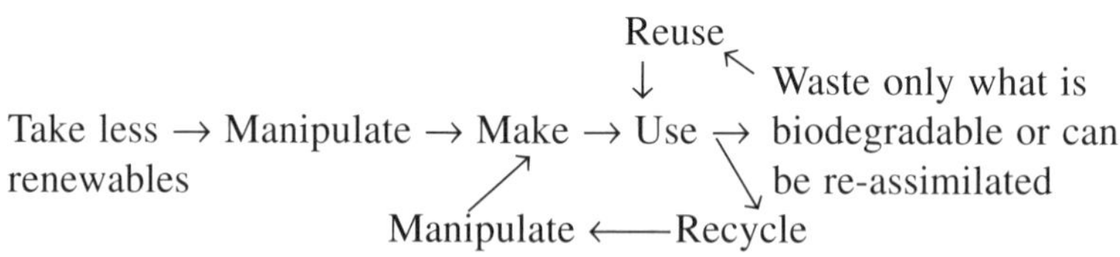

Before reusing and recycling can become economic the main economic hurdles to overcome are the laws of supply and demand and economies of scale. To do this we need to change the technical paradigm. Henry Ford leveraged his success selling cars to devise more efficient methods of production. As a consequence he was able to sell his cars more cheaply, increasing sales, providing more money for innovation, which reduced costs even further and so on. Ford was able to sell more at lower prices and yet make more money by achieving economies of scale. The laws of economics rely on positive feedback loops. Industrial economies of scale tend to increase value linearly, while the laws of supply and demand would dictate that exponentially more is sold or used the lower the price for the same quality. It can however take years before these laws kick in. For example, during the first ten years, Microsoft's profits were negligible. They started to rise in

1985 and then exploded. The experience of Federal Express was similar. The same applies to fax machines and the internet which similarly languished for some time before becoming ubiquitous.

The question is whether the world can wait for an explosion in the recycling business to take place. There is a desperate need to achieve sustainability quickly. What factor or factors are missing? What will make it happen as a matter of profitable economics rather than policy? The trouble is that right now it just costs too much to reuse and recycle for these processes to be driven by economics alone. As a consequence government intervention in the form of regulation (Germany, some other countries and some states in some countries, e.g., South Australia in Australia to some extent) and subsidies (most of the rest of the developed world) are required for what is a desirable social outcome. How can re-using and recycling move beyond the desirable, subsidized by tax dollars to being supported by the push and drag of sound economics? The problem is that in many cases it is more expensive to reuse and recycle than to use newly extracted resources. There would be a rapid turnaround in the sustainability industry if this hurdle could be overcome so that it was cheaper to reuse or recycle.

There are two main costs involved in re-using and recycling. The costs of sorting waste streams and transporting sorted recyclable materials back to a location in which they can be reused.

The second law of thermodynamics (the law of entropy) was formulated in the middle of the 19th century by Clausius and Thomson. Like most natural processes, waste streams tend to follow this law in that wastes at the point of elimination from the techno-process tend to be all mixed up. Disorder is prevalent for two main reasons; things are made with mixed materials and the waste collection process tends to mix them up even more.

Current technical paradigms underlying the techno-process generally require separate inputs. Costs are incurred and waste generated in separating what is required from the balance of material as nature itself rarely concentrates. As mentioned earlier, one study found that around 93% of materials used in production do not end up in saleable products but in waste (Von Weizsäcker, *et al.*, 1997). Re-using and recycling is even more uneconomic because the cost of un-mixing increasingly complex waste streams is prohibitive. After recycling is completed there is the cost of returning the materials back to manufacturers who can use them. Simultaneously dealing with the disassembly/sorting constraints, scale cost, material problems and transport issues during recycling are critical cost challenges.

I once had the pleasure of a discussion with Edward de Bono, the inventor of the words 'lateral thinking' about the TecEco cements I invented. He said that what was needed for market success was a 'killer' application, an application that just could not help but succeed. To get over the law of economies of scale and to make the sorting of wastes economic so that they can be used as low cost inputs for the techno-process new technical paradigms are required. These new paradigms involve redefining the way materials are made, used and recycled:

- including intelligence in materials
- a 'killer' application for unsorted wastes
- adding value to wastes by redefining them as inputs to new materials.

7.7.1 Managing waste efficiently

Others have realized the economies of reusing the output of one process or enterprise in another before they become further mixed up at the tip face. By applying modern intelligent data management techniques databases are being maintained that match wastes from one process or enterprise with the needs of another. This type of industrial integration of 'ecology' is to be encouraged. But what if we could take the concept a lot further? What if we could get over the identification problem and confound the law of entropy at the tip face?

The means to very efficiently sort wastes may just lie in the silicon chip. The cost and size of silicon with embedded memory and intelligence are both falling exponentially. Silicon chips already have a diverse range of uses. For example, they are being used in paint by car manufacturers for identification purposes and one was recently put in the ear of my dog for the same reason. Silicon chips will one day be as plentiful as what they could be embedded in. They will tell us the cost at the check-out, the manufacturer, warranty details who the owner is and what waste stream a robot should put them into when eventually wasted. The only economic hurdle that would remain would be the efficient transportation back to manufacturing points of these waste streams.

The intelligent sorting of materials gets over the entropy law problem but transport is still an issue. What if wastes could be utilized depending on their class of physical properties rather than their chemistry as aggregates or fillers for concretes. Concretes are used everywhere, over two tonnes per person on the planet per annum was the figure mentioned earlier. Utilized in vast volumes waste materials that offered useful broadly defined physical properties such as light weight, tensile strength, insulating capacity, strength or thermal capacity would add value to composites.

There are many wastes that are just too costly to further sort into specific waste streams such as many plastics. There are also waste streams such as mine tailings, furnace sands, quarry dusts and the like for which no particular use could otherwise be found. Glasses tend to share in common many properties as do plastics, wood, ceramics and so on. Glasses are brittle, tough and abrasion resistant. Plastics are generally light, insulating and have tensile strength. What if it did not matter if glasses were mixed up with glasses and plastics with other plastics? The solution is to use these materials in composites for their physical properties rather than for their chemical composition. The problem then becomes one of finding a potentially cheap, un-reactive but strong binder with the right rheology for use by, for example, robots of the future (see page 307).

Plastics, epoxies and other inorganic binders are just too expensive. The choice for durability and cost is a mineral binder. Ordinary Portland cement concretes are not suitable because they are too reactive to use with a wide range of wastes as aggregates or fillers. The breakthrough has been the development of a wide range of blended calcium-magnesium binders with a low long-term pH developed by TecEco that are internally much drier and that therefore do not react with wastes.

TecEco technologies also offer significant abatement or in the case of eco-cements, sequestration. A major advantage of the TecEco technology over all other sequestration and abatement proposals is that the technology itself is viable even without a value being placed on abatement and sequestration. The TecEco technology is but one very important example of where redefining materials can make a big difference to sustainability by, for example, providing composites that utilize wastes. There are other ways improvements in materials will make a big difference.

Gold is a very 'heavy' and potentially toxic metal. Ever heard of it going to landfill? Gold is not wasted because it has too high a value. By underpinning the value for carbon artificially created by the Kyoto treaty with a real value we can solve the problem of global warming and climate change as, driven by economics, people will invent all sorts of ways to capture the gas out of the air. It is the same with other wastes that are detrimental to the environment – we need to find ways of incorporating them as re-inputs to materials in the techno-process, thereby adding value.

7.8 Sustainable materials for the built environment

In the last few years tremendous progress has been made improving the lifetime energy performance of buildings and the architects and engineers who have led the way are to be applauded for suggesting that many products can be used, recycled, and used again without losing any material quality – in cradle-to-cradle cycles (McDonough and Braungart, 2002). Fundamental changes are necessary to materials if we are to achieve real improvements in the value/impact ratio loosely referred to as 'sustainability'. The way forward will be utilizing new technical paradigms defined by innovative new materials such as TecEco tec and eco-cements and geopolymers.

Materials are the lasting substances that flow through the techno-process. They are the link between the geosphere-biosphere and techno-sphere and hence everything between and defining the take and waste. To reduce the impact of the techno-process that describes the flow of these materials from take to waste, it is fundamental that we think about the materials we use to construct our built environment and the molecules they are made of. With the right materials technology, because of its sheer size the built environment could reduce the take from the geosphere-biosphere and utilize many different wastes including carbon dioxide.

Materials used to construct the built environment should, as well as the required properties have low embodied energies, low lifetime energies, and low greenhouse gas emissions when considered on a whole of life cycle basis. They should also preferably be made from renewable resources and easily be recycled or reassimilated by the geosphere-biosphere. The choice of materials in construction has a huge impact on many value properties including weight, embodied energies, fuel related and chemical emissions, lifetime energies, user comfort and health, use of recycled wastes, durability, recyclability and the properties of wastes returned to the geosphere-biosphere. Fundamental changes are necessary to achieve real sustainability and if these are to occur without economic disruption, as the materials we use control the sustainability of the systems we proliferate, the materials paradigm we live in will also have to change. Given the enormous materials flows involved, the obvious place to improve sustainability is the built environment. The materials used determine net emissions, the impact of extraction, how they can be reused and the effects of wastage on earth systems. For the construction industry to progress much further, the impact of the techo-process on the environment must be reduced and materials that are more sustainably made and that deliver greater sustainability in use are required. Materials science will increase in importance as the race to develop materials for the future gets under way. Many materials need a serious makeover or reinventing! Consider some of the ways in which this could occur.

7.8.1 Lighter weight materials

Lighter materials mean lighter buildings which in turn means fewer materials, lower embodied energies and construction costs. For example, materials that are light in weight are not generally strong. It should be technically possible to combine these properties without increasing costs or embodied energies as, for example, with eco-cement formulations for sawdust bricks.

7.8.2 Embodied energies and emissions

Materials have embodied process energy and emissions and, in the case of calcined materials, associated chemical emissions. Without considering chemical releases the amount of carbon dioxide emitted is directly proportional to the energy used during manufacture (the embodied process energy). This is because over 95% of the world's power is generated from fossil fuels. The embodied process energy per unit mass of materials used in buildings (and hence CO_2 emitted) varies enormously from about two gigajoules per tonne for concrete to hundreds of gigajoules per tonne for aluminium. Because some materials such as concrete have associated chemical releases of carbon dioxide, the differing properties of materials, differing quantities required to perform the same task and different design requirements, using these values alone to determine preferred materials to reduce emissions is inappropriate.

A reduction in process energy during manufacture implies greater economic efficiency. Ways of making existing materials or new materials with the same or better functionality and properties but with lower embodied energies are therefore more cost affective. Examples include the new TecEco tec-kiln which combines calcining and grinding and can be heated using solar, wind or waste energy. As a consequence the manufacture of TecEco cements involves much less process energy. Furthermore grinding is only 1–2% efficient and most of the energy required ends up as heat reducing energy required for calcining by around 30%.

Materials such as concrete and timber, having lower embodied energy intensities and hence emissions per tonne, are used for construction in very large quantities whereas the materials with high energy content and emissions, such as stainless steel and aluminium, are used in smaller quantities. The average suburban house would contain in the order of 600–1000 Gigajoules of embodied energy. The most used material is concrete followed by ceramics (brick). Because such huge quantities are used however, by far the most embodied energy (and hence emissions) for the average building is in concrete. Masonry and ceramics (together as a group) follow. Because so much concrete is used in construction generally and because concrete also has associated chemical releases, the effect of using sustainably made tec and eco-cements instead of concrete, masonry and ceramics could cause the embodied energy in an average house in Australia to drop by more than 250–300 Gigajoules and emissions in the order of 15–18 tonnes CO_2

Geopolymers potentially also have very low embodied energies at least until all the fly ash in the world is utilized. Although the production of metakaolin or kandoxi involves process energies, they are generally only used in better quality geopolymers. The best publicly understood example of embodied energies and emissions are those associated with the manufacture of ordinary Portland cement.

7.8.3 Lifetime energies

The physical properties of materials such as thermal capacity, insulating ability and opacity have a substantial effect on lifetime energies. It is true to say that good design is to some extent based on best utilizing the properties of the materials used. The properties of many materials are too focused. It is possible to develop new materials with more than one property currently considered as conflicting. For example, materials that are good insulators do not generally have a high heat capacity. These properties are not necessarily mutually exclusive and combining insulating and heat capacity has huge potential for reducing the lifetime energies of buildings.

7.8.4 Heat-absorbing or releasing materials

Phase change materials are becoming available such as waxes or encapsulated glaubers salts that have very high heat capacity and therefore increase the heat capacity of the composites in which they are placed. Other materials such as some zeolites have a very high absorption energy for water and can work in a similar way. Using these new materials and TecEco cement composites, materials with both high heat or cold capacity and insulting capability could be made.

7.8.5 Using waste in new materials

Many wastes such as cellulose fiber from wood, plastics, straw, sugarcane bagasse, kenaf, hemp and guayule, have physical properties such as light weight and tensile strength that would make them suitable for use in composites with cheap mineral binders. One of the problems holding up the development of such composites has been the high alkalinity of Portland which causes weakness as a result of internal reaction. TecEco tec and eco-cements are nowhere near as alkaline and for this reason, and because they stick very well to other materials, they will be pivotal in the quest to utilize more wastes, converting them to resources.

7.8.6 Healthy materials

Some materials are more comfortable to live with than others. Combined with good design they provide the right temperature, noise levels and other creature comforts required. It is also essential that buildings are healthy to live in. Unfortunately, many new materials such as modern varnishes, glues in plywoods and plastic finishes give off gases that not only smell but that are potentially carcinogenic or hormonal in action. Many traditional paints such as lime wash and those based on casein do not have this problem. Modern paints are now being reinvented to be more healthy.

Carbonating concretes and mortars such as those based on lime or the new eco-cement require the material in which they are used to be able to 'breathe' to carbonate properly. Allowing walls to breathe makes buildings healthier to live in. Ever walked up rocks on a beach on a hot day? The materials we use have a strong influence on our comfort levels on a wider scale. Recently in Tokyo and other large cities with badly designed concrete, the 'hot island phenomenon,' whereby the city itself is several degrees warmer than surrounding areas has become a major issue.

7.8.7 Using recycled materials

Supplementary cementitious materials like fly ash and ground vitrified blast furnace slag are increasingly being used in concrete. The use of other recycled wastes for their physical properties is lagging. The use of supplementary cementious materials reduces the need for concrete and is therefore more sustainable.

Geopolymers are very simply described as low-temperature silica-alumina glasses in which, instead of heat, alkalis are used to break down silicates that incorporate metallic cations when they reform. They are most suitable for use with wastes that contain silica and alumina and provide a high-temperature stable environment for encapsulation of problem elements such as heavy metals. TecEco tec-cements have many of the properties of geopolymers. TecEco cements generally are suitable to incorporate wastes mainly because they are chemically benign and because they stick so well to other materials. They are described by:

- lower reactivity (less water, lower pH)
- reduced solubility of heavy metals (lower pH)
- greater durability
- increased density and impermeability
- increased homogeneity
- no bleed water
- less attack by salts in ground or sea water, and
- increased dimensional stability with less cracking.

The use of cementitious binders to bind all sorts of wastes to make composites for constructing the built environment makes sense. The public must however be assured there is no risk of coming into contact with heavy metals. In a TecEco tec-cement Portland cement brucite matrix:

- OPC takes up lead, some zinc and germanium.
- Brucite (Fig. 7.19) and hydrotalcite are both excellent hosts for toxic and hazardous wastes.
- Heavy metals not taken up in the structure of Portland cement minerals or trapped within the brucite layers end up as hydroxides with minimal solubility.

Brucite, one of the main minerals formed in TecEco tec-cements has an ideal

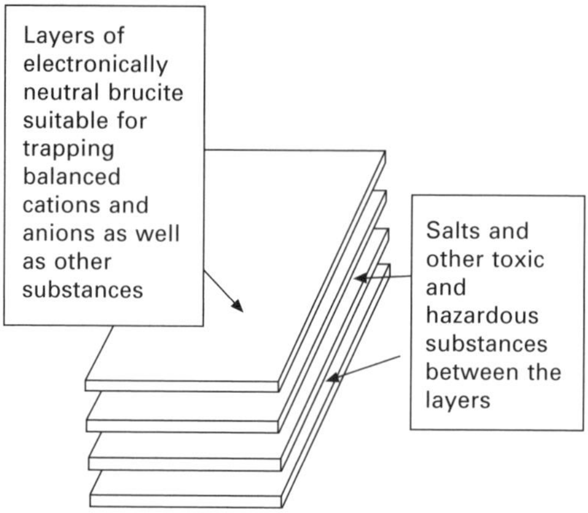

7.19 Structure of brucite.

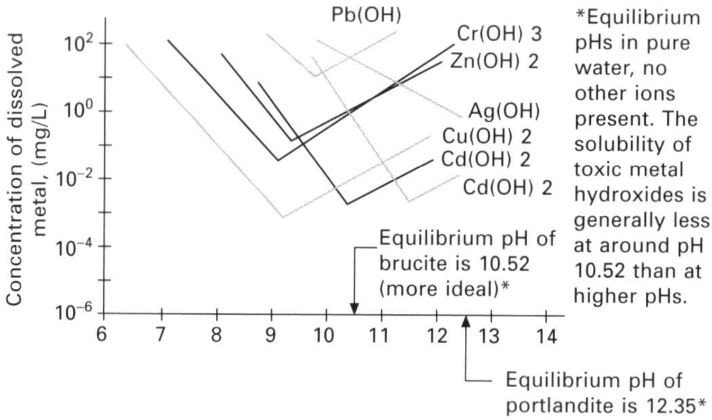

7.20 pH conditions in TecEco cements.

structure for capturing balanced cation anion pairs and has an equilibrium pH at a suitable lower level to minimize the solubility of heavy metals (Fig. 7.20).

7.8.8 More durable materials

Materials that remain useful for longer are required. The time is past when the likes of Henry Ford could reduce the quality of materials and components made from them deliberately to sell more vehicles more often. The less often materials or things from them are made the less energy consumed and lower emissions and other detrimental outcomes. Imagine if functionality and service were purchased instead of energy and things. This was a theme in the book *Natural Capitalism* and an example given was that instead of buying electricity one would purchase heating and lighting (Hawken *et al.*, 2000).

I recently had the pleasure of entertaining in Tasmania the representative of a Brazilian company, Magnesita S.A. He told me that his company no longer just sold refractory bricks. They were paid on the basis of downtime experienced by their steel-producing clients. With better quality bricks and less downtime they were paid more. Geopolymers and TecEco cements are much more durable mainly because the minerals in them are thermodynamically very stable or less soluble.

7.8.9 Recycled materials

Materials and the things that they are used to make should be engineered with their next use in mind so that they never exit the techno-process and become waste. An example already used in the construction of the built environment is modular formwork. Materials will be required that are either biodegradable or easily recycled within the techno-sphere. Examples of materials that are currently not recycled that could be are bricks and blocks. Bonded together using softer carbonating eco-cement or lime mortars these would be easier to clean and thus recycle as they were in the past.

7.9 Creating more sustainable production: eco-cements

Concrete is the biggest material flow on the planet[11] and therefore offers significant opportunities for improvements in sustainability. Other than eco-cements and carbonating lime mortars that carbonate and therefore have a clear advantage and tec-cements that perform well because less is used, there are a number of other novel cements with intrinsically lower energy requirements and CO_2 emissions than conventional Portland cements that have been developed including high belite (C_2S) and calcium sulfoaluminate (C_4A_3S) types as shown in Table 7.2.

From Table 7.2 it can be seen that uniquely carbonating lime mortars and TecEco eco-cements are by far the most sustainable in terms of CO_2 and can even be net carbon sinks. Using a building material that is CO_2 neutral or even sequesters carbon makes a lot of sense – after all, that is what nature has been doing for the past 3–7 billion years depending on your opinion as to the age of the earth. Utilizing wastes also make sense. The potential for keeping the planet in the way that we can survive on it is enormous.

Because of the high efficiency of CO_2 capture by magnesium oxide, and for a number of other technical reasons, TecEco propose the partial substitution of Portland cement with magnesia (MgO) which in eco-cements in porous materials fully carbonates.

[11] Previously mentioned at over two tonnes per person on the planet per annum.

Table 7.2 CO_2 released by de-carbonation during the manufacture of cements and components*

Compound	CO_2 released through decarbonation in producing 1 tonne (tonnes CO_2/tonne compound)	CO_2 potentially recaptured in a porous concrete or mortar – tonnes CO_2 per tonne compound	Net emissions (if no capture – tonnes CO_2 per tonne compound)	Net emissions (if capture for MgO and CaO only – tonnes CO_2 per tonne compound)	Example of cement type
MgO	1.09	1.09	0	–1.09 (net sequestration)	Eco-cement mortar
CaO	0.78	0.78	0	–0.78 (net sequestration)	Carbonating lime mortar
C_3S	0.578	0.289	0.289	Not feasible technically yet	Alite cement
C_2S	0.511	0.255	0.256	Not feasible technically yet	Belite cement
C_3A	0.594	0	0.594	Not feasible technically yet	Tri calcium aluminate cement
PC	0.54	0.27 (variable)	0.27	Not feasible technically yet	Portland cement
1PC:2MgO	0.99	0.817	0.173	–0.817 (net sequestration)	Eco-cement with no pfa
1PC:2MgO:3pfa†	0.445	0.367	0.077	–0.367 (net sequestration)	Eco-cement with pfa
1PC:2pfa^2	0.27	0.137	0.137	Only feasible for the MgO component	Very high fly ash cement
0.05MgO:95PC:2pfa	0.18	0.092	0.092	Only feasible for the MgO component	Tec-cement assuming 1/3 (0.334%) less binder required
C_4A_3S	0.216	0	0.216	Not feasible technically yet	Calcium sulfoaluminate cement

* Geopolymers are unique and are not included in the above table because they fall into the class of chemical cements.
† Assuming no emissions for pfa or that they are accounted for in the power industry.

$MgCO_3 \rightarrow MgO + \downarrow CO_2$ – efficient low temperature calcination and capture

$MgO + \downarrow CO_2 + 3H_2O \rightarrow MgCO_3.3H_2O$ – sequestration as building material

Understanding the magnesium thermodynamic cycle depicted in Fig. 7.21 is very important in relation to understanding the massive potential of eco-cements for sustainability. There are several important features of this cycle; it is a cycle and relatively speaking it does not take much energy to make it go around and around. Better still, this can be done at relatively low temperatures. Calcining can therefore be carried out in a closed system and the CO_2 captured. Grinding and calcining at the same time produces more uniform product, allows the capture of CO_2 and utilizes waste heat from grinding.[12] Eco-cements in a relatively porous matrix such as a concrete block, porous road pavement or mortar complete the cycle, gaining strength by carbonating. The CO_2 required comes out of the surrounding air.

The implications of using TecEco cements in the context of mineral sequestration to solve the problem of global warming are discussed in section 7.9.1. There are other sustainability advantages as well:

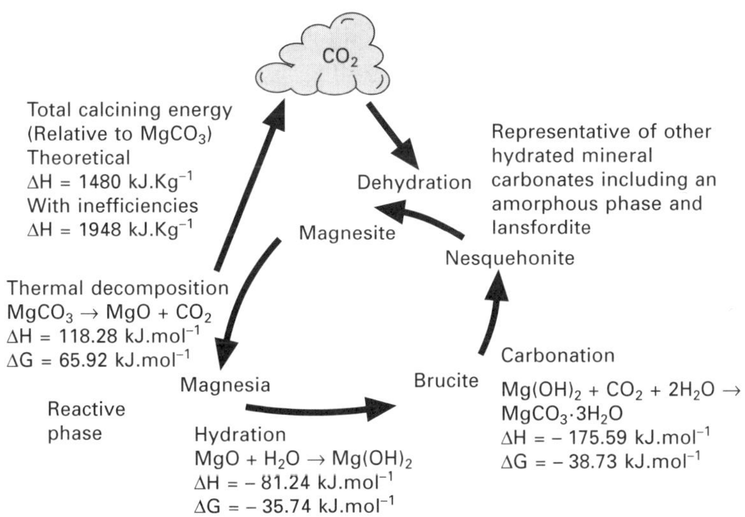

7.21 The magnesium thermodynamic cycle.

[12] Around 98% of the energy that goes into grinding ends up as heat. Some 30% of the energy in a conventional cement plant goes into the grinding process.

- Magnesium has a strong affinity for water in solution and does not lose it readily during carbonation. The result is that solid hydrated carbonates like nesquehonite are formed that are 83 mass% water and CO_2! – cheap sustainable binder!
- Magnesium carbonates are generally fibrous and acicular and therefore add microstructural strength.
- The long-term pH is much lower than Portland cement concretes. Combined with the fact that magnesium minerals seem to stick well to other materials, the result is that a high proportion of wastes can be included.
- TecEco cements are generally much more durable. Materials that last longer are much more sustainable.

The underlying chemistry is very similar, however, eco-cements are potentially superior to lime mortars because:

- The calcination phase of the magnesium thermodynamic cycle takes place at a much lower temperature
- Magnesium minerals are generally more fibrous and acicular than calcium minerals and hence a lot stronger.
- Water forms part of the binder minerals that make the cement component go further.
- Magnesium hydroxide in particular, and to some extent the carbonates, are less reactive and mobile and thus much more durable.
- A less reactive environment with a lower long term pH is achieved.
- Magnesium having a low molecular weight, proportionally a much greater amount of CO_2 is captured.

The low molecular weight of magnesium results in a higher CO_2 content per tonne as in the calculations below.

$$\frac{CO_2}{MgCO_3} = \frac{44}{84} = 52\%$$

$$\frac{CO_2}{CaCO_3} = \frac{44}{101} = 43\%$$

7.9.1 Sequestration processes

Alkali metals like calcium and magnesium release a large amount of chemically bound CO_2 when their oxides are made from their carbonates. If this chemically released CO_2 could be captured during manufacture and reabsorbed during setting, forming a strong binder, there would be huge opportunities for safe sequestration in the built environment with no possibility of leakage or detrimental effects on a wider scale.

The capture of CO_2 at source during the manufacturing process is easier for the calcination of magnesium carbonates than any other carbonate mainly because the process occurs at relatively low temperatures. TecEco Pty. Ltd. own intellectual property in relation to a new tec-kiln in which grinding and calcining[13] can occur at the same time in the same vessel for higher efficiencies and easy capture of CO_2. Provided sufficient uses can be found for pure CO_2 produced during manufacture whereby it is also permanently sequestered, a system for sequestration on a massive scale using carbonates as building materials is very promising. Possibilities for alternative permanent disposal are in materials such as plastics or deep underground where CO_2 reacts with country rock forming more carbonate.

By utilizing the new TecEco technologies and intelligent design, there is no reason why buildings could not be much more sustainable than current state of the art – not only with low embodied energies and emissions but low lifetime energies as well (Fig. 7.22). Using either forsterite or serpentine as inputs, or magnesite precipitated from seawater the tec-kiln technology previously mentioned provides a method of calcining the magnesium carbonate produced using solar derived intermittent energy or waste energy from other sources. The magnesium oxide (MgO) produced can be used to directly

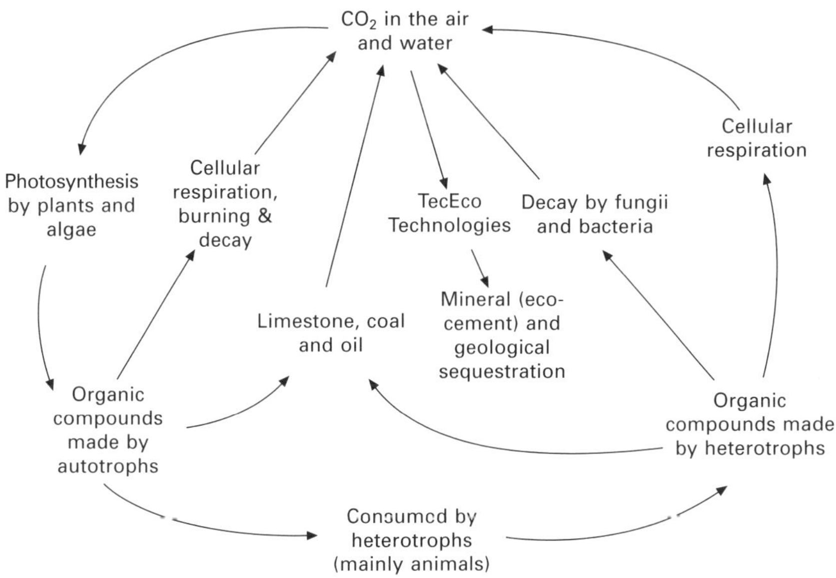

7.22 The carbon cycle as TecEco would like to modify it (adapted from Kimball (2004)).

[13] Calcining in the context of this document refers to the heating of limestone or magnesite to drive off CO_2 and produce the oxide.

sequester more CO_2 in a scrubbing process or to sequester carbon as hydrated magnesium carbonates in the built environment. Using tec-kiln technology magnesium fixes carbon dioxide and therefore acts as a concentrator (Fig. 7.23). Figure 7.24 is a process diagram showing combined mineral and eco-cement sequestration.

The idea of capturing CO_2 as carbonate in the built environment mimics what has in fact naturally been occurring for millions of years.[14] Carbonates formed in seawater are the natural, large-scale, long-term sink for carbon dioxide, however the process takes over 1000 years to equilibrate. Good evidence of the enormous volumes of CO_2 that have been released from the interior of the earth during many volcanic episodes over the last few billion years is the high percentage (7%) of the earth's surface covered in rocks such as limestone, dolomite and magnesite.

If carbon dioxide is captured during the calcining process for the manufacture of reactive magnesia then it can either be geologically sequestered or recaptured as eco-cements in porous cementitious materials for the built environment. Eco-cements gain strength with the formation of magnesium carbonates including lansfordite, nesquehonite and an amorphous phase mineral all of which, because of their generally acicular shape, add microstructural as well as innate strength as binders.

Oil has remained trapped in strata for millions of years and on this basis it is argued that carbon dioxide pumped down to push it up would also remain trapped. Given the fact that on average the pH of the earth is the same as that of seawater (8.2) then some neutralization with the formation of immobile carbonates is also expected.

Kinetically, the carbonation of eco-cements in the built environment will only proceed rapidly in porous materials such as bricks, block, pavers, mortars, porous pavement, etc., which, fortunately, make up a large proportion of building materials. All these materials, including mortars, require the use of appropriate course aggregates for carbonation to occur efficiently. The effect of the substitution of Portland cement with MgO in a simple concrete brick formulation containing 15% cement with and without capture of CO_2 during manufacture of magnesia is depicted in Fig. 7.25, from which it can be deduced that almost two-thirds of the CO_2 in concrete is potentially abated if substituted by eco-cement in porous products such as bricks, blocks, pavers, mortars, etc.

With CO_2 capture or the inclusion of organic fiber materials and fillers for strength and insulation, cementitous building materials that act as net carbon sinks are feasible. Organic fibers may include wood fiber, straw, sugarcane bagasse, kenaf, hemp and guayule. Organic fillers could include much sawdust

[14] There have been at least seven other epochs of global warming easily discernible from the geological evidence.

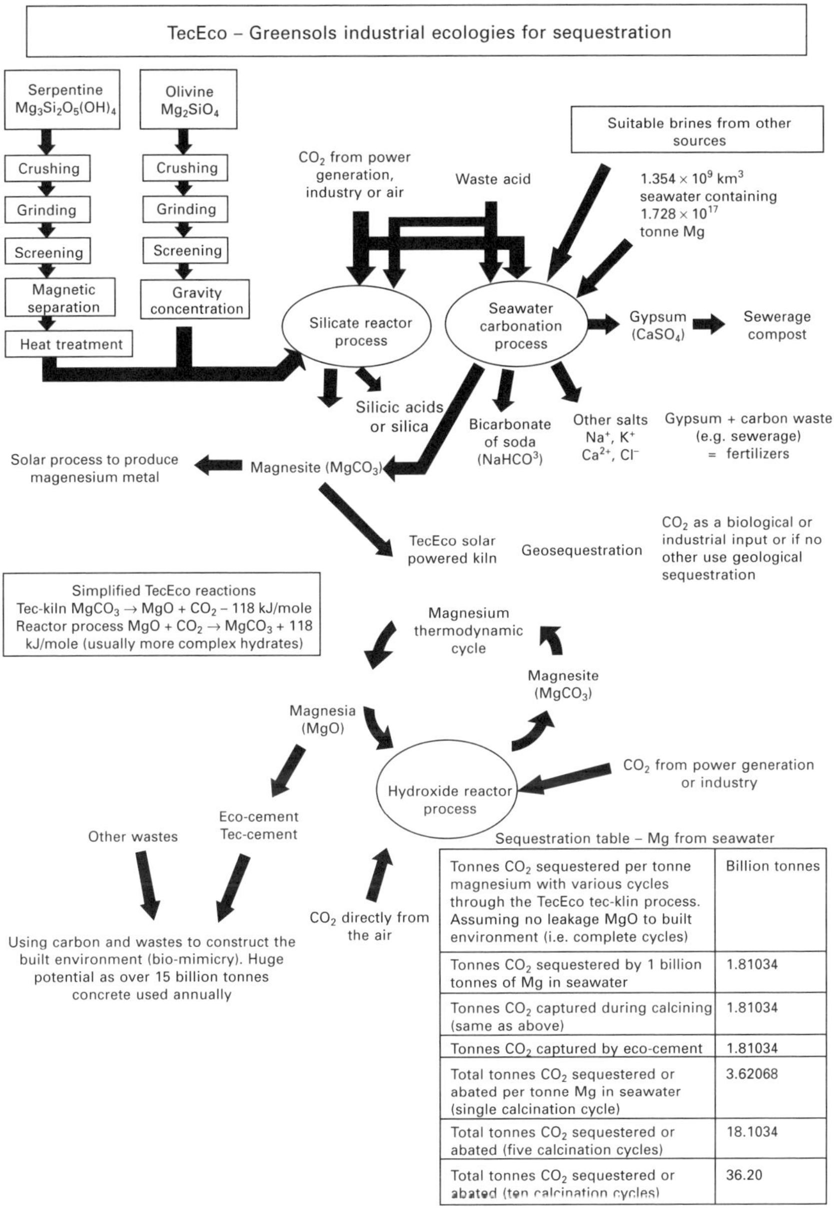

Tonnes CO_2 sequestered per tonne magnesium with various cycles through the TecEco tec-klin process. Assuming no leakage MgO to built environment (i.e. complete cycles)	Billion tonnes
Tonnes CO_2 sequestered by 1 billion tonnes of Mg in seawater	1.81034
Tonnes CO_2 captured during calcining (same as above)	1.81034
Tonnes CO_2 captured by eco-cement	1.81034
Total tonnes CO_2 sequestered or abated per tonne Mg in seawater (single calcination cycle)	3.62068
Total tonnes CO_2 sequestered or abated (five calcination cycles)	18.1034
Total tonnes CO_2 sequestered or abated (ten calcination cycles)	36.20

7.23 The TecEco process for saving the planet – how TecEco technology can turn global warming around by connecting mineral and geological sequestration.

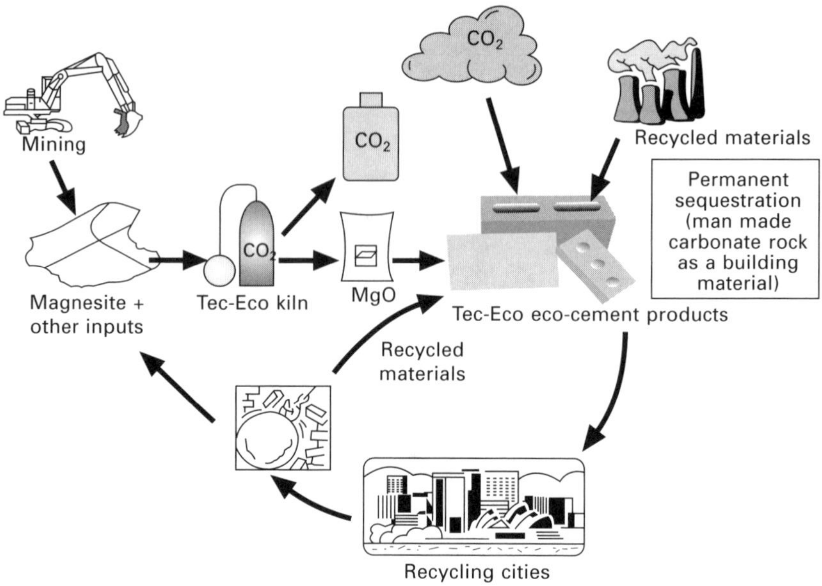

7.24 TecEco cements for sustainable cities.

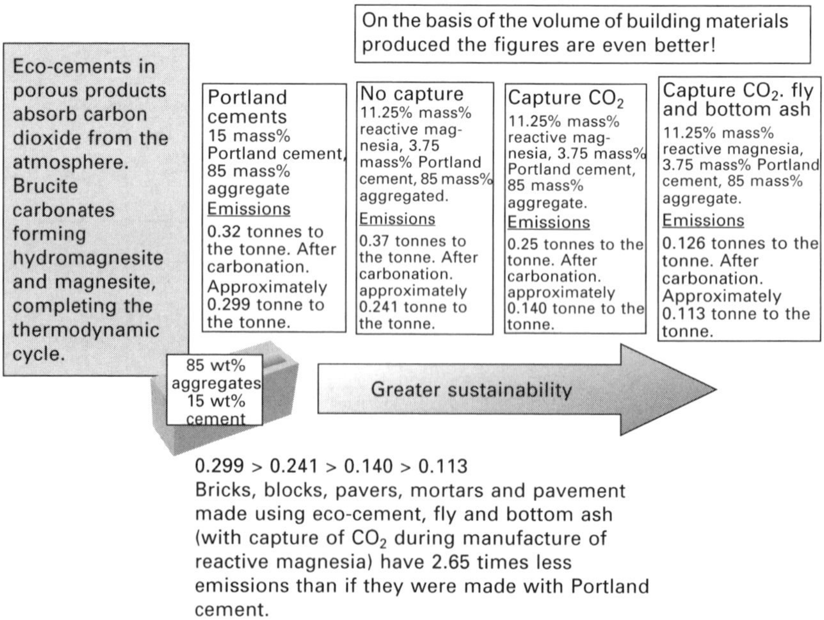

7.25 Abatement in a concrete brick containing 15% eco-cement.

currently going to waste. The above explanation is simplistic. As the energy considerations are complex, readers are directed to the website of TecEco at www.tececo.com where further papers are available.

7.9.2 The practicalities of sequestration

The deposition of carbonate sediments is a slow process and involves long periods of time. Ways of accelerating sequestration using carbonates include geological sequestration and mineral carbonation. Although promoted by the petroleum industry as a means of extracting remaining reserves of oil 'there are significant fundamental research needs that must be addressed before geologic formations can be widely used for carbon sequestration' (NETL, 2004). Mineral carbonation, the reaction of CO_2 with non-carbonated minerals to form stable, benign mineral carbonates, has been identified as a possible safe, long-term option for storing carbon dioxide by many authors (NETL, 2004; Seifritz, 1990; Lackner *et al.*, 1995; Dahlin, *et al.*, 2000; O'Connor *et al.*, 2000; Fauth, *et al.*, 2001).

Mineral sequestration is a process and a number of universities and research organizations around the world are working on it. Although there still some kinetic issues the process is workable. TecEco have demonstrated that eco-cement concretes are suitable for block making and that they sequester considerable amounts of carbon dioxide. The only component of the entire process that is not proven is the TecEco kiln technology. It will take about half a million Australian dollars to do so but this is small change for saving the planet. The cost of mineral sequestration even without the advantages added by TecEco technology is predicted to be quite reasonable.

Assuming thermodynamically efficient processes a cost including rock mining, crushing and milling, of around $US 20/tonne of CO_2 are suggested for mineral sequestration starting with forsterite or serpentine rocks. For a 66% efficient power plant this would add less than 1 cent to the cost of a kilowatt hour of electricity (Yegulalp, *et al.*, 2001). A new process that extracts magnesite from brines (see Fig. 7.21 on page 321) and uses TecEco technology to drive the magnesium thermodynamic cycle (see Fig. 7.21 on page 321) has recently been promoted by John Harrison and Professor Chris Cuff that promises to reduce the cost of sequestration on a massive scale.

A combined process involving TecEco tec-kiln technology would sequester several times more carbon and involve the eventual production of eco-cement concrete components that also sequester carbon to create the built environment and would be therefore potentially very profitable even if there were an even lower price for carbon of say less than $10. The Kyoto protocol will also encourage the development of other technologies whereby CO_2 becomes a resource and as a result the process will eventually be supported by an economically driven price for CO_2. The use of TecEco eco-cements concretes would also be favoured as magnesite, the raw material, would not have to be mined. An added advantage would be the permanent disposal of carbon dioxide with no possibility for leakage.

7.9.3 Other sustainable binders

Tec-cements

TecEco tec-cements include a much lower proportion of reactive magnesia and reduce emissions by further extending supplementary materials and reducing water whereby greater strength is obtained for less total binder. Further details are available from the TecEco website.

Calcium sulfoaluminate cements

Calcium sulfoaluminate cements were developed in China during the 1970s at the China Building Materials Academy and were intended for the manufacture of self-stressing concrete pipes due to their expansive properties. Sulfoaluminate cements contain as main phases belite (C2S), tetracalcium trialuminate sulfate (C4A3 $\bar{S}$bar) and gypsum. The main mineral formed is ettringite. They have very low emissions associated with their manufacture and various blends with OPC and pozzolans are being trialed with some success.

Geopolymers

Geopolymers utilize silicon and aluminium containing raw materials like fly ash and metakaolin. As metakaolin has to be manufactured and therefore has an embodied energy, the objective, particularly for lower quality geopolymers, is to use as much fly ash as possible or fly ash only. Global supplies of ground granulated blast furnace slag are starting to run out and eventually fly ash will also become a scarce resource and cost more than it already does. It then becomes debatable whether the coal industry or geopolymer industry should bear the associated carbon tax. Research into geopolymers is continuing and obstacles such as placement problems associated with extreme viscosity of the material and the critical quantitative addition of chemicals and water required are being investigated.

7.10 Making sustainability profitable

In the past it was considered that economic development was linked to growth in the use of resources and energy and population growth. Most now understand that change is itself a stimulant for economic growth. We therefore have nothing to fear from the process of innovative change towards more sustainable outcomes.

Some businesses have pillaged resources and have not embraced sustainable development and the concept of natural capital (Hawken *et al.*, 2000). Changing the way these businesses think is a matter of communicating to them that they can do something for themselves and the environment through greater resource use efficiency and reduced energy consumption. Sustainable development is about investing and strengthening socially and environmentally responsible businesses whilst at the same time reducing negative social or ecological consequences through awareness programs that slowly change the way in which more recalcitrant businesses operate.

To be acceptable in markets technology usually improves. For example, fridges today are many times more efficient than fridges of 20 years ago – even without CFC refrigerants! Not only are products rapidly becoming more efficient, so too are manufacturing processes. The biggest factor in increasing economic growth and raising living standards over time in the past has been the economy's ability to produce more out of less, i.e., to become more productive. Productivity is good economics. Productive companies reduce inputs for the same output quality, volume and cost. They are much more efficient.

Mother nature was an economist. She is frugal, efficient, productive and integrated and by observing climax natural ecologies we can learn much about productivity and efficiency. Increases in productivity not only bring greater profits, they reduce the demand on natural resources. Integrating productivity and efficiency across enterprises whereby communities get more out of less makes even more sense and can further reduce total take and waste impacts. The same applies on a global scale.

'Environmental protection is a market, not a cost' was the message of the keynote speaker, James B. Quinn, at the recent forum on sustainable development at Parliament House Canberra, Australia (Quinn, 1999). Amory Lovins of the Rocky Mountain Institute puts it simply when he says that green technology 'will happen, and happen rapidly – because it's profitable' (Cook, 2000). Sustainability is good business. The main reason is simply a cost/value issue – recycling wastes and minimizing inputs reduces input costs. In relation to construction, lower embodied and lifetime energies for buildings means that owning and living in them will cost less.

When trying to solve a problem of immense proportions a good strategy is to first assess the solution, the means of applying it and the outcome from the effort of applying it. Put simply the input/outcome ratio or 'bang for buck' is important for success. The obvious place, that seems to have been missed by just about everybody, to focus sustainability efforts is the built environment. It is our footprint on the globe. Given the size of the built environment there are huge opportunities for improving the techno-process and whilst doing so solving pollution and climate change problems. The built environment is a major proportion of the techno-sphere and our lasting legacy on the planet. In this dominant proportion of all materials flows unsustainable practices abound evidenced by the high volume of wastage going to landfill.

Most of what we take, manipulate and make that we do not consume immediately, goes into the materials with which we construct the built environment or 'techno-sphere'. Buildings and infrastructure probably account for around 70% of all materials flows.[15] Of this 'Buildings account for 40% of the materials and about a third of the energy consumed by the world economy. Combined with eco-city design principles, green building technologies therefore have the potential to make an enormous contribution to a required 50% reduction in the energy and material intensity of consumption in the post-modern world.' (Rees, 1999)

> In 1999, construction activities contributed over 35% of total global CO_2 emissions – more than any other industrial activity. Mitigating and reducing the impacts contributed by these activities is a significant challenge for urban planners, designers, architects and the construction industry, especially in the context of population and urban growth, and the associated requirement for houses, offices, shops, factories and roads. (UNEP, 2001)

Since the wealthiest 25% of the human population consume 80% of the world's economic output (Bruntland, 1987), approximately 64% of the world's economic production/consumption and pollution is associated with cities in rich countries. Only 12% is tied to cities in the developing world (Rees, 1999). In short, 'half the people and three-quarters of the world's environmental problems reside in cities, and rich cities, mainly in the developed North, impose by far the greater load on the ecosphere and global commons' (Rees, 1997).

Australia is one of the biggest energy consumers per capita in the world and has tripled consumption in the past 20 years with 94% of our current energy consumption from fossil fuels (2001). According to the *Human Settlements Theme Report, State of the Environment Australia 2001* (CSIRO, 2001),

> Carbon dioxide (CO_2) emissions are highly correlated with the energy consumed in manufacturing building materials. On average, 0.098 tonnes of CO_2 are produced per gigajoule of embodied energy of materials used in construction. The energy embodied in the existing building stock in Australia is equivalent to approximately ten years of the total energy consumption for the entire nation. Choices of materials and design principles have a significant impact on the energy required to construct a building. However, this energy content of materials has been little considered in design until recently, despite such impacts being recognized for over 20 years.

[15] TecEco estimate arrived at by including infrastructure with buildings.

To date urbanization has been considered as having serious negative implications for global sustainability. The impact of new materials such as TecEco eco-cement could reverse this as cities would become giant carbon sinks and repositories of what were formerly wastes. The impact of tec and eco-cements is discussed in section 7.9. Cities have dense concentrations of people, the juxtaposition of many industries provides real opportunities to reduce energy and material through-put by integration beyond the enterprise level.

We use easily obtained fossil fuels to heat and power our complex civilization, yet all the energy we will ever need is produced by our sun[16] and directly available as well as from natural forces of, wind, water and wave that are indirectly derived from the sun's energy reaching the earth. Sustainability embraces creating new from used, and requires minimal impact on natural resources such as old growth forest, coal or oil. Eventually it will mean living with the massive solar flux. As can be seen from Table 7.3 fossil fuel energy is only a very small fraction of total energy on the planet.

Table 7.3 Different energy fluxes on the planet*

10^{13} Watts	Total	Heat	Wind	Evaporation	Photosynthesis
Solar	12100	8000	37	4000	4
Earth heat	3.2				
Tidal	0.3				
World techno-process energy demand	1.5				

* Estimated from various sources.

To date the main practical emphasis has been on designing buildings with low lifetime energies. Little effort has been made to reduce the impact of materials on lifetime energies, embodied energies and emissions. There is huge scope for sequestration and conversion of waste to resource given the massive size of the materials flows involved in the built environment. With the right materials technology, because of its sheer size, the built environment could reduce the take from the geosphere-biosphere and utilize many different wastes including carbon dioxide. If sustainable development could be achieved then our cities could support large numbers of people while limiting their impact on the natural environment.

[16] Every half-hour more solar energy strikes the Earth than is released globally by the burning of all fossil fuels in a year.

7.10.1 Entrepreneurs and innovation

Entrepreneurs (I consider myself to be one) are those rare optimistic individuals who conceive new business opportunities and who take on the risks required to convert those ideas into reality. They seek opportunities, both for themselves and for the communities they belong to, and therefore are an important part of any society. As agents of change, entrepreneurs take on the responsibility of identifying new commercial ventures, incubating ideas and championing their adoption, assembling the resources needed to bring these ideas to commercial reality and, finally, launching and growing business ventures.

In a market-based economy, environmental entrepreneurs will play a critical role in the adoption of green business practices. They constitute the main 'pull' factor that entices architects and engineers to go green, as opposed to the 'push' factors of government regulation and stakeholder/lobby group pressure. All good teams need good leaders and much co-operative teamwork is required by the inhabitants of earth if we are to survive the long-term future.

Consider why entrepreneurs do things. Perhaps it is because many of them have a mission in life other than just making money. According to Sean Covey, author of the *7 Habits of Highly Effective Teens*, 'life is a mission, not a career. A career is a profession. A mission is a cause. A career asks, "What's in it for me?" A mission asks, "How can I make a difference?" Martin Luther King's mission was to ensure civil rights for all people. Gandhi's mission was to liberate 300 million Indians. Mother Teresa's mission was to clothe the naked and feed the hungry' (Covey, 1999). Many entrepreneurs have a passion to develop an idea or invention, run a better business – that is their mission; the money sometimes follows.

Governments around the world should do all in their power to encourage environmental entrepreneurs to take a greater and more rewarding role in markets as by doing so, new technologies will doubtless emerge resulting in less emissions and more efficient sequestration.

7.10.2 The role of government

The public has a perception that governments will do something about sustainability. They should be disappointed as in the past most effort has been mostly orientated towards mapping the extent of the problem, getting political mileage out of it and to some extent reducing energy consumption. Unfortunately, the democratic system has a fatal flaw. The outlook of politicians and thus governments is usually not much beyond the next election. As a consequence policy is generally extremely short sighted and too directly connected to the needs of the here and now rather than those of the future. Totalitarian governments can, in contrast move much more quickly such as demonstrated by the announcement by the Chinese central government of a move away from polluting clay bricks.

Global problems are not just the concern of one or two countries but all people on the planet. World federalists believe we need a system of democratic global governance on top of (not instead of) national governments. Such a system would provide enforceable legal mechanisms for resolving conflicts and safeguarding the environment. Perhaps they have a point.

In spite of the two UN 'Habitat' conferences on urban prospects,[17] cities have not been given serious attention in the mainstream sustainability debate. For example the World Conservation Strategy of 1980, which first used the term 'sustainable development', paid little attention to accelerating urbanization. The Brundtland report (Bruntland, 1987) did discuss the issue, but the main emphasis was on the 'urban crisis in developing countries' (UNCHS, 2004b).

> The population of cities is increasing much more than country areas and cities have a network of linkages that extend far beyond their boundaries. Sustainable urban development requires consideration of the carrying capacity of the entire ecosystem supporting such development, including the prevention and mitigation of adverse environmental impacts occurring outside urban areas. The unsafe disposal of waste leads to the degradation of the natural environment; aquifers, coastal zones, ocean resources, wetlands, natural habitats, forests and other fragile ecosystems are also affected. (UNCHS, current)

It is essential that governments realize that urban settlement has much promise for human development and the protection of the world's natural resources. Dense concentrations of people and many industries, provides real opportunities to reduce energy and material through-put. Recycling based on the chemical composition of wastes is much more feasible in big cities not faced with transport issues for return to manufacturers. High population densities render rapid transport systems economic. Industrial wastes can be treated economically because of the economies of scale and the agglomeration economies of having many similar industries together. Supporting moves towards greater sustainability in construction is a newly recognized direction for governments and will require further changes in the regulatory regime. Innovation is essential for change and change must occur, particularly with construction materials for greater sustainability.

In western countries the existing framework in which entrepreneurs can raise capital to bring to reality their creativity is market driven with little government assistance or intervention. For example, at the present time in Australia, S 708 of the Corporations Act 2001 regarding fund-raising is still

---

[17] Vancouver in 1976 and Istanbul in 1996.

far too restrictive as is the attitude of many venture capitalists – indeed the word 'venture' has been misapplied. To change this, market related inducements are required and arguably this is the role of government. To help create a new set of sustainable market segments it is encouraging that some governments are already introducing market drivers. In Australia the NSW government seems to be leading the way, and has introduced a framework for the recognition of carbon sequestration certificates. Hopefully the Australian Federal Government takes up the challenge and is not tardy in introducing a national system. Globally we have Kyoto and now it is a reality Australia and the US should join. During the Clinton era the USA recognized the need to prime the pump with research. The present Bush regime is doing little.

The Australian Greenhouse Office have actively purchased abatement. Unfortunately, however, there has not been significant effort to incubate new sustainable technologies. Minor exceptions are some cleaner production initiatives under the Greenhouse Challenge Program. The UK seem to be leading the field with Prime Minister Tony Blair keen to develop three million new homes in the south west of Britain using more sustainable technology. The point is that sustainable development can be achieved in cities and that they could support large numbers of people with limited impact on the natural environment. Governments generally need to realize this.

Sustainability should be a cornerstone for all government policy and expenditure. Governments should also facilitate economic systems that encourage sustainability such as performance standards, carbon trading and deposit legislation. A change towards acceptance of products that are environmentally sustainable is occurring and will accelerate with the re-education activity of governments around the world. For sustainability the take and waste components of the techno-process need to be reduced and to what is renewable and preferably biodegradable.

Various policy options are being experimented with such as:

1. Research and development funding priorities
2. Procurement policies
   Government in most countries is more than one-third of the economy and can strongly influence change through:
   - life cycle purchasing policy
   - funding of public projects and housing linked to sustainability such as recycling.
3. Intervention policies
   - building codes including mandatory adoption of performance specification
   - requiring the recognition and accounting for externalities
   - extended producer responsibility (EPR) legislation
   - mandatory use of minimum standard materials that are more sustainable
   - mandatory eco-labelling.

4. Taxation and incentive policies
   - direct or indirect taxes, bonuses or rebates to discourage/encourage sustainable construction, etc.
   - a national system of carbon taxes
   - an international system of carbon trading
5. Sustainability education

Consider building codes, research and development funding and policies of encouraging recycling and legislating for extended user responsibility as they are most relevant to the materials theme of this chapter.

Building codes

The emphasis has been on lifetime rather than embodied energies because potential lifetime energy reductions through good design are significant. Most OECD countries have set up energy efficiency standards for new dwellings and service sector buildings: this includes all European countries, Australia, Canada, the USA, Japan, Korea and New Zealand. Some non-OECD countries outside Europe have also established mandatory or voluntary standards for service buildings and Singapore and the Philippines were among the first. To date building codes have not encouraged the use of more energy efficient materials, in spite of the huge impact of materials not only on embodied energies but lifetime energies as well.

Research and development funding

The early North American and to some extent European approach was to 'prime the pump' and research global warming. The early Australian approach to sustainability was much cruder and involved the outright purchase of abatement with little research. To date, materials science, which this paper demonstrates is fundamental, has received little funding globally. As of early 2004 European priorities were genomics and biotechnology for health, information society technologies, nanotechnologies, intelligent materials and new production processes, aeronautics and space, food safety and health risks, sustainable development and global change, citizens and governance in the European knowledge-based society. In the USA Federal funding priorities include nanotechnology, defense, and aeronautics. Geological sequestration has also been a priority in many countries as it is associated with the petroleum industry and doubts have been raised as to the transparency of funding (Wilson, 2004).

To date Australian research priorities have not included materials science, however, the Australian research funding priorities for 2004–2005 may well lead the way globally as, included under the heading 'Frontier Technologies for Building and Transforming Australia', are advanced materials. 'Advanced materials for applications in construction, communications, transport, agriculture and medicine' include ceramics, organics, biomaterials, smart material and fabrics, composites, polymers and light metals (ARC, 2004).

Recycling

Recycling involves a series of activities by which wastes are collected, sorted, processed and converted into raw materials and used in the production of new products. Recycling is carried out by individuals, volunteers, businesses and governments. For high value waste, recycling is profitable and undertaken by business, usually by buying back wastes, but not so as the value declines. Governments generally have recognized the importance of recycling but have gone about the introduction of recycling through councils and local authorities in completely the wrong way. As the hazards of discarded wastes do not correlate with their value, many wastes are recycled by the authority of legislation or power of producer organizations.

Instead of being forced upon us using good taxpayer dollars for 'feel good' reasons, much more effort should have been put into the development of technologies to make the process economic. With market forces driving recycling much more would occur much more efficiently. The problem is to make the process of recycling less 'feel good' and more economic so it is driven by market forces. The solution is to use materials not just for their chemical composition but for their physical property. TecEco cements open this possibility.

Accounting for externalities – the true cost of the techno-process

> ...the exponential growth curve of cost associated with negative impacts or 'externalities' such as climate change, salinity, acid sulphate soils, river system degradation, or general pollution has, up until now, been a legacy for future generations to deal with. For decades the cash price of goods and services has been artificially deflated, with much of the real cost being outsourced on to the environment.
>
> The costs, however, that are backing up on us – bush fires, dust storms, floods, soil erosion, salinity, changes in disease patterns, hurricanes and cyclones – can often be attributed, at least in part, to climate change. An integral part of the dilemma we have is the denial that anything truly threatening is happening. (Wain, 2004a)

Accounting standards that recognize the value of natural capital (Hawken *et al.*, 2000) are required so that the true costs of techno-processes that extract and waste are borne by those who gain the benefit of doing so.

Extended producer responsibility (EPR)

EPR incorporates negative externalities from product use and end-of-life in product prices. Producers are made responsible for environmental effects over the entire product life cycle so that the cost of compliance cannot be shifted to a third party and must therefore be incorporated into product prices. Examples of EPR regulations include emissions and fuel economy standards (use stage) and product take-back requirements (end of life) such as deposit legislation, and mandatory returns policies which tend to force design with disassembly in mind.

Producers are made responsible for collecting and recycling end-of-life products. Waste-management costs are shifted to those most capable of reducing disposal costs by changing designs for recyclability, longevity, reduced toxicity, and limited volume of waste generated. Disposal costs are reflected in product prices so consumers can make more informed decisions. The above solutions all involve a cost. What if benefits could be incorporated?

What governments should be doing and why

Governments appear not to understand or see the benefit of fostering change to the techno process whereby they could substantially alleviate the problems of the planet. If some do, the evidence that they have done little about it is the low priority for materials research. According to Paul Zane Pilzer 'technology is the major determinant of wealth as it determines the nature and supply of physical resources' (Pilzer, 1990). Why then is it that so little government research funding is to change the technical paradigm for reducing, re-using and recycling materials? Materials are, after all, a major part of the flow of resources in the techno-process and fundamental for sustainability. Governments need to focus on fundamental research that leads to change in the technology factor. As Pilzer's first law states: 'By enabling us to make productive use of particular raw materials, technology determines what constitutes a physical resource' (Pilzer, 1990). Pilzer goes on further to explain that definitional technologies are those that enable us to make use of particular resources. Wastes are potentially a huge resource. Improvements in recovery and utilization technologies will one day make them of significant value.

Fortunately, some governments such as the EU are starting to research how we could live more sustainably on the planet. I am not, however, aware of any country or group of countries that prioritize the development research into materials as a way of reducing the take and waste in the techno-process, maximizing utility and making reuse and recycling more profitable. It is important that governments all over the world co-operate to bring about sustainability, take an active role and recognize their responsibility to seek sustainability as a cornerstone to all government expenditure and policy.

They must facilitate economic systems that encourage sustainability such as carbon trading and EPR. The challenge is to achieve 'common good' without a disproportionate impact on taxation.

7.10.3 The role of professionals

There are a large number of professionals, professions and industries involved in the value chain through the techno-process. The approach must not only be holistic in terms of what is done but holistic in that it involves a wide range of people. With particular reference to the construction business, the challenge is to understand materials and implement them in more sustainable ways.

Architects (including specifiers)

Architects and their specifiers dictate what sort of space we live in, with what material and how that space is constructed. The have been responsible for much of the reduction in lifetime energies of buildings so far achieved. Architects will:

- prescribe the sorts of spaces and the configurations of those spaces which directly affect how we interact with others inside and outside them
- decide what materials the spaces are built with and how they affect our health and the health of the environment
- decide how the spaces relate one to another and to the outside world and determine if the spaces are separate entities or connected, whether they look inward or outward, whether they encourage interaction or isolation
- prescribe how we furnish our spaces. Indeed, architects are the arbiters of taste in buildings rather like fashion houses direct what we wear.

Many new materials offer interesting architectural features taking them beyond mere product. The ability to design materials as products will change the relationship between scientists, architects and engineers. 'The architect becomes a coordinator between various kinds of embedded intelligences, applying their knowledge to create something society needs, instead of the usual passive role of an architect selecting given products' (Griff, 2004).

Engineers

Engineers hang the dreams of architects together as structural reality in 3D space. They turn complex ideas into reality and are therefore involved in some of the most innovative and sustainable new structures in the world.

Town planners

Town planners build cities. They decide where to place suburbs and how suburbs will look in relation to houses, transport, facilities, activities, etc. Town planners oversee the development of an entire town or region and must understand the needs of the environment, local environment, economy and population and how to interact those needs. In this role, and as the people who approve building construction, town planners are pivotal if we are to have more sustainable communities.

7.11 Conclusion

Materials are the key to sustainability in the built environment and innovative new materials will allow architects and engineers to build structures that have greater value as they are more pleasing to use, live in or look at, healthier for us and much more sustainable. Huge quantities of materials are used. Their choice profoundly affects many value properties relevant to sustainability including weight, embodied energies, fuel related and chemical emissions, lifetime energies, user comfort and health, use of recycled wastes, durability, recyclability and the properties of wastes returned to the geosphere-biosphere.

A holistic approach to sustainability in which all things possible are done is most likely to work and the built environment offers tremendous opportunities for the cultural and technical changes required. Exciting new benchmark technologies such tec and eco-cements offer a way forward and of solving the two greatest problems on the planet of global warming and waste. The direction is clear, technology can help us change the techno process. By doing so the process becomes more economic and thus self propelled with less government intervention.

Finding 3 under the heading 'Transport and Urban Design' of the recent ISOS conference in Australia applies globally. It stated in part. '…The Federal Government should promote Australian building innovations (e.g. eco-cement) that contribute global solutions towards sustainability; provide more sustainable city innovation R&D funds; and re-direct some housing and transport funds towards sustainable cities demonstration projects.[18] Technology can make it possible to achieve a far greater measure of sustainability, to economically reduce, reuse and recycle. The potential multipliers from spending on research and development are huge. As Fred Pearce reported in *New Scientist Magazine* (Pearce, 2002), 'There is a way to make our city streets as green as the Amazon rainforest.'

---

[18] ISOS Conference, 14th November Canberra, ACT, Australia communique downloadable from http://www.isosconference.org.au/entry.html.

7.12 Appendix: suggested policies for governments for a sustainable built environment

Many countries in the world such as Canada, the UK and the US have targeted research as means of priming the pump to address sustainability, climate action and other environmental ramifications of urban development. In spite of much rhetoric, the reverse has been the case in Australia. Research funding is driven by market forces and abatement, for example, blatantly market driven with the Australian Greenhouse Office doing little more than buying emissions reductions. A genuine national carbon trading system would be better than the fragmented one that is currently evolving and the Federal government should as a priority introduce such a system to add confidence to the carbon dollar. Now that Kyoto is a reality, joining would seem appropriate.

It makes sense economically to reduce emissions (given scale, emissions = energy = money (and with carbon trading, more money)). Market forces alone will not, however, be enough to drive important research into materials, argued in this chapter as fundamental to sustainability. More subtle government intervention and funding is required. In January 2002, four priority areas were identified in Australia for ARC funding: Complex/Intelligent Systems, Genome/Phenome Research, Nano- and Bio-materials and Photon Science and Technology. As of January 2004, the funding agencies have still not caught up with the importance of funding materials science as the key to the changes in our usage patterns required for sustainability.

The mechanisms available to the Commonwealth Government to bring about change towards sustainability, particularly in the built environment where ecologically sustainable patterns of settlement are required, include:

- Research and development funding priorities to include the funding of research development and deployment of new materials technologies.
- Procurement policies
  Government in Australia is more than one/third of the economy and can strongly influence change through
  - life cycle purchasing policy
  - funding of public projects and housing linked to sustainability.
- Intervention policies
  - building codes including mandatory adoption of performance specification
  - deposit legislation on not just packaging, but a wide range of goods including pallets and other commonly thrown away recyclables in construction
  - encouragement of cradle-to-cradle loops
  - mandatory use of minimum standard materials that are more sustainable.
- Taxation and incentive policies

  - direct or indirect taxes, bonuses, grants or rebates to discourage/encourage sustainable construction, etc.
  - a national system of carbon taxes
  - an international system of carbon trading
- Sustainability education.

In the context of the built environment these mechanisms are described further below.

Research and development funding priorities to include the funding of research development and deployment of new materials technologies

National priorities for research need to be brought up to date with the urgent requirements for climate action and should include materials science, energy and sustainable technologies. Such changes should urgently be reflected in ARC funding policy.

To 'prime the pump' and get urgent research development and deployment happening, direct funding as a matter of national importance of promising new materials science breakthroughs should be considered. TecEco is one of the leading companies in the race for more sustainable materials and would be prepared to negotiate a share in royalty income with the government.

Procurement policies

Life cycle purchasing policy

The development of purchasing procedures that are governed by life cycle performances will be a strong tool for the government to assess and adopt materials that will have an overall minimal impact on the environment, society and economics of the nation.

Funding of public projects and housing linked to sustainability

Prescriptive specifications detail the material to be used for a project. If the Commonwealth Government specified the use of new more sustainable materials such as TecEco's magnesium-based cements in appropriate construction projects (following verification research), it would know that certain measures to bring about urban development reform were being met. In order to bring this about it will be necessary to employ sustainable champions. Akin to a quantity surveyor, such champions would have the role of keeping the project on track to ensure that material selection and procedures have minimal detrimental impacts.

Intervention policies

Building codes including mandatory adoption of performance specification

Current outdated formula-based standards supported by the industry because of vested interests and fear of change should be required to follow the modern trend of performance-based specifications giving suppliers criteria to adhere to. As a matter of urgency the Commonwealth Government could adopt performance-based specifications for cement/concrete that:

- restrict the amount of embodied energy
- restrict the associated emissions of greenhouse gases
- detail the durability and other characteristics of the finished product.

Formula-based standards are holding up the development of much better composite materials directed towards existing market demand for more than just strength. Other desirable properties such as durability, insulating capacity, light weight and so on can only be introduced by innovation currently throttled by formula-based standards. Given the volume of materials flows the sustainability outcomes of a shift to performance-based standards will be significant.

National deposit legislation

The wastage of brand new materials on a building site is appalling and in Australia around 15%. The quantity of used materials from building and construction going to landfill in Australia is around 40% (Atkinson, 2003) and higher in Europe with greater building churn. In the context of construction deposit legislation covering items commonly thrown away that could be recycled may be in order. A novel idea and world first would be to include bricks, blocks, glass, metal roofing iron and many other recyclable building products. Savings in landfill costs alone would make the proposal worthwhile Facilities are already available in most cities to accept such waste. The new TecEco cements will help get over some problems such as gypsum in recycled building aggregates and sodium leaching out of glass cullet.

Encouragement of cradle-to-cradle loops (McDonough and Braungart, 2002)

To encourage recycling and remanufacturing the concept of service should be encouraged. For example, if heating and lighting were supplied instead of electricity, utilities would make sure the most efficient systems available were installed. Returning the responsibility for products including their disposal or reuse back to manufacturers has merit in that recycling is encouraged and quality assured. Technical recycling is kept within the technosphere and biological recycling less likely to be tainted.

Mandatory use of minimum standard materials that are more sustainable

As the need is urgent it may be necessary to ensure that the changes required come about to develop legislation requiring the mandatory use of more sustainable materials such as the calcium magnesium cement blends proposed by TecEco.

Taxation and incentive policies

Direct or indirect taxes, grants, bonuses or rebates to discourage/ encourage sustainable construction, etc.

Taxes and rebates are a powerful tool in the hands of governments to direct expenditure in perceived desirable directions.

A national system of carbon taxes

The proliferation of various state-run systems such as in Queensland and NSW should be superseded as soon as possible by worthwhile national carbon trading legislation secured by agreement with the states and ideally as part of the Kyoto protocol. Such a step may be difficult to implement, but should not cost the public purse much. A significant outcome would be the leverage to market forces already present that will bring about quantum improvements in sustainability at little overall net cost, but much required reallocation of capital.

An international system of carbon trading

Joining Kyoto would internationalize and legitimize carbon trading and would be worth it for the carbon credits attached to multiple cycle TecEco sequestration technology if for no other reason.

Sustainability education

Emphasis needs to be given to education across all disciplines in relation to sustainability. The sustainability of the built environment is influenced by a wide range of professionals including investors, developers, local governments, clients, accountants and many more. An understanding of sustainability needs to be fostered from a young age in everyone. It should be mandatory for engineers to include more materials science in their courses as many have little understanding of the environmental impact of the materials they use. The education of professionals into the consequences of using different materials and the benefits of using sustainable materials is a high priority and should be linked to higher education funding.

Conclusion

The role of governments is to ensure that public need is addressed even if the public do not understand that need. Action is required to address climate change and the more subtle detrimental effects of sprawling urban settlement. Policies and actions are required that do not impinge too heavily on budgets and several such legislative and policy opportunities to bring about the changes required have been suggested in this appendix.

If all Australian governments committed to building social, environmental and economic sustainability into every element of governance, we could be an example to the rest of the world. The manufacture and use of low energy, low emissions cements has been demonstrated by TecEco to be technically feasible and should be encouraged as a matter of national importance utilizing policies such as those suggested.

7.13 References

AAAS (2004). American Association for the Advancement of Science Population and Environment Atlas, AAAS American Association for the Advancement of Science, USA (http://atlas.aaas.org/index.php?part=1&sec=waste).

Anon (2004). *Annual Amount and Composition of Waste Consigned to Landfill*, South Australian Department of Environment and Heritage.

ARC (2004). National Research Priorities and Associated Priority Goals, Australian Research Council, Australia (http://www.arc.gov.au/pdf/2004_designated_national_research_priorities_&_ associate.pdf).

Atkinson, M. (2003). *Keynote Address*, Energy Efficiency Conference.

Batterham, R. (2004). Employment, Workplace Relations and Education. *Senate-References*, Hansard Senate.

Benyus, J. M. (1997). *Biomimicry, Innovation Inspired by Nature*, HarperCollins Books.

Brown, P. (2004). Melting ice: the threat to London's future, *The Guardian*, 14 July.

Bruntland, G. H., ed. (1987), *Our Common Future*, Report of the World Commission on Environment and Development (WCED), Oxford University Press.

Campbell, C. (2002). *How Long Can Oil Last Sunday Business Post*, Ireland.

Campbell, C. (2004). Oil and Gas Liquids, 2004 Depletion Scenario, Association for the Study of Peak Oil, Ireland (http://www.peakoil.net/uhdsg/Default.htm).

c-f-c.com (2004). Chlorofluorocarbons – CFCs, CFC Star Tec Inc., USA (http://www.c-f-c.com/supportdocs/cfcs.htm).

CICA (2002). Industry as a partner for sustainable development. Confederation of International Contractors Association, France.

CMDL (2003). *Carbon Cycle-Greenhouse Gases*, Climate Monitoring and Diagnostics Laboratory (CMDL), National Oceanic and Atmospheric Administration (NOAA), USA (www.cmdl.noaa.gov).

Commoner, B. (1972). The Environmental Costs of Economic Growth. *Chemistry in Britain* **8**(2): 52–65.

Cook, B. (2000). New designs on natural capitalism. Hong Kong General Chamber of Commerce.

Covey, S. (1999). *The 7 Habits of Highly Effective Teams*, Glasgow, Simon & Schuster.

CSIRO (2001). *Human Settlements Theme Report, State of the Environment Australia 2001*, Australian Government Department of Environment and Heritage, Australia.

Dahlin, D. C., O'Connor, W. K., Nilsen, D.N., Rush, G.E., Walters, R.P. and Turner, P.C. (2000). *A method for permanent CO_2 sequestration: Supercritical CO_2 mineral carbonation*, 17th Annual International Pittsburgh Coal Conference, USA.

DEH (2004). Annual Amount and Composition of Waste Consigned to Landfill (2004). Department of Environment and Heritage, South Australia (http://www.environment.sa.gov.au/reporting/human/waste/landfill.html).

Doyle, A. (2004). *Reuters, UK.*

Durie, R. A. (1998). Invited keynote address: Australian Greenhouse Gas Mitigation Responses, *Sustainable Energy and Environmental Technology Conference*, Gold Coast, Australia.

EPA (2003). Office of Solid Waste Strategic Planning Document 2003–2008. Environmental Protection Agency, USA.

Fauth, D. J., Baltrus, J. P., Soong, Y., Knoer, J.P., Howard, B.H., Graham, W. J., Maroto-Vales, M.M. and Andrésen, J.M. (2001). Carbon Storage and Sequestration as Mineral Carbonates. *Prepr. Symp. Am. Chem. Soc., Div. Fuel Chem* **46(1)**: 278.

Gelbspan, R. (2004). Toward a Global Energy Transition. In R. Gelbspan, *Boiling Point*, Basic Books, USA.

Gismondi, M. and Rees, W. E. (2004). 'Aurora on Line with William Rees', Aurora website of Interviews with Leading Thinkers and Writers, Athabasca University, Canada (http://aurora.icaap.org/talks/rees.html).

GISS (2002). Global Trends: 2002 Summation. *Goddard Institute for Space Studies*, USA.

GISS (2005). Global Temperature Anomaly. Goddard Institute for Space Studies, USA (http://www.giss.nasa.gov/data/update/gistemp/graphs/Fig.A.pdf).

Griff, A. (2004). Material Matters, *LA Architect*. Los Angeles, L A Architect, 30–35.

GRRN (2000). Wasting and Recycling in the United States 2000. Grass Roots Recycling Network, USA (www.grrn.org).

Harrison, P. (1992). *The Third Revolution, Environment, Population and a Sustainable World* London, Penguin Books.

Haughton, R. (2004). Understanding the Global Carbon Cycle, Woods Hole Institute, USA.

Hawken, P., Lovins, A. and Lovins, L.H. (2000). *Natural Capitalism: Creating the Next Industrial Revolution*. Earthscan Publishers Limited, UK.

Hendriks, C. A., Worrell, E., Price, L., Martin, N. and Ozawa Meida, L. (2002). *Emission Reduction of Greenhouse Gases from the Cement Industry*, International Energy Agency (IEA), USA.

Herbertson, J. (2004). 'Environmental drivers and emerging directions', AASMIC Inaugural Conference, Association for the Advancement of Sustainable Materials in Construction, Australia (www.aasmic.org).

Herbertson, J. and Green, M. (2003). Research data, The Natural Step, Environmental Institute, Australia (http://all.naturalstep.org).

Herbertson, J. and Sutton, P. (2002). Foundations of Sustainable Resource Processing. *Green Processing 2002 Conference Proceedings*, Australian Institute of Mining and Metallurgy (AUSIMM), Australia.

Hillman, M. and Fawcett, T. (2004). *How We Can Save the Planet*, London, Penguin Books.
IEA (2003). Key World Energy Statistics, International Energy Agency, France (http://www.iea.org/bookshop/add.aspx?id=144).
Johannes, R.E. (1992). *Worlds of the Lagoon: Fishing and Marine Lore in the Palau District*, Micronesia. University of California Press, USA.
Kimball, J. W. (2004). Kimball's Biology Pages, (http://users.rcn.com/jkimball.ma.ultranet/BiologyPages/C/CarbonCycle.html).
Lackner, K., Wendt, C., and Sharp, D.H. (1995). 'Carbon dioxide disposal in carbonate minerals' *Energy*. **20**: 1153–1170.
Langenfields, R. (2003). Oxygen measurements yield greenhouse clues, CSIRO, Australia (http://www.csiro.au/index.asp?id=OxygenMeasurements&type=mediaRelease).
Malthus, T. (1798). *An Essay on the Principle of Population, as it Affects the Future Improvement of Society with Remarks on the Speculations of Mr. Godwin, M. Condorcet, and Other Writers*, London, UK.
Marland, G., Boden, T., Brenkert, A. and Andres, R.J. (1999). Historical and projected global greenhouse gas emissions. In J. Kinsman, C.V. Mathai, M. Baer, E. Holt, and M. Trexler (eds), *Global Climate Change: Science, Policy and Mitigation/Adaptation Strategies*. Air & Waste Management Association, Pittsburgh, pp. 37–42.
Matear, R. (2002). Deep Ocean Losing Oxygen, CSIRO, Australia (http://www.csiro.au/index.asp?id=proceanoxygen&type=mediaRelease).
McDonough, W. and Braungart, M. (2002). *Cradle to Cradle: Remaking the Way We Make Things*, North Point Press, USA.
MIRO (2002). RC114 – Targeted Research Action on Waste Minimisation and Recycling, Mining Industry Research Organisation (MIRO), UK (http://www.miro.co.uk/projects/arcprojs/rc114.htm).
NETL (2004). Chemical and Geologic Sequestration of Carbon Dioxide, National Energy Technology Laboratory (NETL), USA (http://www.netl.doe.gov/products/r&d/annual_reports/2001/cgscdfy01.pdf).
O'Connor, W. K., Dahlin, D. C., Nilsen, D. N., Walters, R. P. and Turner, P. C (2000). Carbon Dioxide Sequestration by Direct Mineral Carbonation: Results from Recent Studies and Current Status *Proceedings of the 25th International Technical Conference on Coal Utilization & Fuel Systems*. National Energy Technology Laboratory, USA, 153–164 (www.netl.doe.gov)
Pearce, F. (1997). The Concrete Jungle Overheats. *New Scientist* (2097): 14.
Pearce, F. (2002). Green Foundations. *New Scientist* **175**(2351): 39–40.
Pilzer, P. Z. (1990). *Unlimited Wealth – The Theory and Practise of Economic Alchemy*, Crown Publishers, UK.
Quinn, P. J. B. (1999). *Keynote address, Forum on Sustainable Development*, Forum on Sustainable Development, Parliament House, Canberra.
Rees, W. E. (1997). Is 'Sustainable City' an oxymoron? *Local Environment* **2**: 303–310.
Rees, W. E. (1999). The Built Environment and Ecosphere: A Global Perspective. *Building Research and Information* **27**(4/5): 206–220.
Sanjour, W. (2001). Royal Roads University Waste Audit Report. Royal Road University, Canada.
Schaller, D. (2004). Beyond Sustainability: From Scarcity to Abundance, *BioInspire Newsletter*. Denver, USA.
Seifritz, W. (1990). *Nature* **345, 486**.
Seligman, M. E. P. (1990). *Learned Optimism: How to Change Your Mind and Your Life*, New York, Pocket Books – Simon & Schuster, Inc.

Sharp, L. (2003). *Green Campuses, The Road from Little Victories to Systemic Transformation*. Harvard University Press, USA

Steffen, W. and Tyson, P. (2001). Global Change and the Earth System: A planet under pressure, International Geosphere-Biosphere Programme (IGBP), Sweden (http://www.igbp.kva.se/cgi-bin/php/frameset.php).

Sutton, P. (2002). Greenleaps newsletter.

The Sunday Tasmanian (2001). Hobart, Tasmania.

The Mercury (2004). Hobart, 3, Tasmania.

Tucker, S. (2000). CSIRO on line brochure, CSIRO, Australia (http://www.dbce.csiro.au/ind-serv/brochures/embodied/embodied.htm).

Tucker, S. (2002). CSIRO Department of Building Construction and Engineering. J. Harrison, pers. comm.

UN (2002). World Population Prospects: The 2002 Revision Population Database. UN Publications, USA.

UNCHS (2004a). Global Trends, United Nations Human Settlements Programme, USA (http://www.unchs.org/habrdd/global.html).

UNCHS (2004b). The Habitat Agenda, United Nations Human Settlements Programme, USA.

University of Oregon (2004). *Electronic Universe Educational Server*. University of Oregon (http://zebu.uoregon.edu/1998/es202/l14.html).

UNEP (2001). *Energy and Cities: Sustainable Building and Construction Summary of Main Issues*, IETC Side Event at UNEP Governing Council, Meeting in Nairobi, Kenya United Nations Environment Programme, USA (www.unep.org).

UNESCO (2004). New Role of Education, United Nations Educational, Scientific and Cultural Organisation, USA.

USGS (2000). Mineral Commodity Summary – Cement. US Geological Survey, USA (www.usgs.gov).

USGS (2004). Mineral Commodity Summary – Cement. US Geological Survey, USA (www.usgs.gov).

Van Oss, H. (2004). Mineral Commodity Summary – Cement.

Van Oss, H., Hendriks, K., and Thomas, D. (2003). Cement Statistics, US Government Survey.

Vitousek, P. M., Mooney, H. A., Lubchenco, J. and Melillo, J.M. (1997). Human Domination of Earth's Eco-Systems. *Science* **277**(July): 494–99.

Von Weizsäcker, E. U., Lovins, A. B., Ulrich, E. and Lovins, H. (1997). *Factor four: Doubling Wealth, Halving Resource Use*, London, Earthscan.

Wain, F. (2004a). *The Canberra Times*. Canberra, Australia.

Wain, F. (2004b). The real cost and implications of collateral damage. *The Canberra Times*, Australia

Wilson, A. (1993). Cement and Concrete: Environmental Considerations. *Environmental Building News* **2**(2).

Wilson, N. (2004). Call for public input to carbon plan. *The Weekend Australian* (May 1–2, 2004).

WRI (1998). WRI, 1998–99 Database Diskette, World Resources Institute (http://www.wri.org).

WRI (2004). World Resources 2002–2004, World Resources Institute (http://www.wri.org).

WWF (2004). WWF's Living Planet Report, World Wildlife Fund (http://www.panda.org/news_facts/publications/general/livingplanet/index.cfm).

Yegulalp, T. M., Lackner, K. S., and Ziock, H.J. (2001). A Review of Emerging Technologies for Sustainable Use of Coal for Power Generation. *The International Journal of Surface Mining, Reclamation and Environment* **15**(52–68): 58.

Index